Shinsaku Fujita

Symmetry and Combinatorial Enumeration in Chemistry

With 54 Figures

Springer-Verlag

Berlin Heidelberg New York
London Paris Tokyo
Hong Kong Barcelona
Budapest

Dr. Shinsaku Fujita
Research Laboratories, Ashigara
Fuji Photo Film Co., Ltd.
Minami-Ashigara, Kanagawaken, 250-01 Japan

ISBN-13: 978-3-540-54126-4   e-ISBN-13: 978-3-642-76696-1
DOI: 10.1007/978-3-642-76696-1

The use of registered names, trademarks, etc. in this publication does not imply, even in the absence of a specific statement, that such names are exempt from the relevant protective laws and regulations and therefore free for general use.

Typesetting: Camera ready by author

51/3140-543210 – Printed on acid-free paper

Preface

This book is written to introduce a new approach to stereochemical problems and to combinatorial enumerations in chemistry. This approach is based on group theory, but different from conventional ways adopted by most textbooks on chemical group theory. The difference stems from their starting points: conjugate subgroups and conjugacy classes.

The conventional textbooks deal with linear representations and character tables of point groups. This fact implies that they lay stress on conjugacy classes; in fact, such group characters are determined for the respective conjugacy classes. This approach is versatile, since conjugacy classes can be easily obtained by examining every element of a group. It is unnecessary to know the group-subgroup relationship of the group, which is not always easy to obtain. The same situation is true for chemical enumerations, though these are founded on permutation groups. Thus, the Pólya-Redfield theorem (1935 and 1927) uses a cycle index that is composed of terms associated with conjugacy classes.

On the other hand, the alternative approach which I take is based on permutation representations and tables of marks. The latter concept has once been discussed in Burnside's excellent textbook (1911) but unjustly neglected for a long time. The tables of marks have been omitted even from voluminous books on the group theory, except Oshima's textbook (1954), Sheehan's paper (1968) and Hässelbarth's one (1985). In this approach, conjugate subgroups are of much importance; the marks are given to the respective representatives of conjugate subgroups. The tables of marks can be regarded as counterparts of character tables. The determination of such conjugate subgroups requires the group-subgroup relationship that is generally difficult of approach. However, this fact is no longer a drawback in the light of our target, because this relationship is also necessary to our target of understanding stereochemistry.

I have adopted the second approach, since this is suitable to our stereochemical problems that have discrete nature. The task of solving them, however, necessitates some new elements that have never been served by the conventional methods. Thus, I have tried to integrate point-group and permutation-group theories. Throughout this book, I have put special emphasis on coset representations, which are transitive permutation representaitons. In order to treat the stereochemical problems comprehensively, I have introduced several new concepts: the SCR (set-of-coset-representation) notation, allowed and forbidden coset representations, chirality fittingness, subduction of coset representations, proligands, promolecules and so forth. I have further extended them to create another set of concepts, *e.g.*, unit subduced cycle indices (USCIs), subduced cycle indices (SCIs), partial cycle indices (PCIs), and elementary superposition. All of these concepts are useful to solve combinatorial enumerations. Although examples in this book are adopted from chemistry only, the method developed here can be utilized to solve any other combinatorial enumerations.

This book may serve as a first introduction for students and researchers intending to begin their works in this field. In particular, I hope that the book would give them some comprehensive perspective other than those given by conventional textbooks.

I am indebted to Mr. Sosuke Hanai for a number of valuable comments on the draft manuscript.

Minami-Ashigara, Kanagawa, Japan
January 1991   Shinsaku Fujita

Contents

Chapter 1

Introduction

Group theory is now an essential tool for chemists. Thus, there have appeared a vast number of pedagogical articles on its applications to chemistry.[1]–[9] In addition, we can enrich our knowledge by means of excellent textbooks on this topic.[10]–[17]

Group theory has been applied to quantum-chemical problems that are concerned with molecular symmetry, *e.g.*,

- symmetry adapted functions for molecular orbital theory,

- ligand field theory, and

- molecular vibrations.

These applications are mainly based on group-representation theory. In this theory, a linear representation of a point group and its reduction into irreducible representations play significant roles, where character tables of point groups are used for assigning such an irreducible representation to an orbital. The textbooks on chemical group theory have already discussed these issues in details; hence, these are not discussed in the present book. However, we should here pay attention to methodology presumed in the applications.

The applications imply a continuous molecular model in which a set of wave functions are superposed to realize a molecule. In other words, the concepts of bonds and atoms become secondary. This approach succeeds to the physical tradition and is against the chemical convention.

On the other hand, there exist at least two fields that require group-theoretical investigations,

- combinatorial enumeration of chemical compounds and reactions, and

- stereochemistry,

which are the main objectives of this book. These applications require another molecular model in which atoms and bonds have primary meanings. In other words, we are grounded on three-dimensional structural formulas. This model has discrete nature, which seems to be suitable for analysis by permutation-group theory.

Accordingly, permutation-group theory is applied to enumeration problems of molecules and reactions.[18] Since the original introduction of Pólya's theorem (1935),[1] chemistry has supplied many problems for verifying the theorem.[19, 20] Such results have been reviewed extensively in 1970s.[21] The concept of double cosets has been applied to enumeration of rearrangements of phosphorus complexes.[22] Methods based on tables of marks have recently been developed for enumerating graphs[23] and compounds,[24] in which their weights (molecular formulas) and symmetries are taken into consideration. Tables of marks were originally described in Burnside's famous textbook[25]; however, they have attracted little attention of mathematicians as well as of chemists. This is because such tables of marks have been published for a limited number of groups in comparison with the character tables above. In the present book, we prepare mark tables for representative (point) groups and use them thoroughly for solving our problems.

Stereochemistry is concerned with chirality and achirality of molecules as well as their interconversion.[26]–[28] Since these concepts are related to molecular symmetry, stereochemical problems are investigated by means of group theory, especially, of point-group theory. Important terms in stereochemistry such as topicities are thus defined by group-theoretical consideration.[29] However, a point group is frequently insufficient to manipulate a discrete molecular model, since it is concerned with a continuous molecular model. For remedying this disadvantage, the concept of framework group has been proposed.[30] An alternative solution[31] is based on local (site) symmetries, which come from crystallographical literatures.[32] Algebraic aspects of chirality phenomena have been discussed on the basis of permutation groups.[33, 34]

The short history above implies that chemical applications of point-group theory and those of permutation-group theory have developed separately, although they are closely related to each other. This book aims at integrating the two theories in order to gain a deeper insight into stereochemistry and chemical enumeration. The missing link for the integration is the concept of "coset representations".

Throughout the course of this book, the following correspondence is significant:

$$
\begin{array}{ccccc}
\text{sets of} & & \text{orbits} & & \text{coset} \\
\text{equivalent} & \overset{\text{chem.}}{\longleftrightarrow} & \text{(equivalence} & \overset{\text{math.}}{\longleftrightarrow} & \text{representations}. \\
\text{ligands} & & \text{classes)} & &
\end{array}
$$

A given molecule is composed of sets of equivalent ligands (or atoms), each of which is regarded as an orbit. This one-to-one correspondence can be referred

---

[1] We refer to this theorem as the Pólya-Redfield theorem, since Redfield discovered this independently.

to as a chemical correspondence. Each orbit in turn corresponds to an approriate coset representation in one-to-one fashion. The latter is a mathematical correspondence. Thereby, we can state that each set of equivalent ligands is governed by a coset representation. This fact means that the properties of the ligand set can be determined by examining those of the corresponding coset representation.

The first half of this book (Chapters 2 to 12) involves discussions on stereochemical problems in the light of the present approach. This half is subdivided into three parts.

Chapters 2 to 4 deal with mathematical foundations. Before we get to work on integrating point-group and permutation-group theories, Chapters 2 and 3 are devoted to presenting essential items of the respective theories. Chapter 4 is a short introduction to axioms and theorems of group theory. These chapters can be skipped if the readers are familiar to group theory.

In Chapters 5 to 8, we discuss coset representations and their chemical applications. These discussions clarify our basic idea that combines the two theories. Thus, we present several new concepts, such as SCR notations (notations based on sets of coset representations), local symmetries, forbidden coset representations, and chirality fittingness, all of which are derived from coset representations.

Chapters 9 to 12 are concerned with subduction of coset representations and its qualitative applications to stereochemistry. Among the key concepts of the present approach, we discuss there the following topics: unit subduced cycle indices with and without chirality fittingness (USCIs and USCI-CFs), desymmetrization lattices, prochirality redefined, and topicities and stereogenicity revisited.

The second half (Chapters 13–21) is concerned with enumeration problems that have been solved in the light of the present USCI approach. This half is also subdivided into three parts.

Chapters 13 and 14 describe the Pólya-Redfield theorem and related topics for an introduction to combinatorial enumerations. We emphasize the importance of obligatory minimum valencies in chemical applications, since any atoms have their own valencies. For this purpose, we take account of orbits of a domain explicitly.

In Chapters 15 to 18, we discuss quantitative applications of unit subduced cycle indices to enumerations of chemical compounds with achiral ligands. In particular, generating functions of new type are obtained by the successive processes, unit subduced cycle indices (USCIs) $\longrightarrow$ subduced cycle indices (SCIs) ($\longrightarrow$ partial cycle indices (PCIs)) $\longrightarrow$ a cycle index (CI). In addition, we introduce the concept of elementary superposition, which provides us with an alternative method of solving enumeration problems without depending on generating functions.

Chapters 19 to 21 contain topics on enumeration of compounds with achiral and chiral ligands, which requires the use of unit subduced cycle indices with chirality fittingness.

Chapters 6 to 19 (except Chapters 13 and 17) are based on the original papers printed in the first pages of the respective chapters. They have however been rewritten thoroughly in order to hold consistency throughout the present book. In particular, we have substituted many illustrative examples for original ones so that the contents of this book become more familiar to organic chemists as well as to theoretical chemists. In several chapters, we have added theorems and their proofs for formulating our problems in a precise manner. Chapters 17, 20 and 21 have been reprinted from the original papers after slight modification, since they are quite suitable for exemplifying the USCI approach.

Appendices A to E are the first collection of the data concerning the USCI approach and are useful to solve stereochemical and combinatorial problems.

Bibliography

[1] J. E. White, *J. Chem. Educ.*, **44**, 128 (1967).

[2] J. Donahue, *J. Chem. Educ.*, **46**, 27 (1969).

[3] a) M. Orchin, H. H. Jaffé, *J. Chem. Educ.*, **47**, 246 (1970). b) M. Orchin, H. H. Jaffé, *J. Chem. Educ.*, **47**, 327 (1970). c) M. Orchin, H. H. Jaffé, *J. Chem. Educ.*, **47**, 510 (1970).

[4] D. I. Ford, *J. Chem. Educ.*, **49**, 336 (1972).

[5] D. P. Strommen, E. R. Lippincott, *J. Chem. Educ.*, **49**, 341 (1972).

[6] E. L. Burrows, M. J. Clark, *J. Chem. Educ.*, **51**, 87 (1974).

[7] M. Herman, J. Lievin, *J. Chem. Educ.*, **54**, 596 (1977).

[8] G. D. Nigam, *J. Chem. Educ.*, **60**, 919 (1983).

[9] S. Fujita, *J. Chem. Educ.*, **63**, 744 (1986).

[10] F. A. Cotton, *Chemical Applications of Group Theory*, 2nd Ed., Wiley-Interscience, New York (1971).

[11] H. H. Jaffé, M. Orchin, *Symmetry in Chemistry*, Wiley, Chichester (1965).

[12] L. H. Hall, *Group Theory and Symmetry in Chemistry*, McGraw-Hill, New York (1969).

[13] D. M. Bishop, *Group Theory and Chemistry*, Clarendon, Oxford (1973).

[14] S. F. A. Kettle, *Symmetry and Structure*, Wiley, Chichester (1985).

[15] M. F. C. Ladd, *Symmetry in Molecules and Crystals*, Ellis Horwood, Chichester (1989).

[16] D. C. Harris, M. D. Bertolucci, *Symmetry and Spectroscopy*, Dover, New York (1989).

[17] I. Hargittai, M. Hargittai, *Symmetry through the Eyes of a Chemist*, VCH, Weinheim (1986).

[18] J. Hinze (Ed.), *The Permutation Group in Physics and Chemistry* (Lecture Notes in Chemistry, Vol. 12), Springer-Verlag, Berlin-Heidelberg-New York (1979).

[19] G. Pólya, R. C. Read, *Combinatiorial Enumeration of Groups, Graphs and Chemical Compounds*, Springer, New York-Berlin-Heidelbelg (1987).

[20] N. L. Biggs, E. K. Lloyd, R. J. Vilson, *Graph Thory 1736–1936*, Oxford University Press, Oxford (1976).

[21] A. T. Balaban (Ed.), *Chemical Application of Graph Theory*, Academic, London (1976).

[22] E. Ruch, D. J. Klein, *Theor. Chim. Acta*, **63**, 447 (1983).

[23] J. Sheehan, *Can. J. Math.*, **20**, 1068 (1968).

[24] W. Hässelbarth, *Theor. Chim. Acta*, **67**, 339 (1985).

[25] W. Burnside, *Theory of Groups of Finite Order*, 2nd Ed., Cambridge Univ. Press, Cambridge (1911).

[26] K. Mislow, *Introduction to Stereochemistry*, Benjamin, New York (1965).

[27] E. L. Eliel, *Stereochemistry of Carbon Compounds*, McGraw-Hill, New York (1962).

[28] M. Nakazaki, *Kagaku to Taisyosei (Chemistry and Symmetry)*, Nankodo (1976).

[29] a) K. Mislow, M. Raban, *Top. Stereochem.*, **1**, 1 (1967). b) K. Mislow, J. Siegel, *J. Am. Chem. Soc.*, **106**, 3319 (1984).

[30] J. A. Pople, *J. Am. Chem. Soc.*, **102**, 4615 (1980).

[31] R. L. Flurry, *J. Am. Chem. Soc.*, **103**, 2901 (1981).

[32] T. Hahn (Ed.), *International Tables for Crystallography*, Vol. A, Kluwer Academic, Dordracht (1989).

[33] a) E. Ruch, *Acc. Chm. Res.*, **5**, 49 (1972). b) E. Ruch, *Angew. Chem. Intern. Ed.*, **16**, 65 (1977).

[34] C. A. Mead, *Top. Curr. Chem.*, **49**, 1 (1974).

Chapter 2

Symmetry and Point Groups

2.1 Symmetry Operations and Elements

Symmetry and (point) groups have been recognized as essential concepts for chemists and there have appeared several excellent textbooks on these topics.[1]–[6] In the light of these textbooks, we can obtain fundamental knowledge on symmetry and group theory. In this section, we revisit a minimum set of concepts concerning symmetry and groups by using an allene molecule ($\mathbf{D}_{2d}$) as an example.

In order to discuss molecular symmetry, we should first introduce *symmetry operations*. A symmetry operation is defined as an operation that moves a given molecule (or generally, a three-dimensional object) from an initial state to another, where the two states cannot be differentiated from each other. Such symmetry operations are classified as follows.

$$\left\{\begin{array}{l} \text{rotations (proper rotations)} \\[1em] \text{rotoreflections (improper rotations)} \left\{\begin{array}{l} \text{reflections} \\ \text{inversion} \\ \text{rotoreflections (narrow)} \end{array}\right. \end{array}\right.$$

We use the following italicized or Greek symbols according to Schönflies' notation:

I	identity	σ_v	reflection in a vertical plane
C_n	rotation by $2\pi/n$	σ_d	reflection in a diagonal plane
σ	reflection	i	inversion
σ_h	reflection in a horizontal plane	S_n	rotoreflection ($S_n = \sigma_h C_n$)

If necessary, we distinguish several symmetry operations of the same kind by sequential subscripts in parentheses *e.g.*, $\sigma_{d(1)}$ and $\sigma_{d(2)}$.

An appropriate set of symmetry operations fixes a point, a line, or a plane. Thereby, such fixed points, lines and planes are referred to as *symmetry elements*. For example, a set of rotations (C_n, C_n^2,..., C_n^{n-1}, C_n^n ($= I$)) fixes a line, which is called a C_n-axis or an n-fold axis. This is one of such symmetry elements. Note

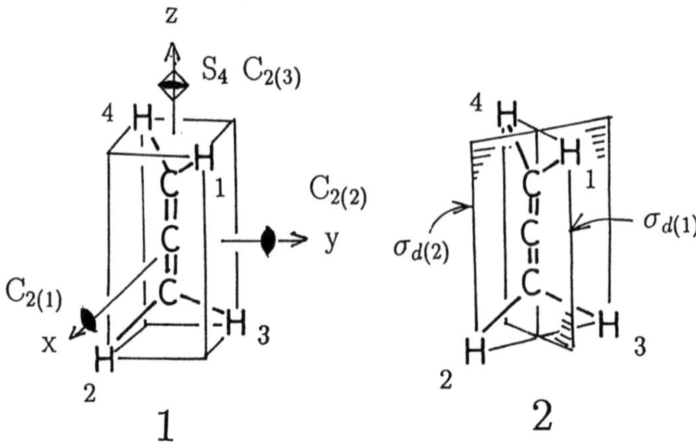

Figure 2.1: Symmetry elements of an allene molecule

that we use a roman capital for designating the axis. For most molecules, there is usually one axis that has higher rotational symmetry than the others. This axis is called the *principal axis*, which is usually selected to be identical with the z-axis. The terms "horizontal" and "vertical" for the reflection planes are concerned with the principal axis.

In this section, we examine the symmetry elements and operations of an allene molecule. Figure 2.1 illustrates all of the symmetry elements. The left figure (**1**) depicts three two-fold axes, which are denoted by roman characters $(C_{2(1)}, C_{2(2)}, C_{2(3)})$. These axes are associated with rotations of π $(= 180°)$ around the x-, y- and z-axes. These operations are designated by the italicized symbols $(C_{2(1)}, C_{2(2)},$ and $C_{2(3)})$. These operations are (proper) rotations.

The $C_{2(3)}$-axis is contained in an S_4-axis, which is associated with a rotoreflection (S_4). The S_4 operation is a combined operation that consists of a rotation (C_4) and a reflection (σ_h); i.e., $S_4 = \sigma_h C_4$. The S_4-axis produces four symmetry operations, S_4, S_4^2 $(= C_{2(3)})$, S_4^3, S_4^4 $(= I)$. The S_4-axis is the principal axis of allene.

The right structure (**2**) shows two mirror planes in the allene molecule; $\sigma_{d(1)}$ and $\sigma_{d(2)}$. We use these symbols to denote the mirror planes as well as reflection operations by these planes; however, this usage may produce no confusion. These planes contain the principal S_4-axis. The subscript d (diagonal) stems from the fact that each of them bisects the angle between the two-fold axes $(C_{2(1)}$ and $C_{2(2)})$ perpendicular to the principal axis.

We have 8 symmetry operations generated by the above procedure: $\mathbf{D}_{2d} = \{I,$ $C_{2(1)}$, $C_{2(2)}$, $C_{2(3)}$, $\sigma_{d(1)}$, $\sigma_{d(2)}$, S_4, and $S_4^3\}$. All of these operations keep the allene molecule invariant. This means that the initial state of the molecule cannot be distinguished from the final state generated by each of the operations. We here define the multiplication of two symmetry operations as successive application of these operations. For example, the first application of the $C_{2(1)}$ operation and the second of the $C_{2(3)}$ operation are regarded as the multiplication of these operations, which is designated by the expression, $C_{2(3)}C_{2(1)}$. Since this multiplication produces the $C_{2(2)}$ operation, we have $C_{2(3)}C_{2(1)} = C_{2(2)}$.

The set $(\mathbf{D}_{2d})$ of the 8 symmetry operations gives a multiplication table (Table 2.1). This table is closed in itself and has no duality of elements in each row and in each column. Hence, the set satisfies the axioms of group theory; in other words, the set is a group.[1] We refer to this group as $\mathbf{D}_{2d}$ point group. We use the term *point group* in the sense that it fixes at least one point.

Table 2.1: Multiplication table of $\mathbf{D}_{2d}$ group

		The 1st operation							
		I	$C_{2(1)}$	$C_{2(2)}$	$C_{2(3)}$	$\sigma_{d(1)}$	S_4^3	S_4	$\sigma_{d(2)}$
	I	I	$C_{2(1)}$	$C_{2(2)}$	$C_{2(3)}$	$\sigma_{d(1)}$	S_4^3	S_4	$\sigma_{d(2)}$
	$C_{2(1)}$	$C_{2(1)}$	I	$C_{2(3)}$	$C_{2(2)}$	S_4	$\sigma_{d(2)}$	$\sigma_{d(1)}$	S_4^3
	$C_{2(2)}$	$C_{2(2)}$	$C_{2(3)}$	I	$C_{2(1)}$	S_4^3	$\sigma_{d(1)}$	$\sigma_{d(2)}$	S_4
The 2nd	$C_{2(3)}$	$C_{2(3)}$	$C_{2(2)}$	$C_{2(1)}$	I	$\sigma_{d(2)}$	S_4	S_4^3	$\sigma_{d(1)}$
operation	$\sigma_{d(1)}$	$\sigma_{d(1)}$	S_4^3	S_4	$\sigma_{d(2)}$	I	$C_{2(1)}$	$C_{2(2)}$	$C_{2(3)}$
	S_4^3	S_4^3	$\sigma_{d(1)}$	$\sigma_{d(2)}$	S_4	$C_{2(2)}$	$C_{2(3)}$	I	$C_{2(1)}$
	S_4	S_4	$\sigma_{d(2)}$	$\sigma_{d(1)}$	S_4^3	$C_{2(1)}$	I	$C_{2(3)}$	$C_{2(2)}$
	$\sigma_{d(2)}$	$\sigma_{d(2)}$	S_4	S_4^3	$\sigma_{d(1)}$	$C_{2(3)}$	$C_{2(2)}$	$C_{2(1)}$	I

Strictly speaking, we should distinguish the concept of "the symmetry of a molecule" from the corresponding concept of "point group". However, there is a close relation between these concepts, *i.e.*, the symmetry of a molecule $\longleftrightarrow$ a combination of symmetry elements $\longleftrightarrow$ a set of symmetry operations $\equiv$ a point group. Chemists' convention permits such an expression that a molecule belongs to an appropriate point group. Thus, we can state that allene belongs to $\mathbf{D}_{2d}$ point group.

---

[1] For the axioms of group theory, see Chapter 4.

2.2 Conjugacy Classes in Point Groups

A conjugate element of an element a ($\in$ G) is defined as sas^{-1}, where $s \in$ G. Conversely, a is a conjugate element of sas^{-1}. A conjugacy class (C) is represented by

$$C = \{sas^{-1} \mid \forall s \in G\}$$

for a given element (a). Of course, the element (a) itself belongs to the conjugacy class (C). The group G is partitioned into several conjugacy classes. The identity element (I) itself constructs a conjugacy class in any case.

In the case of the $\mathbf{D}_{2d}$ group, the eight symmetry operations are partitioned into five conjugacy classes, *i.e.*, $C_1 = \{I\}$, $C_2 = \{C_{2(1)}, C_{2(2)}\}$, $C_3 = \{C_{2(3)}\}$, $C_4 = \{S_4, S_4^3\}$, and $C_5 = \{\sigma_{d(1)}, \sigma_{d(2)}\}$.

2.3 Subgroups of Point Groups

A subset (**H**) of a group **G** that forms a group under the multiplication of **G** is called a *subgroup* of **G**. This is denoted by the symbol $\mathbf{H} \leq \mathbf{G}$. An identity group $\{\,I\,\}$ and **G** itself are always subgroups of **G**. In this section, we list the subgroups of the $\mathbf{D}_{2d}$ group.

$\mathbf{C}_2$ **Point Groups.** By starting from the 8 symmetry operations, we have a $\mathbf{C}_2$ point group, *i.e.*, $\mathbf{C}_2 = \{\,I, C_{2(3)}\,\}$. This set is verified to be a group by constructing a multiplication table (Table 2.2), where the subscript (3) is abbreviated for generality's sake. Other $\mathbf{C}_2$ groups can be selected from the 8 symmetry operations,

Table 2.2: Multiplication table of $\mathbf{C}_2$ group

		The 1st operation	
		I	C_2
The 2nd	I	I	C_2
operation	C_2	C_2	I

i.e., $\mathbf{C}_2' = \{I, C_{2(1)}\}$ and $\mathbf{C}_2'' = \{I, C_{2(2)}\}$. These also have the same muliplication table as Table 2.2.

$\mathbf{C}_s$ **Point Groups.** By selecting a mirror plane from the 8 symmetry operations, we have a $\mathbf{C}_s$ point group, *e.g.*, $\mathbf{C}_s = \{I, \sigma_{d(1)}\}$. This is verified by constructing a multiplication table (Table 2.3). For general discussion, we use the symbol (σ) instead of $\sigma_{d(1)}$. Another $\mathbf{C}_s$ group is selected from the 8 symmetry operations, *i.e.*, $\mathbf{C}_s' = \{I, \sigma_{d(2)}\}$,

If the symbol C_2 in Table 2.2 is replaced by a letter A and if σ in Table 2.3 is also replaced by A, these tables become identical with each other. This fact

Table 2.3: Multiplication table of C_s group

		The 1st operation	
		I	σ
The 2nd	I	I	σ
operation	σ	σ	I

indicates that the two groups have the same structure as abstract groups. In other words, they are isomorphic.

S_4 Point Group. The symmetry operation (S_4) is a kind of improper rotation (rotoreflection). If we select $\{I, S_4, C_{2(3)}, S_4^3\}$ from the $\mathbf{D}_{2d}$ group, we have an $\mathbf{S}_4$ point group. The multiplication table for $\mathbf{S}_4$ group is shown in Table 2.4, where the subscript (3) of $C_{2(3)}$ is omitted. The $\mathbf{S}_4$ group is cyclic, since all of the elements

Table 2.4: Multiplication table of S_4 group

		The 1st operation			
		I	S_4	C_2	S_4^3
	I	I	S_4	C_2	S_4^3
The 2nd	S_4	S_4	C_2	S_4^3	I
operation	C_2	C_2	S_4^3	I	S_4
	S_4^3	S_4^3	I	S_4	C_2

are represented by S_4^n, where n is an integer.

C_{2v} Point Group. We select a $\mathbf{C}_2$-axis and two of such mirror planes that intersect each other at the $\mathbf{C}_2$-axis in a perpendicular fashion. This selection gives a $\mathbf{C}_{2v}$ point group. For example, we have the $\mathbf{C}_{2v}$ group, $\mathbf{C}_{2v} = \{I, C_{2(3)}, \sigma_{d(1)}, \sigma_{d(2)}\}$ as a subgroup of the $\mathbf{D}_{2d}$ group. If we change the subscrips for generality, we have $\mathbf{C}_{2v} = \{I, C_2, \sigma_{v(1)}, \sigma_{v(2)}\}$, whose multiplication table is shown in Table 2.5. It should

Table 2.5: Multiplication table of C_{2v} group

		The 1st operation			
		I	C_2	$\sigma_{v(1)}$	$\sigma_{v(2)}$
	I	I	C_2	$\sigma_{v(1)}$	$\sigma_{v(2)}$
The 2nd	C_2	C_2	I	$\sigma_{v(2)}$	$\sigma_{v(1)}$
operation	$\sigma_{v(1)}$	$\sigma_{v(1)}$	$\sigma_{v(2)}$	I	C_2
	$\sigma_{v(2)}$	$\sigma_{v(2)}$	$\sigma_{v(1)}$	C_2	I

be noted that the subscripts d and v depend upon subgroups selected.

$\mathbf{D_2}$ **Point Group.** The $\mathbf{D_2}$ subgroup cantains all of the proper rotations of the $\mathbf{D}_{2d}$ group. Thus, the subgroup consists of the three C_2 operations, *i.e.*, $\mathbf{D_2} = \{I, C_{2(1)}, C_{2(2)}, C_{2(3)}\}$, whose multiplication table is shown in Table 2.6.

Table 2.6: Multiplication table of $\mathbf{D_2}$ group

		The 1st operation			
		I	$C_{2(1)}$	$C_{2(2)}$	$C_{2(3)}$
	I	I	$C_{2(1)}$	$C_{2(2)}$	$C_{2(3)}$
The 2nd	$C_{2(1)}$	$C_{2(1)}$	I	$C_{2(3)}$	$C_{2(2)}$
operation	$C_{2(2)}$	$C_{2(2)}$	$C_{2(3)}$	I	$C_{2(1)}$
	$C_{2(3)}$	$C_{2(3)}$	$C_{2(2)}$	$C_{2(1)}$	I

The $\mathbf{D_2}$ group is isomorphic to the $\mathbf{C}_{2v}$ group.

2.4 Conjugate and Normal Subgroups of Point Groups

A *conjugate subgroup* of $\mathbf{H}$ ($\leq \mathbf{G}$) is defined as a set represented by

$$s\mathbf{H}s^{-1} = \{sas^{-1} \mid \forall a \in \mathbf{H}\},$$

where the symbol s is a given element of $\mathbf{G}$.

Among the three $\mathbf{C_2}$ groups of $\mathbf{D}_{2d}$, $\mathbf{C}_2' = \{I, C_{2(1)}\}$ and $\mathbf{C}_2'' = \{I, C_{2(2)}\}$ are conjugate. They are not conjugate to the remaining $\mathbf{C_2} = \{I, C_{2(3)}\}$. The two $\mathbf{C}_s$ groups, $\mathbf{C}_s = \{I, \sigma_{d(1)}\}$ and $\mathbf{C}_s' = \{I, \sigma_{d(2)}\}$, are conjugate to each other.

It is worthwhile discussing a geometrical meaning of conjugate subgroups. For example, if we choose the S_4 operation for the s operation, we have $S_4 \mathbf{C}_2' S_4^{-1}$ as being conjugate to the $\mathbf{C}_2'$ group. According to Table 2.1, we have $S_4 \mathbf{C}_2' S_4^{-1} = S_4 \mathbf{C}_2' S_4^3 = \mathbf{C}_2''$. Although we do not discuss an exact meaning, we should here point out that the operation by S_4 and S_4^{-1} appearing in this equation represents a coordinate transformation. First, we choose the x, y, and z-axes as found in the left structure of Fig. 2.1. An anti-clockwise rotoreflection of the coordinate system by $\pi/2$ around the z-axis (*i.e.* S_4) affords another system. When we compare the initial state and the ending one, we find a transformation: $\mathbf{C}_2' \rightarrow \mathbf{C}_2''$. Thus, conjugate subgroups are equivalent with respect to an appropriate coordinate transformation.

A subgroup $\mathbf{H}$ ($\leq \mathbf{G}$) is called a *normal subgroup* if $s\mathbf{H}s^{-1} = \mathbf{H}$ for $\forall s \in \mathbf{G}$. In other words, a normal subgroup is conjugate to itself. For example, $\mathbf{C_2} = \{I, C_{2(3)}\}$ is a normal subgroup of the $\mathbf{D}_{2d}$ group.

2.5 Non-Redundant Set of Subgroups for a Point Group

In the preceding section, we have examined the subgroups of the $\mathbf{D}_{2d}$ group. We summarize them briefly:

$$C_1 = \{I\}$$
$$C_2 = \{I, C_{2(3)}\}$$
$$C_2' = \{I, C_{2(1)}\} \qquad C_2'' = \{I, C_{2(2)}\}$$
$$C_s = \{I, \sigma_{d(1)}\} \qquad C_s' = \{I, \sigma_{d(2)}\}$$
$$S_4 = \{I, S_4, C_{2(3)}, S_4^3\}$$
$$C_{2v} = \{I, C_{2(3)}, \sigma_{d(1)}, \sigma_{d(2)}\}$$
$$D_2 = \{I, C_{2(1)}, C_{2(2)}, C_{2(3)}\} \quad \text{and}$$
$$D_{2d} = \{I, C_{2(1)}, C_{2(2)}, C_{2(3)}, \sigma_{d(1)}, \sigma_{d(2)}, S_4, \text{ and } S_4^3\} \ .$$

Since conjugate subgroups are considered to be equivalent, we select a representative from every set of such conjugate subgroups. A set of such representatives is called a *(non-redundant) set of subgroups* (SSG). With respect to the $\mathbf{D}_{2d}$ group, we obtain an SSG:

$$SSG_{\mathbf{D}_{2d}} = \{\mathbf{C}_1, \mathbf{C}_2, \mathbf{C}_2', \mathbf{C}_s, \mathbf{S}_4, \mathbf{C}_{2v}, \mathbf{D}_2, \mathbf{D}_{2d}\}.$$

Figure 2.2 (left) illustrates the group-subgroup relationship of the SSG.

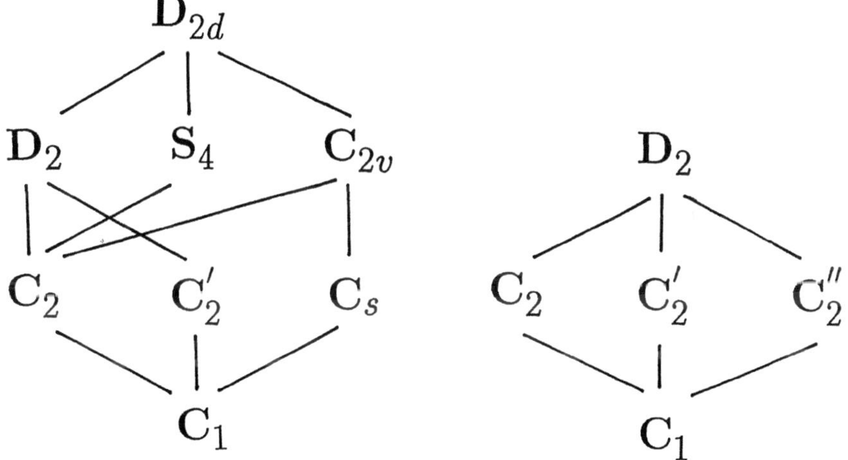

Figure 2.2: Group-subgroup lattices of $\mathbf{D}_{2d}$ and $\mathbf{D}_2$ groups

On the other hand, Fig. 2.2 (right) depicts the group-subgroup lattice of $\mathbf{D}_2$ group. The SSG of the $\mathbf{D}_2$ group is represented by

$$SSG_{\mathbf{D}_2} = \{\mathbf{C}_1, \mathbf{C}_2, \mathbf{C}_2', \mathbf{C}_2'', \mathbf{D}_2\}.$$

It should be noted that the three C_2 groups are not conjugate in the series of the D_2 point group. Let us consider the difference between the conjugate subgroups of the D_{2d} group and those of the D_2 group. The D_{2d} group is considered to be created by adding a $\sigma_{d(1)}$ plane to the D_2 group. The addition of the mirror plane makes the $C_{2(1)}$- and the $C_{2(2)}$-axes equivalent. Thereby, the C_2'- and C_2''-subgroups become conjugate when we consider the D_{2d} group. As a result, either of them is selected as a member of the SSG. However, the $C_{2(3)}$-axis is superimposed to neither of the other two-fold axes. This means that the C_2 is a distinct member of the SSG.

The construction of the D_{2d} group by adding the $\sigma_{d(1)}$ plane to the D_2 group is represented formally as follows.

$$\begin{aligned}
D_{2d} &= D_2 + D_2\sigma_{d(1)} \\
&= \{I, C_{2(1)}, C_{2(2)}, C_{2(3)}\} + \{\sigma_{d(1)}, C_{2(1)}\sigma_{d(1)}, C_{2(2)}\sigma_{d(1)}, C_{2(3)}\sigma_{d(1)}\} \\
&= \{I, C_{2(1)}, C_{2(2)}, C_{2(3)}\} + \{\sigma_{d(1)}, S_4^3, S_4, \sigma_{d(2)}\}
\end{aligned}$$

This formulation is related to a coset decomposition, as described in Chapter 3. The order of the symmetry operations in Table 2.1 stems from the order appearing in the formulation.

For the C_{2v} group, we have

$$SSG_{C_{2v}} = \{C_1, C_2, C_s, C_s', C_{2v}\},$$

where the C_s and C_s' subgroups are no longer conjugate to each other. Since the D_{2d} group is alternatively considered to be generated by adding a dihedral $C_{2(1)}$-axis to the C_{2v} group, *i.e.*,

$$\begin{aligned}
D_{2d} &= C_{2v} + C_{2v}C_{2(1)} \\
&= \{I, C_{2(3)}, \sigma_{d(1)}, \sigma_{d(2)}\} + \{C_{2(1)}, C_{2(3)}C_{2(1)}, \sigma_{d(1)}C_{2(1)}, \sigma_{d(2)}C_{2(1)}\} \\
&= \{I, C_{2(3)}, \sigma_{d(1)}, \sigma_{d(2)}\} + \{C_{2(1)}, C_{2(2)}, S_4^3, S_4\}.
\end{aligned}$$

As a result of the addition of the $C_{2(1)}$-axis, the two mirror planes ($\sigma_{d(1)}$ and $\sigma_{d(2)}$) become equivalent. This means that the C_s and the C_s' are conjugate within the D_{2d} group.

Bibliography

[1] F. A. Cotton, *Chemical Applications of Group Theory*, 2nd Ed., Wiley-Interscience, New York (1971).

[2] G. Burns, *Introduction to Group Theory with Applications*, Academic, New York (1977).

[3] H. H. Jaffé, M. Orchin, *Symmetry in Chemistry*, Wiley, Chichester (1965).

[4] S. F. A. Kettle, *Symmetry and Structure*, Wiley, Chichester (1985).

[5] M. F. C. Ladd, *Symmetry in Molecules and Crystals*, Ellis Horwood, Chichester (1989).

[6] I. Hargittai, M. Hargittai, *Symmetry through the Eyes of a Chemist*, VCH, Weinheim (1986).

Chapter 3

Permutation Groups

3.1 Permutations and Cycles

Consider a set $\Delta = \{ \delta_1, \delta_2, \ldots, \delta_{|\Delta|} \}$. A one-to-one mapping from Δ to Δ is called a *permutation*. The number $|\Delta|$ is called the *degree* of the permutation. Since we here take accout only of such mappings, the elements of Δ may be any objectives. For simplicity's sake, we consider a set of positive integers, $\Delta = \{1, 2, 3, 4\}$.

When the set $\{1, 2, 3, 4\}$ is transformed into $\{2, 4, 1, 3\}$ (*i.e.* $1{\to}2$, $2{\to}4$, $3{\to}1$, and $4{\to}3$), this transformation is expressed by the symbol,

$$\begin{pmatrix} 1 & 2 & 3 & 4 \\ 2 & 4 & 1 & 3 \end{pmatrix} \quad \text{or} \quad (1\ 2\ 4\ 3).$$

The latter expression, which is called a *cycle*, denotes a sequential transformation, $1 \to 2 \to 4 \to 3 \to 1$. A cycle of length r is called an r-cycle. Thus, the cycle $(1\ 2\ 4\ 3)$ is a 4-cycle. In general, a permutation is represented by a product of cycles. For example, we have

$$\begin{pmatrix} 1 & 2 & 3 & 4 \\ 3 & 4 & 1 & 2 \end{pmatrix} = (1\ 3)(2\ 4).$$

The right-hand side is obtained by sequential transformations, $1 \to 3 \to 1$ and $2 \to 4 \to 2$. The multiplication of two permutations is defined as successive application of these permutations. The result of the multiplication is called a product. For example,

$$\underbrace{\begin{pmatrix} 1 & 2 & 3 & 4 \\ 3 & 4 & 1 & 2 \end{pmatrix}}_{2nd} \underbrace{\begin{pmatrix} 1 & 2 & 3 & 4 \\ 2 & 4 & 1 & 3 \end{pmatrix}}_{1st} = \begin{pmatrix} 1 & 2 & 3 & 4 \\ 4 & 2 & 3 & 1 \end{pmatrix},$$

This equation corresponds to the successive permutations, $\{1,2,3,4\} \overset{1st}{\to} \{2,4,1,3\} \overset{2nd}{\to} \{4,2,3,1\}$.

The identity corresponds to the permutation,

$$\begin{pmatrix} 1 & 2 & 3 & 4 \\ 1 & 2 & 3 & 4 \end{pmatrix}.$$

Let the symbol i^p denote the mapping of the original letter i by p.[1] Then, the corresponding permutation is represented by

$$p = \begin{pmatrix} 1 & 2 & 3 & 4 \\ 1^p & 2^p & 3^p & 4^p \end{pmatrix}.$$

The inverse of the permutation (p) is given by the expression,

$$p^{-1} = \begin{pmatrix} 1^p & 2^p & 3^p & 4^p \\ 1 & 2 & 3 & 4 \end{pmatrix}.$$

In the light of this terminology, the product of two permutations,

$$p = \begin{pmatrix} 1 & 2 & 3 & 4 \\ 1^p & 2^p & 3^p & 4^p \end{pmatrix} \text{ and } q = \begin{pmatrix} 1 & 2 & 3 & 4 \\ 1^q & 2^q & 3^q & 4^q \end{pmatrix},$$

is represented by

$$qp = \begin{pmatrix} 1 & 2 & 3 & 4 \\ 1^q & 2^q & 3^q & 4^q \end{pmatrix} \begin{pmatrix} 1 & 2 & 3 & 4 \\ 1^p & 2^p & 3^p & 4^p \end{pmatrix} = \begin{pmatrix} 1 & 2 & 3 & 4 \\ (1^p)^q & (2^p)^q & (3^p)^q & (4^p)^q \end{pmatrix}.$$

Any permutation can be represented by a product of cycles:

$$p = \begin{pmatrix} 1 & 2 & \cdots & n \\ 1^p & 2^p & \cdots & n^p \end{pmatrix} = (j_1\, j_2\, \cdots\, j_r)(k_1\, k_2\, \cdots\, k_s)\cdots \tag{3.1}$$

Suppose that we construct a permutation tpt^{-1} by starting from p and t. The tpt^{-1} permutation is called a *similarity transformation*. Then, we have a useful lemma.

Lemma 3.1 *Suppose that a permutation is represented by eq. 3.1. Let t be a permutation of the same degree. Then,*

$$tpt^{-1} = (j_1^t\, j_2^t\, \cdots\, j_r^t)(k_1^t\, k_2^t\, \cdots\, k_s^t)\cdots.$$

Proof. Since the cycles in the right-hand side of eq. 3.1 are interchangeable, we can presume $r \geq s \geq \cdots (> 0)$ without losing generality. Then, we have $\{j_1^t, j_2^t, \cdots, j_r^t\} \xrightarrow{t^{-1}} \{j_1, j_2, \cdots, j_r\} \xrightarrow{p} \{j_2, j_3, \cdots, j_r, j_1\} \xrightarrow{t} \{j_2^t, j_3^t, \cdots, j_r^t, j_1^t\}$. This transformation corresponds to the cycle $(j_1^t\, j_2^t\, \cdots\, j_r^t)$. A similar transformation is verified for the set of letters contained in each of the other cycles. Hence, we obtain

$$tpt^{-1} = (j_1^t\, j_2^t\, \cdots\, j_r^t)(k_1^t\, k_2^t\, \cdots\, k_s^t)\cdots.$$

Hence, p and tpt^{-1} have products of cycles having the same length.

---

[1] We also use the symbol $p(i)$ to designate the same mapping.

3.2 Permutation Groups

A group whose elements are permutations is called a *permutation group*. Let us consider the following four permutations:

$$p_1 = \begin{pmatrix} 1 & 2 & 3 & 4 \\ 1 & 2 & 3 & 4 \end{pmatrix} = (1)(2)(3)(4) \qquad p_2 = \begin{pmatrix} 1 & 2 & 3 & 4 \\ 2 & 1 & 4 & 3 \end{pmatrix} = (1\ 2)(3\ 4)$$

$$p_3 = \begin{pmatrix} 1 & 2 & 3 & 4 \\ 3 & 4 & 1 & 2 \end{pmatrix} = (1\ 3)(2\ 4) \qquad p_4 = \begin{pmatrix} 1 & 2 & 3 & 4 \\ 4 & 3 & 2 & 1 \end{pmatrix} = (1\ 4)(2\ 3).$$

Since the set of the permutations $\mathbf{P} = \{p_1, p_2, p_3, p_4\}$ gives the following multiplication table (Table 3.1), the set is concluded to be a group. The number of elements (permutations) in a group is called the *order* of the group; thus, the $\mathbf{P}$ group is a permutation group of order 4. Since these elements are permutations concerning

Table 3.1: Multiplication table of P group

		The 1st permutation			
		p_1	p_2	p_3	p_4
	p_1	p_1	p_2	p_3	p_4
The 2nd	p_2	p_2	p_1	p_4	p_3
permutation	p_3	p_3	p_4	p_1	p_2
	p_4	p_4	p_3	p_2	p_1

a domain $\Delta = \{1, 2, 3, 4\}$, this group is a permutation group of degree 4.[2] This permutation group is isomorphic to the $\mathbf{D}_2$ point group. The multiplication table (Table 3.1) is equivalent to that of $\mathbf{D}_2$ point group if we take account of the correspondence: $I \sim p_1$, $C_{2(1)} \sim p_2$, $C_{2(2)} \sim p_3$, and $C_{2(3)} \sim p_4$. Since the two groups ($\mathbf{P}$ and $\mathbf{D}_2$) have the same structure in an abstract fashion, they are referred to as being isomorphic. This relationship is illustrated in the formula (1), which is a

---

[2] The degree of a permutation group should not be confused with its order. The degree is the size ($|\Delta|$) of the domain on which the permutation group acts; on the other hand, the order is the number of elements (permutations) contained in the group. See Chapter 4.

top view of an allene molecule (see Fig. 2.1 in Chapter 2).

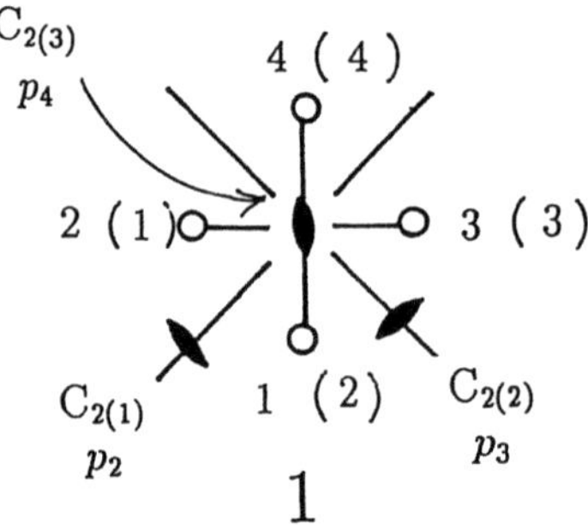

Consider all of the permutations on $\Delta = \{1,2,\ldots,n\}$. As shown later (Section 3.4), they construct a permutation group called a *symmetric group*.[3] The symmetric group of degree n is designated by the symbol $S^{[n]}$.[4] Thereby, any permutation group of degree n is regarded as a subgroup of $S^{[n]}$.

Let us consider a permutation group P of degree n. If we construct tPt^{-1} according to

$$tPt^{-1} = \{tpt^{-1} \mid \forall p \in P, \exists t \in S^{[n]}\},$$

then Lemma 3.1 shows that this transformation affords only interchange of letters contained. Since any p corresponds to tpt^{-1} in one-to-one fashion, it is unnecessary to differentiate between P and tPt^{-1}. In other words, P and tPt^{-1} are equivalent, *i.e.* $P \cong tPt^{-1}$.

For example, we renumber the four positions of 1 as shown in the parentheses, *i.e.*, $1 \to 2$ and $2 \to 1$. This renumbering corresponds to the permutation $t = (1\ 2)(3)(4)$ and yields the following results.

$$tp_1t^{-1} = \begin{pmatrix} 1 & 2 & 3 & 4 \\ 1 & 2 & 3 & 4 \end{pmatrix} = (1)(2)(3)(4) \qquad tp_2t^{-1} = \begin{pmatrix} 1 & 2 & 3 & 4 \\ 2 & 1 & 4 & 3 \end{pmatrix} = (1\ 2)(3\ 4)$$

$$tp_3t^{-1} = \begin{pmatrix} 1 & 2 & 3 & 4 \\ 4 & 3 & 2 & 1 \end{pmatrix} = (1\ 4)(2\ 3) \qquad tp_4t^{-1} = \begin{pmatrix} 1 & 2 & 3 & 4 \\ 3 & 4 & 1 & 2 \end{pmatrix} = (1\ 3)(2\ 4).$$

The resulting permutation group $tPt^{-1} = \{tp_1t^{-1}, tp_2t^{-1}, tp_3t^{-1}, tp_4t^{-1}\}$ is equivalent to P defined above.

3.3 Transitivity and Orbits

Let G be a permutation group acting on the set represented by $\Delta = \{\delta_1, \delta_2, \ldots, \delta_{|\Delta|}\}$. If there exists an element $g \in G$ that satifies $\delta_j = \delta_i^g$ for $\forall \delta_i$ and δ_j, the G

---

[3]Don't confuse a symmetric group with other terms concerning molecular "symmetry".
[4]Don't confuse this symbol with such a symbol as S_n that denotes a point group.

group is called a *transitive permutation group* on Δ; otherwise, **G** is referred to as an *intransitive permutation group.*

When **G** is transitive, the corresponding set (Δ) is referred to as constructing one *orbit*. The group $\mathbf{P} = \{p_1, p_2, p_3, p_4\}$ described in the preceding section is a transitive permutaiton group, because it satisfies this criterion. The set $\Delta = \{1, 2, 3, 4\}$ constructs a single orbit on the action of the **P** permutation group. The members of the orbit are equivalent to each other. This is consistent with the fact that the four positions of the allene skeleton are equivalent.

When **G** is intransitive, the set Δ has two or more orbits. For illustrating an intransitive permutation group, let us examine a subset of the **P** group, *i.e.*, $\mathbf{P}'$ $= \{p_1, p_4\} = \{ \begin{pmatrix} 1 & 2 & 3 & 4 \\ 1 & 2 & 3 & 4 \end{pmatrix}, \begin{pmatrix} 1 & 2 & 3 & 4 \\ 4 & 3 & 2 & 1 \end{pmatrix} \} = \{(1)(2)(3)(4), (1\ 4)(2\ 3)\}$. Obviously, this set constructs a permutation group on $\Delta = \{1, 2, 3, 4\}$. The $\mathbf{P}'$ group is a subgroup of **P**. It is easy to understand that the $\mathbf{P}'$ permutation group is isomorphic to the $\mathbf{C}_2$ point group. If we examine the concrete forms of these permutations carefully, the four integers are partitioned into two sets, *i.e.*, $\{1, 4\}$ and $\{2, 3\}$. Hence, the permutation group ($\mathbf{P}'$) is intransitive; the original set (Δ) is thus divided into the two orbits.

3.4 Symmetric Groups

Let us consider a permutation on $\Delta = \{1, 2, \ldots, n\}$,

$$ p = \begin{pmatrix} 1 & 2 & \ldots & n \\ 1^p & 2^p & \ldots & n^p \end{pmatrix}. $$

The total number of such permutations is $n!$. The set of all of the permutations constructs a group, although the proof is here abbreviated. This group is called a *symmetric group of degree n* and represented by the symbol $\mathbf{S}^{[n]}$. The order of the $\mathbf{S}^{[n]}$ is $n!$. Obviously, the symmetric groups are transitive.

Let us consider the symmetric group of degree 3 ($\mathbf{S}^{[3]}$). The order of $\mathbf{S}^{[3]}$ is $3! = 6$. Thus, we have six permutations as the elements of $\mathbf{S}^{[3]}$:

$$ p_1 = \begin{pmatrix} 1 & 2 & 3 \\ 1 & 2 & 3 \end{pmatrix} = (1)(2)(3), \quad p_2 = \begin{pmatrix} 1 & 2 & 3 \\ 2 & 3 & 1 \end{pmatrix} = (1\ 2\ 3), $$

$$ p_3 = \begin{pmatrix} 1 & 2 & 3 \\ 3 & 1 & 2 \end{pmatrix} = (1\ 3\ 2), \quad p_4 = \begin{pmatrix} 1 & 2 & 3 \\ 1 & 3 & 2 \end{pmatrix} = (1)(2\ 3), $$

$$ p_5 = \begin{pmatrix} 1 & 2 & 3 \\ 2 & 1 & 3 \end{pmatrix} = (1\ 2)(3), \quad p_6 = \begin{pmatrix} 1 & 2 & 3 \\ 3 & 2 & 1 \end{pmatrix} = (1\ 3)(2). $$

These six permutations construct a multiplicaiton table (Table 3.2).

Table 3.2: Multiplication table of $S^{[3]}$ group

		The 1st permutation					
		p_1	p_2	p_3	p_4	p_5	p_6
	p_1	p_1	p_2	p_3	p_4	p_5	p_6
The 2nd	p_2	p_2	p_3	p_1	p_5	p_6	p_4
permutation	p_3	p_3	p_1	p_2	p_6	p_4	p_5
	p_4	p_4	p_6	p_5	p_1	p_3	p_2
	p_5	p_5	p_4	p_6	p_2	p_1	p_3
	p_6	p_6	p_5	p_4	p_3	p_2	p_1

Theorem 3.1 *Let p and q be elements of a symmetric group $S^{[n]}$ group. If the two elements are conjugate to each other, they have products of cycles of the same form. Conversely, if the two elements have products of cycles of the same form, they are conjugate.*

Proof. Any permutation of $S^{[n]}$ group can be represented by a product of cycles:

$$p = \begin{pmatrix} 1 & 2 & \cdots & n \\ 1^p & 2^p & \cdots & n^p \end{pmatrix} = (j_1\ j_2\ \cdots\ j_r)(k_1\ k_2\ \cdots\ k_s)\cdots.$$

Since the cycles in the right-hand side are interchangeable, we can presume $r \geq s \geq \cdots (> 0)$ without losing generality. Let t be any element of $S^{[n]}$. Lemma 3.1 affords

$$tpt^{-1} = (j_1^t\ j_2^t\ \cdots\ j_r^t)(k_1^t\ k_2^t\ \cdots\ k_s^t)\cdots.$$

Therefore, p and tpt^{-1} have the products of cycles of the same form.

Conversely, suppose that the two elements (p and q) have products of cycles of the same form, *i.e.*,

$$p = (j_1\ j_2\ \cdots\ j_r)(k_1\ k_2\ \cdots\ k_s)\cdots$$
$$q = (j_1'\ j_2'\ \cdots\ j_r')(k_1'\ k_2'\ \cdots\ k_s')\cdots.$$

If we select an appropriate permutation,

$$t = \begin{pmatrix} j_1 & j_2 & \cdots & j_r & k_1 & k_2 & \cdots & k_s & \cdots \\ j_1' & j_2' & \cdots & j_r' & k_1' & k_2' & \cdots & k_s' & \cdots \end{pmatrix},$$

we obtain the following transformation: $\{j_1', j_2', \cdots, j_r', \cdots, \} \xrightarrow{t^{-1}} \{j_1, j_2, \cdots, j_r, \cdots,\} \xrightarrow{p} \{j_2, j_3, \cdots, j_r, j_1, \cdots,\} \xrightarrow{t} \{j_2', j_3', \cdots, j_r', j_1', \cdots,\}$. Hence, this transformation affords the expression, $q = tpt^{-1}$.

In order to generalize this theorem, we define cycle structures.

Definition 3.1 *Let p be a permutation on $\Delta = \{1, 2, \ldots, n\}$, as being*

$$p = \begin{pmatrix} 1 & 2 & \ldots & n \\ 1^p & 2^p & \ldots & n^p \end{pmatrix}$$

Suppose that the permutation is divided into the product of ν_n of n-cycles, $\cdots$, ν_2 of 2-cycles, and ν_1 of 1-cycles, where ν_i's ($i = 1$ to n) are non-negative integers. Then, the permutation p is referred to as possessing the cycle structure (ν), which is defined by the formula,

$$(\nu) \equiv (1^{\nu_1} 2^{\nu_2} \cdots n^{\nu_n}),$$

where

$$1\nu_1 + 2\nu_2 + \cdots + n\nu_n = n. \tag{3.2}$$

Such a cycle structure corresponds to the form of a product of cycles in one-to-one fashion. Theorem 3.1 indicates that there is one conjugacy class according to every cycle structure satifying eq. 3.2. Hence, we arrive at the following theorem.

Theorem 3.2 *The total number of the conjugacy classes of $S^{[n]}$ is equal to the number of possible cycle structures.*

There is an alternative way to designate a conjugacy class of a symmetric group.

Definition 3.2 *Let p be a permutation on $\Delta = \{1, 2, \ldots, n\}$, as being*

$$p = \begin{pmatrix} 1 & 2 & \ldots & n \\ 1^p & 2^p & \ldots & n^p \end{pmatrix}.$$

Suppose that the permutation is divided into θ_1-cycle, θ_2-cycle, $\cdots$, and θ_n, where

$$\theta_1 + \theta_2 + \cdots + \theta_n = n \tag{3.3}$$

and

$$\theta_1 \geq \theta_2 \geq \cdots \geq \theta_n \geq 0.$$

Then, the partition $[\theta]$ is defined by the expression,

$$[\theta] = [\theta_1, \theta_2, \cdots, \theta_n].$$

The cycle structure (Def. 3.1) is represented in terms of the partition (Def. 3.2) as being

$$[\theta] = [\overbrace{n, n, \cdots, n}^{\nu_n}, \overbrace{n-1, n-1, \cdots, n-1}^{\nu_{n-1}}, \cdots, \overbrace{1, 1, \cdots, 1}^{\nu_1}, \overbrace{0, 0, \cdots, 0}^{n - \sum_i \nu_i}].$$

$$n \text{ non-negative integers}$$

Theorem 3.3 *Suppose that a conjugacy class of $S^{[n]}$ has a cycle structure represented by*

$$(\nu) = (1^{\nu_1} 2^{\nu_2} \cdots n^{\nu_n}).$$

Then, the number of elements (permutations) contained in the conjugacy class is represented by

$$n_{(\nu)} = \frac{n!}{1^{\nu_1}\nu_1! \, 2^{\nu_2}\nu_2! \cdots n^{\nu_n}\nu_n!}. \tag{3.4}$$

Proof. The number of ways in which n objects are placed on n positions is represented by $n!$. This value contains duplicated counting; ν_i of i-cycles ($i = 1$ to n) should be conunted once. Moreover, each i-cycle ($i = 1$ to n) represents the same cycle if the members are permuted sequentially; hence, i^{ν_i} of i-cycles should be counted once. Therefore, we obtain eq. 3.4.

Example **3.1** For illustrating Theorems 3.2 and 3.3, we examine $S^{[3]}$. In this case, there are three possible cycle structures, *i.e.* (1^3), (3^1) and $(1^2 2^1)$. These cycle structures determine the conjugacy classes of the $S^{[3]}$ group as follows:

$$
\begin{array}{llll}
(1^3) & [1,1,1] & : & (1)(2)(3); & 3!/(1^3 3!) = 1 \\
(3^1) & [3,0,0] & : & (1\ 2\ 3),\,(1\ 3\ 2); & 3!/(3^1 1!) = 2 \\
(1^2 2^1) & [2,1,0] & : & (1)(2\ 3),\,(1\ 2)(3),\,(1\ 3)(2); & 3!/(1^1 1! 2^1 1!) = 3
\end{array}
$$

3.5   Parity

A 2-cycle is called a transposition. Any cycle can be represented by a product of transpositions, *i.e.*,

$$(j_1\ j_2\ \cdots p_r) = (j_1\ j_r)(j_1\ j_{r-1}) \cdots (j_1\ j_3)(j_1\ j_2). \tag{3.5}$$

Since any permutation is represented by a product of cycles, this equation indicates that the permutation is further represented by a product of transpositions. If the permutation is represented by an odd number of transpositions, it is called an *odd permutation*. Otherwise, it is called an *even permutation*. This attribute of the permutation (odd or even) is called *parity*, since this is independent of ways in which the permutation is represented by a product of transpositions,

Equation 3.5 affords the following lemma.

Lemma 3.2 *A cycle $(j_1\ j_2\ \cdots p_r)$ is odd if r is even; and is even if r is odd.*

Since any permutation is a product of cycles, Lemma 3.2 provides the following discriminant.

Theorem 3.4 *When a permutation of* $S^{[n]}$ *has a cycle structure represented by*

$$(\nu) = (1^{\nu_1} 2^{\nu_2} \cdots n^{\nu_n}),$$

the parity of the permutation is determined by

$$D_{(\nu)} = \prod_{i=2}^{n} (-1)^{(i-1)\nu_i} = \begin{cases} +1 & \cdots \quad even \\ -1 & \cdots \quad odd \, . \end{cases} \tag{3.6}$$

This theorem implies that, if p and q ($\in S^{[n]}$) have the same cycle structure (ν), they have the same parity. This can be restated; if they belong to the same conjugacy class, they have the same parity.

3.6 Alternating Groups

Suppose that we select all of the even pemutations from the elements of a symmetry group ($S^{[n]}$). This selection gives an alternating group of degree n ($A^{[n]}$).

Theorem 3.5 *The alternating group of degree n ($A^{[n]}$) is a normal subgroup of the ($S^{[n]}$) group.*

Proof. Let p be any pemutation of $A^{[n]}$. And let t be any permutation of $S^{[n]}$. Then, tpt^{-1} belongs to the same conjugacy class as p. Because two conjugate permutations have the same parity and because p belongs to $A^{[n]}$, tpt^{-1} (for $\forall p \in A^{[n]}$ and $\forall t \in S^{[n]}$) is concluded to be a element of $A^{[n]}$. This means that $A^{[n]}$ is a normal subgroup of $S^{[n]}$.

Theorem 3.6 $|\, S^{[n]} \,| = n!, \quad |\, A^{[n]} \,| = n!/2$

The proof is abbreviated.

Example **3.2** Let us examine the symmetric group $S^{[4]}$, the order of which is calculated to be $4! = 24$. This group is isomorphic to $\mathbf{T}_d$ group as shown in Table 3.3. The alternating group of degree 4 ($A^{[4]}$) is composed of 12 permutations, p_1 to p_{12}. The order of $A^{[4]}$ is calculated to be $4!/2 = 12$.

When we collect permutations having the 1-cycle (4), we obtain a subgroup of $S^{[4]}$, *i.e.*, $\{p_1, p_6, p_{11}, p_{13}, p_{16}, p_{17}\}$. This collection corresponds to the operation that fixes the integer 4. As a result, the subgroup is isomorphic to the symmetric group of degree 3 ($S^{[3]}$).

The subgroup composed of $\{p_1, p_6, p_{11}, p_{13}, p_{16}, p_{17}, p_{19}, p_{20}\}$ is isomorphic to $\mathbf{D}_{2d}$. This correspondence will be discussed in Chapter 5.

Table 3.3: The symmetric group of degree 4

$\mathbf{T}_d$				$\mathbf{S}^{[4]}$
I	$\sim$	p_1	$=$	$(1)(2)(3)(4)$
$C_{2(1)}$	$\sim$	p_2	$=$	$(1\ 2)(3\ 4)$
$C_{2(2)}$	$\sim$	p_3	$=$	$(1\ 3)(2\ 4)$
$C_{2(3)}$	$\sim$	p_4	$=$	$(1\ 4)(2\ 3)$
$C_{3(1)}$	$\sim$	p_5	$=$	$(1)(2\ 4\ 3)$
$C_{3(2)}$	$\sim$	p_6	$=$	$(1\ 2\ 3)(4)$
$C_{3(3)}$	$\sim$	p_7	$=$	$(1\ 3\ 4)(2)$
$C_{3(4)}$	$\sim$	p_8	$=$	$(1\ 4\ 2)(3)$
$C_{3(1)}^2$	$\sim$	p_9	$=$	$(1)(2\ 3\ 4)$
$C_{3(4)}^2$	$\sim$	p_{10}	$=$	$(1\ 2\ 4)(3)$
$C_{3(2)}^2$	$\sim$	p_{11}	$=$	$(1\ 3\ 2)(4)$
$C_{3(3)}^2$	$\sim$	p_{12}	$=$	$(1\ 4\ 3)(2)$
$\sigma_{d(1)}$	$\sim$	p_{13}	$=$	$(1)(2\ 3)(4)$
$\sigma_{d(2)}$	$\sim$	p_{14}	$=$	$(1)(2)(3\ 4)$
$\sigma_{d(3)}$	$\sim$	p_{15}	$=$	$(1)(2\ 4)(3)$
$\sigma_{d(4)}$	$\sim$	p_{16}	$=$	$(1\ 2)(3)(4)$
$\sigma_{d(5)}$	$\sim$	p_{17}	$=$	$(1\ 3)(2)(4)$
$\sigma_{d(6)}$	$\sim$	p_{18}	$=$	$(1\ 4)(2)(3)$
$S_{4(3)}$	$\sim$	p_{19}	$=$	$(1\ 2\ 4\ 3)$
$S_{4(3)}^2$	$\sim$	p_{20}	$=$	$(1\ 3\ 4\ 2)$
$S_{4(1)}$	$\sim$	p_{21}	$=$	$(1\ 3\ 2\ 4)$
$S_{4(1)}^2$	$\sim$	p_{22}	$=$	$(1\ 4\ 2\ 3)$
$S_{4(2)}$	$\sim$	p_{23}	$=$	$(1\ 2\ 3\ 4)$
$S_{4(2)}^2$	$\sim$	p_{24}	$=$	$(1\ 4\ 3\ 2)$

Bibliography

[1] B. Baumslag, B. Chandler, *Theory and Problems of Group Theory*, McGraw-Hill, New York (1968).

[2] M. A. Armstrong, *Groups and Symmetry*, Springer-Verlag, New York Berlin Hidelberg (1988)

[3] T. Ōyama, *Yugen Chikan Gun (Permutation Groups of Finite Order)*, Syōkabō, Tokyo (1981).

[4] T. Inui, Y. Tanabe, Y. Onodera, *Ōyō Gun-Ron (Group Theory and Its Applicaitons in Physics)*, Syōkabō, Tokyo (1976).

Chapter 4

Axioms and Theorems of Group Theory

In Chapters 2 and 3, we have used some concepts of group theory in a practical fashion. In this chapter, we restate the axioms of the group theory and several ralated concepts such as subgroups and cosets in a more logical fashion. Two types of partitions of group elements are discussed to classify them. The one is a partition according to cosets. And the other is a partition based on conjugacy.

4.1 Axioms and Multiplication Tables

A set (G) of elements with a multiplication is defined as a *group*, if the set satisfies the following axioms.

Axiom 4.1 (Axioms of Group Theory)

I. The group is a closed set with respect to the multiplication. This means that, for any two elements a, b of $\mathbf{G}$, there exists a multiplication and that the resulting product ab is also an element of $\mathbf{G}$.

II. The multiplication is associative, that is to say,

$$(ab)c = a(bc) \tag{4.1}$$

for any three elements selected from $\mathbf{G}$.

III. There exists an identity element (I) that satisfies

$$Ia = aI. \tag{4.2}$$

IV. There exists an inverse a^{-1} for any element a selected from $\mathbf{G}$. The inverse also belongs to the set (G) and satisfies

$$a^{-1}a = aa^{-1} = I. \tag{4.3}$$

In the present book, we deal with a group **G** that involves a finite number of elements. The finite number is called the *order* of the group (**G**) and denoted as $|G|$.

Let the symbol C_3 denote a rotation by 120° ($2\pi/3$) around the z-axis. Then, we have rotations by 240° (C_3^2) and by 360° (C_3^3). The latter is equal to an identity operation (I). The resulting set $\mathbf{C}_3 = \{I, C_3, C_3^2\}$ can be easily verified to satisfiy Axiom 4.1. Hence, the $\mathbf{C}_3$ is a group. The multiplication table for $\mathbf{C}_3 = \{I, C_3, C_3^2\}$ is shown in Table 4.1.

Table 4.1: Multiplication table of $\mathbf{C}_3$ group

		The 1st operation		
		I	C_3	C_3^2
The 2nd	I	I	C_3	C_3^2
operation	C_3	C_3	C_3^2	I
	C_3^2	C_3^2	I	C_3

Let us work out a chloroform molecule (Fig. 4.1). The symmetry operations

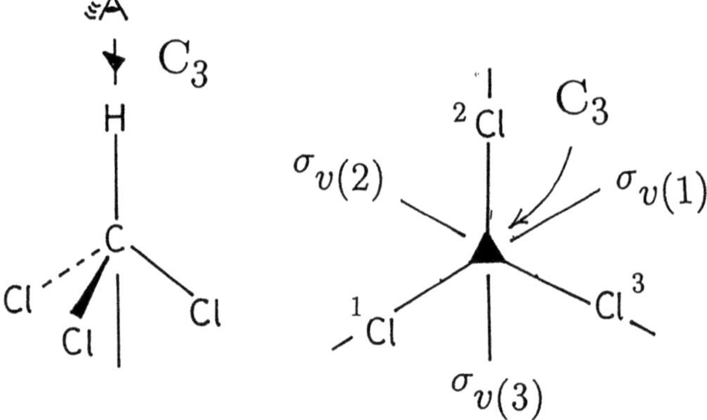

Figure 4.1: Symmetry elements of chloroform

that keep the chloroform molecule invariant are $\mathbf{C}_{3v} = \{I, C_3, C_3^2, \sigma_{v(1)}, \sigma_{v(2)}, \sigma_{v(3)}\}$. The set ($\mathbf{C}_{3v}$) is a group, since it satisfies the axioms of group theory. Table 4.2 is the multiplication table of the $\mathbf{C}_{3v}$ group. It should be noted that Table 4.2 contains several examples in which multiplicaiton of two elements is incommutable. For example, compare $\sigma_{v(2)}\sigma_{v(1)} = C_3$ with $\sigma_{v(1)}\sigma_{v(2)} = C_3^2$.

The $\mathbf{C}_{3v}$ group is isomorphic to the symmetric group of degree 3 ($\mathbf{S}^{[3]}$). This fact is verified geometrically by numbering the three chlorine atoms and by ef-

Table 4.2: Multiplication table of C_{3v} group

		The 1st operation					
		I	C_3	C_3^2	$\sigma_{v(1)}$	$\sigma_{v(2)}$	$\sigma_{v(3)}$
The 2nd operation	I	I	C_3	C_3^2	$\sigma_{v(1)}$	$\sigma_{v(2)}$	$\sigma_{v(3)}$
	C_3	C_3	C_3^2	I	$\sigma_{v(2)}$	$\sigma_{v(3)}$	$\sigma_{v(1)}$
	C_3^2	C_3^2	I	C_3	$\sigma_{v(3)}$	$\sigma_{v(1)}$	$\sigma_{v(2)}$
	$\sigma_{v(1)}$	$\sigma_{v(1)}$	$\sigma_{v(3)}$	$\sigma_{v(2)}$	I	C_3^2	C_3
	$\sigma_{v(2)}$	$\sigma_{v(2)}$	$\sigma_{v(1)}$	$\sigma_{v(3)}$	C_3	I	C_3^2
	$\sigma_{v(3)}$	$\sigma_{v(3)}$	$\sigma_{v(2)}$	$\sigma_{v(1)}$	C_3^2	C_3	I

fecting each symmetry operation. Thereby, we have the following correspondence. Note that $\mathbf{S}^{[3]} = \{p_1, p_2, \ldots, p_6\}$,

$$I \sim p_1 = (1)(2)(3) \qquad C_3 \sim p_2 = (1\ 2\ 3) \qquad C_3^2 \sim p_3 = (1\ 3\ 2)$$
$$\sigma_{v(1)} \sim p_4 = (1)(2\ 3) \qquad \sigma_{v(2)} \sim p_5 = (1\ 2)(3) \qquad \sigma_{v(3)} \sim p_6 = (1\ 3)(2)$$

4.2 Subgroups

A subgroup ($\mathbf{H}$) of a group $\mathbf{G}$ is defined as a subset of $\mathbf{G}$ that froms a group under the multiplication of $\mathbf{G}$. The symbol ($\mathbf{H} \leq \mathbf{G}$) denotes this relationship. Obviously, the identity group $\{I\}$ is a subgroup of any $\mathbf{G}$; and $\mathbf{G}$ itself is always a subgroup of $\mathbf{G}$.

Let us examine Table 4.2 for the C_{3v} group. This table contains a block that is identical with the multiplication table of the C_3 group (Table 4.1). This fact indicates that C_3 is a subgroup of C_{3v}, *i.e.*, $C_3 \leq C_{3v}$. Other subgroups of C_{3v} are $\mathbf{C}_s = \{I, \sigma_{v(1)}\}$, $\mathbf{C}'_s = \{I, \sigma_{v(2)}\}$, and $\mathbf{C}''_s = \{I, \sigma_{v(3)}\}$.

Theorem 4.1 *A subset ($\mathbf{H}$) of a group ($\mathbf{G}$) is a subgroup of $\mathbf{G}$, if and only if ab^{-1} belongs to $\mathbf{H}$ for $\forall a \in \mathbf{H}$ and $\forall b \in \mathbf{H}$.*

Proof. Suppose that $\mathbf{H}$ is a subgroup of $\mathbf{G}$ and that $a \in \mathbf{H}$ and $b \in \mathbf{H}$. Obviously, b^{-1} belongs to $\mathbf{H}$. Because of Axiom 4.1(I), $ab^{-1} \in \mathbf{H}$.

Conversely, suppose that

$$ab^{-1} \in \mathbf{H} \quad \text{for} \quad \forall a, \forall b \in \mathbf{H}. \tag{4.4}$$

This equation holds when b is equal to a. Hence, we have $I = aa^{-1} \in \mathbf{H}$, which indicates the existence of an identity element in $\mathbf{H}$. Since $I \in \mathbf{H}$ and $b \in \mathbf{H}$, eq. 4.4 affords $Ib^{-1} \in \mathbf{H}$. That is to say, $b^{-1} \in \mathbf{H}$. This shows the existence of an inverse element in $\mathbf{H}$. For $a \in \mathbf{H}$ and $b^{-1} \in \mathbf{H}$, eq. 4.4 provides $a(b^{-1})^{-1} \in \mathbf{H}$. That is to say, $ab \in \mathbf{H}$. As a result, the subset ($\mathbf{H}$) is a group, since it satisfies Axioms 4.1(I) to (IV). Therefore, $\mathbf{H}$ is a subgroup of $\mathbf{G}$.

Let us exemplify Theorem 4.1 by using $\mathbf{C}_{3v}$ and the subgroup $\mathbf{C}_3$. From the data of Tables 4.1 and 4.2, we easily obtain $I^{-1} = I$, $C_3^{-1} = C_3^2$, and $C_3^{-2} = C_3$. These inverses belong to the $\mathbf{C}_3$ subgroup. Hence,

$$\mathbf{C}_3 I^{-1} = \mathbf{C}_3 I = \{II, C_3 I, C_3^2 I\} = \{I, C_3, C_3^2\} = \mathbf{C}_3, \tag{4.5}$$
$$\mathbf{C}_3 C_3^{-1} = \mathbf{C}_3 C_3^2 = \{IC_3^2, C_3 C_3^2, C_3^2 C_3^2\} = \{C_3^2, I, C_3\} = \mathbf{C}_3, \tag{4.6}$$
$$\text{and } \mathbf{C}_3 C_3^{-2} = \mathbf{C}_3 C_3 = \{IC_3, C_3 C_3, C_3^2 C_3\} = \{C_3, C_3^2, I\} = \mathbf{C}_3. \tag{4.7}$$

These equations verify Theorem 4.1.

4.3   Cosets

In order to investigate properties of a group, the concept of *coset* is of essential importance. This concept is concerned with a number of useful tools to clarify the structure of a given group.

Definition 4.1 (Coset) *Let $\mathbf{H}$ be a subgroup of a finite group $\mathbf{G}$. We select an element (t) of $\mathbf{G}$ that does not belong to $\mathbf{H}$. Suppose that we collect all products (at) for $\forall a \in \mathbf{H}$. Then the resulting subset of $\mathbf{G}$, i.e.,*

$$\mathbf{H}t = \{at \mid \forall a \in \mathbf{H}\}, \tag{4.8}$$

is called a (right) coset *concerning $\mathbf{H}$.*[1] *The element (t) is called a* representative *of the coset.*

Let us again work out $\mathbf{C}_3$ and $\mathbf{C}_{3v}$ groups. When we select an identity (I) as a representative, the resulting coset is $\mathbf{C}_3$ itself. That is

$$\mathbf{C}_3 I = \{I, C_3, C_3^2\} = \mathbf{C}_3.$$

This equation is identical with eq. 4.5. Equations 4.6 and 4.7 indicate that the same coset is produced, whichever representative $(C_3$ or $C_3^2)$ is selected. Another representative, $\sigma_{v(1)}$, provides a coset represented by

$$\mathbf{C}_3 \sigma_{v(1)} = \{I\sigma_{v(1)}, C_3 \sigma_{v(1)}, C_3^2 \sigma_{v(1)}\} = \{\sigma_{v(1)}, \sigma_{v(2)}, \sigma_{v(3)}\}.$$

Theorem 4.2 *Let $\mathbf{H}$ be a subgroup of $\mathbf{G}$. We select a coset $\mathbf{H}t$.*

1)   If $s \in \mathbf{H}t$, then $st^{-1} \in \mathbf{H}$.

---

[1]In a similar way, we can define a left coset as $t\mathbf{H}$. In general, the right and the left coset are not identical with each other. See Section 4.6. In the present book, we mainly employ right cosets. Hence, we use the term *coset* for designating a right coset, unless otherwise stated.

2) *If $st^{-1} \in$ H, then $s \in$ Ht.*

Proof.

1) Since $s \in$ H, we can select $\exists a \in$ H, which satisfies $s = at$. Hence, $st^{-1} = a \in$ H.

2) Since $st^{-1} \in$ H, we select an element (a) that satisfies $st^{-1} = a \in$ H. Hence, $s = at$. Because of the definition of a coset, $at \in$ Ht. These facts indicate $s \in$ Ht.

Let us examine the coset, $C_3\sigma_{v(1)} = \{\sigma_{v(1)}, \sigma_{v(2)}, \sigma_{v(3)}\}$, wherein H $= C_3$ and $t = \sigma_{v(1)}$. In this case, Theorem 4.2 corresponds to the following equations,

$$(C_3\sigma_{v(1)})\sigma_{v(1)}^{-1} = C_3(\sigma_{v(1)}\sigma_{v(1)}^{-1}) = C_3$$

and

$$(C_3\sigma_{v(1)})\sigma_{v(1)}^{-1} = \{\sigma_{v(1)}\sigma_{v(1)}^{-1}, \sigma_{v(2)}\sigma_{v(1)}^{-1}, \sigma_{v(3)}\sigma_{v(1)}^{-1}\} = \{I, C_3.C_3^2\} = C_3.$$

Theorem 4.3 *Let* H *be a subgroup of* G. *We select a coset* Ht.

1) *If $r \in$ Ht, and $s \in$ Ht, then $rs^{-1} \in$ H.*

2) *If $rs^{-1} \in$ H and $s \in$ Ht, then $r \in$ Ht.*

Proof.

1) We can select $\exists a$ and $\exists b$ from H to satisfy $r = at$ and $s = bt$ because of the presumption. Hence, we have $rs^{-1} = at(bt)^{-1} = att^{-1}b^{-1} = ab^{-1}$. This means that $rs^{-1} \in$ H, because of Theorem 4.1.

2) We can select $\exists a \in$ H, which satisfies $rs^{-1} = a$, because $rs^{-1} \in$ H. This provides $r = as$. From the relation of $s \in$ Ht, we can select $\exists b$ from H that satisfies $s = bt$. Hence, we obtain $r = as = abt$. Since a and b $(\in$ H) satisfy $ab \in$ H, this equation indicates that $r = abt \in$ Ht.

Theorem 4.4 *Let* H *be a subgroup of* G *and* Ht *be a coset. We select an element of s $(\in$ H) other than t and construct another coset,* Hs $= \{bs \mid \forall b \in$ H$\}$. *Then,* Hs *is identical with* Ht.

Proof. We select an element (a) from H that satisfies $s = at$. This provides $bs = bat$ for $\forall b \in$ H. Since a and b are elements of H, the product ba also belongs to H. From the relationship $\{\forall b\} =$ H, we obtain $\{\forall ba\} =$ H. This means that

$$Hs = \{bs \mid \forall b \in H\} = \{bat \mid \forall ba \in H\} = Ht.$$

Hence, $\mathbf{H}s$ is identical with $\mathbf{H}t$.

Theorem 4.4 indicates that any element of a coset may be selected as a representative of the coset. This is exemplified by using $\mathbf{C}_3 \leq \mathbf{C}_{3v}$. The multiplication table (Table 4.2) yields

$$
\begin{aligned}
\mathbf{C}_3\sigma_{v(1)} &= \{I\sigma_{v(1)}, \mathbf{C}_3\sigma_{v(1)}, \mathbf{C}_3^2\sigma_{v(1)}\} = \{\sigma_{v(1)}, \sigma_{v(2)}, \sigma_{v(3)}\}, \\
\mathbf{C}_3\sigma_{v(2)} &= \{I\sigma_{v(2)}, \mathbf{C}_3\sigma_{v(2)}, \mathbf{C}_3^2\sigma_{v(2)}\} = \{\sigma_{v(2)}, \sigma_{v(3)}, \sigma_{v(1)}\}, \\
&\text{and} \\
\mathbf{C}_3\sigma_{v(3)} &= \{I\sigma_{v(3)}, \mathbf{C}_3\sigma_{v(3)}, \mathbf{C}_3^2\sigma_{v(3)}\} = \{\sigma_{v(3)}, \sigma_{v(1)}, \sigma_{v(2)}\}.
\end{aligned}
$$

These equations indicate that the cosets, $\mathbf{C}_3\sigma_{v(1)}$, $\mathbf{C}_3\sigma_{v(2)}$, and $\mathbf{C}_3\sigma_{v(3)}$, are identical with each other.

Theorem 4.5 *Let $\mathbf{H}t_1$ and $\mathbf{H}t_2$ be two cosets of $\mathbf{G}$ concerning a subgroup $\mathbf{H}$. If $\mathbf{H}t_1$ and $\mathbf{H}t_2$ are different from one another, they have no common elements.*

Proof. Assume that there exists a common element(s), *i.e.*, $s \in \mathbf{H}t_1$ and $s \in \mathbf{H}t_2$. Hence, there are appropriate elements, a and b,$(\in \mathbf{H})$ that satisfy $s = at_1$ and bt_2. When we select any $ct_1 \in \mathbf{H}t_1$, we arrive at the relationship, $ct_1 = ca^{-1}at_1 = ca^{-1}s = ca^{-1}bt_2$. Since c, a^{-1}, and b belong to $\mathbf{H}$, we have $ca^{-1}b \in \mathbf{H}$. It follows that $ct_1 = (ca^{-1}b)t_2 \in \mathbf{H}$. This equation holds for all $ct_1 \in \mathbf{H}t_1$. Hence, we obtain $\mathbf{H}t_1 \subseteq \mathbf{H}t_2$. Similarly, we arrive at $\mathbf{H}t_2 \subseteq \mathbf{H}t_1$. These equations result in $\mathbf{H}t_2 = \mathbf{H}t_1$.

Theorem 4.5 affords a foundation to classify the elements of a finite group $\mathbf{G}$ by a subgroup $\mathbf{H}$. First, we select a representative (t_1) from $\mathbf{G}$ and construct a coset $\mathbf{H}t_1$. Next, we choose another representative (t_2) from $\mathbf{G}$ that does not belong to $\mathbf{H}t_1$ and construct a second coset $\mathbf{H}t_2$. When we repeat this process to cover all of the elements of $\mathbf{G}$, we end up with the following partition.[2]

$$
\mathbf{G} = \mathbf{H}t_1 + \mathbf{H}t_2 + \ldots + \mathbf{H}t_{r-1} + \mathbf{H}t_r = \sum_{i=1}^{r} \mathbf{H}t_i, \tag{4.9}
$$

where t_1 is equal to I. This classification is called a *(right) coset decomposition* of $\mathbf{G}$ by $\mathbf{H}$. There exist several ways to select a set of representatives (a *transversal*),

$$
\{t_1, t_2, \ldots, t_{r-1}, t_r\}.
$$

However, Theorem 4.5 assures that the coset decomposition represented by 4.9 is determined uniquely.

Theorem 4.6 (Lagrange) *The order of a subgroup $\mathbf{H}$ of a finite group $\mathbf{G}$ is a divisor of the order of the group.*

---

[2] We use a plus sign $(+)$ in place of a union $(\cup)$ if combined sets are disjoint. We also use $\sum$ for designating a "set summation" $(\bigcup_{i=1}^{r} \mathbf{H}t_i)$ if the sets are disjoint.

Proof. Note that a coset decomposition (eq. 4.9) holds for this case. When the order of $\mathbf{H}$ is denoted as $|\mathbf{H}|$, the order of each coset is equal to $|\mathbf{H}|$. Hence, $|\mathbf{H}|\, r = |\mathbf{G}|$.

Example **4.1** Let us examine the $\mathbf{C}_{3v}$ group. This group contains six symmetry operations, *i.e.*,

$$\mathbf{C}_{3v} = \{I, C_3, C_3^2, \sigma_{v(1)}, \sigma_{v(2)}, \sigma_{v(3)}\}.$$

We select a $\mathbf{C}_3$ group as a subgroup, *i.e.*,

$$\mathbf{C}_3 = \{I, C_3, C_3^2\}.$$

When we choose I and $\sigma_{v(1)}$ as representatives, we obtain

$$\mathbf{C}_{3v}I = \{I, C_3, C_3^2\} \ \text{ and } \ \mathbf{C}_3\sigma_{v(1)} = \{I\sigma_{v(1)}, C_3\sigma_{v(1)}, C_3^2\sigma_{v(3)}\} = \{\sigma_{v(1)}, \sigma_{v(2)}, \sigma_{v(3)}\}$$

by using the multiplication table (Table 4.2). Therefore, the corresponding coset decomposition is represented by

$$\mathbf{C}_{3v} = \mathbf{C}_3 + \mathbf{C}_3\sigma_{v(1)}.$$

4.4 Equivalence Relations

In order to study the properties of a group, we must classify the elements of the group in terms of a distinct criterion. The criterion is based on an equivalence relation, which is defined to manipulate a set in general.

Definition 4.2 (Equivalence relation) *Suppose that a relation represented by the symbol ($\sim$) is defined between arbitrary two elements of a given set $\mathbf{G}$, and that there are only two cases between any two elements (a and b), i.e., $a \sim b$ and $a \not\sim b$. If the relation satifies*

(1) $a \sim a$ (reflective)

(2) $a \sim b$ then $b \sim a$ (symmetric) and

(3) $a \sim b$, $b \sim c$, then $a \sim c$ (transitive),

then this relation is called an equivalence relation. *When $a \sim b$, we say that a is equivalent to b.*

The equivalence relation is one of fundamental concepts in any branches of mathematics. We can construct a subset by collecting equivalent elements of a given set. This subset is named an *equivalence class*.

An equivalence relation concerning a set provides us with a tool to classify the elements of the set into several equivalence classes. The classification is called a *partition* in terms of the equivalence relation.

This section shows that cosets are a kind of such equivalence classes. When we compare Theorems 4.2 to 4.4, we arrive at the idea that the relation $ab^{-1} \in \mathbf{H}$ can be used as an equivalence relation (Fig. 4.2).

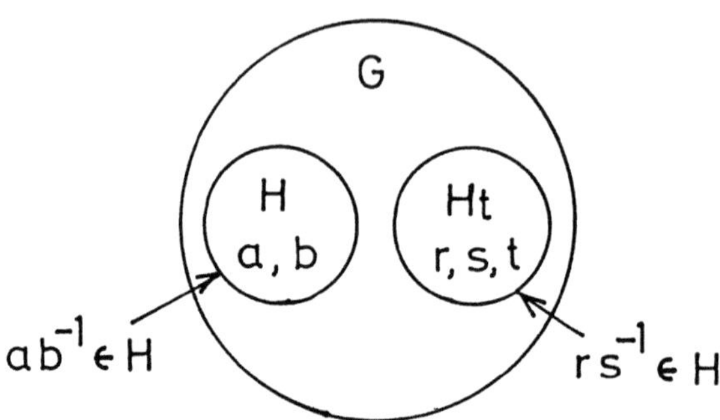

Figure 4.2: Cosets as equivalence classes

Theorem 4.7 (Cosets as equivalence classes) *Let* $\mathbf{H}$ *be a subgroup of a group* $\mathbf{G}$ *of finite order. If the relation* $(a \sim b)$ *is defined between a pair of elements* a, b *($\in \mathbf{G}$) to mean that* $ab^{-1} \in \mathbf{H}$, *then* $a \sim b$ *is an equivalence relation on* $\mathbf{G}$.

Proof. The relation satisfies Definition 4.2 as follows.

1) $aa^{-1} = I \in \mathbf{H}$. Hence, $a \sim a$.

2) Suppose $a \sim b$; *i.e.*, $ab^{-1} \in \mathbf{H}$. Then $\mathbf{H} \ni (ab^{-1})^{-1} = ba^{-1}$. Hence, $b \sim a$.

3) Suppose $a \sim b$ and $b \sim c$; that is, $ab^{-1} \in \mathbf{H}$ and $bc^{-1} \in \mathbf{H}$. Hence,

$$ac^{-1} = a(b^{-1}b)c^{-1} = (ab^{-1})(bc^{-1}) \in \mathbf{H}.$$

This means that $a \sim c$.

Theorem 4.7 indicates that a (right) coset by $\mathbf{H}$ ($\leq \mathbf{G}$) is an equivalence class. Hence, the coset decomposition represented by eq. 4.9 is a partition of the group ($\mathbf{G}$).

4.5 Conjugacy Classes

A conjugate element of an element a ($\in$ **G**) is defined as an element (sas^{-1}), where $s \in$ **G**. Conversely, the a is a conjugate element of the sas^{-1}. The following theorem can be easily proved.

Theorem 4.8 *Consider two elements, sas^{-1} and tat^{-1}, which are conjugate to an element (a). Then, sas^{-1} and tat^{-1} are conjugated to each other.*

Theorem 4.9 *We consider a relation $a \sim b$ to be $b = sas^{-1}$ for an appropriate $s \in$ **G**. This relation is an equivalence relation.*

Proof. The relation satisfies Definition 4.2 as follows.

1) $a = IaI^{-1}$ and $I \in$ **G**. Hence, we have $a \sim a$.

2) Suppose $b = sas^{-1}$. Then, $(s^{-1})b(s^{-1})^{-1} = s^{-1}(sas^{-1})(s^{-1})^{-1} = a$. Since s^{-1} is an element of **G**, we have $b \sim a$.

3) Suppose $b = sas^{-1}$ and $c = sbs^{-1}$. Then, $c = s(sas^{-1})s^{-1} = (ss)a(ss)^{-1}$. Since ss is an element of **G**, this means that $a \sim c$.

Theorems 4.8 and 4.9 afford a foundation to classify the elements of the group $(\mathbf{G})$.

Definition 4.3 (Conjugacy classes) *A conjugacy class (C) is defined as*

$$\mathsf{C} = \{sas^{-1} \mid \forall s \in \mathbf{G}\}$$

*for a given element (a) of **G**.*

Of course, the element (a) itself belongs to C. Then, the group $(\mathbf{G})$ is partitioned into several conjugacy classes. Such conjugacy classes are regarded as equivalence classes in terms of Theorem 4.9.

Example 4.2 Let us again examine the $\mathbf{C}_{3v}$ group. The multiplication table (Table 4.2) affords the following results.

$$III = I, C_3 I C_3^{-1} = I, C_3^2 I C_3^{-2} = I,$$

$$\sigma_{v(1)} I \sigma_{v(1)}^{-1} = I, \sigma_{v(2)} I \sigma_{v(2)}^{-1} = I, \sigma_{v(3)} I \sigma_{v(3)}^{-1} = I;$$

$$IC_3 I = C_3, C_3 C_3 C_3^{-1} = C_3, C_3^2 C_3 C_3^{-2} = C_3,$$

$$\sigma_{v(1)} C_3 \sigma_{v(1)}^{-1} = C_3^2, \sigma_{v(2)} C_3 \sigma_{v(2)}^{-1} = C_3^2, \sigma_{v(3)} C_3 \sigma_{v(3)}^{-1} = C_3^2;$$

$$IC_3^2 I = C_3^2, C_3 C_3^2 C_3^{-1} = C_3^2, C_3^2 C_3^2 C_3^{-2} = C_3^2,$$

$$\sigma_{v(1)}C_3^2\sigma_{v(1)}^{-1} = C_3, \sigma_{v(2)}C_3^2\sigma_{v(2)}^{-1} = C_3, \sigma_{v(3)}C_3^2\sigma_{v(3)}^{-1} = C_3;$$

$$I\sigma_{v(1)}I = \sigma_{v(1)}, C_3\sigma_{v(1)}C_3^{-1} = \sigma_{v(3)}, C_3^2\sigma_{v(1)}C_3^{-2} = \sigma_{v(2)},$$

$$\sigma_{v(1)}\sigma_{v(1)}\sigma_{v(1)}^{-1} = \sigma_{v(1)}, \sigma_{v(2)}\sigma_{v(1)}\sigma_{v(2)}^{-1} = \sigma_{v(3)}, \sigma_{v(3)}\sigma_{v(1)}\sigma_{v(3)}^{-1} = \sigma_{v(2)};$$

$$I\sigma_{v(2)}I = \sigma_{v(2)}, C_3\sigma_{v(2)}C_3^{-1} = \sigma_{v(1)}, C_3^2\sigma_{v(2)}C_3^{-2} = \sigma_{v(3)},$$

$$\sigma_{v(1)}\sigma_{v(2)}\sigma_{v(1)}^{-1} = \sigma_{v(3)}, \sigma_{v(2)}\sigma_{v(2)}\sigma_{v(2)}^{-1} = \sigma_{v(2)}, \sigma_{v(3)}\sigma_{v(2)}\sigma_{v(3)}^{-1} = \sigma_{v(1)};$$

$$I\sigma_{v(3)}I = \sigma_{v(3)}, C_3\sigma_{v(3)}C_3^{-1} = \sigma_{v(2)}, C_3^2\sigma_{v(3)}C_3^{-2} = \sigma_{v(1)},$$

$$\sigma_{v(1)}\sigma_{v(3)}\sigma_{v(1)}^{-1} = \sigma_{v(2)}, \sigma_{v(2)}\sigma_{v(3)}\sigma_{v(2)}^{-1} = \sigma_{v(1)}, \sigma_{v(3)}\sigma_{v(3)}\sigma_{v(3)}^{-1} = \sigma_{v(3)}.$$

Therefore, the $\mathbf{C}_{3v}$ is partitioned into three conjugacy classes, *i.e.*, $C_1 = \{I\}$, $C_2 = \{C_3, C_3^2\}$, and $C_3 = \{\sigma_{v(1)}, \sigma_{v(2)}, \sigma_{v(3)}\}$.

It is worthwhile discussing a geometrical meaning of a conjugacy class. For example, the expression, $C_3\sigma_{v(1)}C_3^{-1} = \sigma_{v(3)}$, is explained by Fig. 4.3. Although we do not discuss the exact meaning, we should here point out that the operation by C_3 and C_3^{-1} appearing in the left-hand side represents a coordinate transformation. First, we choose the x, y, and z-axes as found in (A). An anti-clockwise rotation of the coordinate system by $2\pi/3$ around the z-axis affords another system shown by (B). When we compare A and B, we find that $\sigma_{v(1)} \longrightarrow \sigma_{v(3)}$, $\sigma_{v(2)} \longrightarrow \sigma_{v(1)}$, and $\sigma_{v(3)} \longrightarrow \sigma_{v(2)}$. These results correspond to the equations, *i.e.*,

$$C_3\sigma_{v(1)}C_3^{-1} = \sigma_{v(3)}, C_3\sigma_{v(2)}C_3^{-1} = \sigma_{v(1)}, \text{ and } C_3\sigma_{v(3)}C_3^{-1} = \sigma_{v(2)}.$$

Thus, a conjugacy class is a set of symmetry operations that are equivalent with respect to coordinate transformations. The results obtained here afford a convention to find a conjugacy class geometrically. For example, the $\sigma_{v(1)}$ plane is moved by succesive C_3 operations as follows;

$$\sigma_{v(1)} \xrightarrow{C_3} \sigma_{v(3)} \xrightarrow{C_3} \sigma_{v(2)} \xrightarrow{C_3} \sigma_{v(1)}.$$

Thus, the three reflection planes belong to the same conjugacy class.

4.6   Conjugate and Normal Subgroups

Definition 4.4 (Conjugate subgroups) *A conjugate subgroup of* $\mathbf{H}$ *(*$\leq \mathbf{G}$*) is defined as a set[3] represented by*

$$s\mathbf{H}s^{-1} = \{sas^{-1} \,|\, \forall a \in \mathbf{H}\},$$

where s is a given element of $\mathbf{G}$.

---

[3]This set can be easily proved to be a group (the proof will be an exercise for readers).

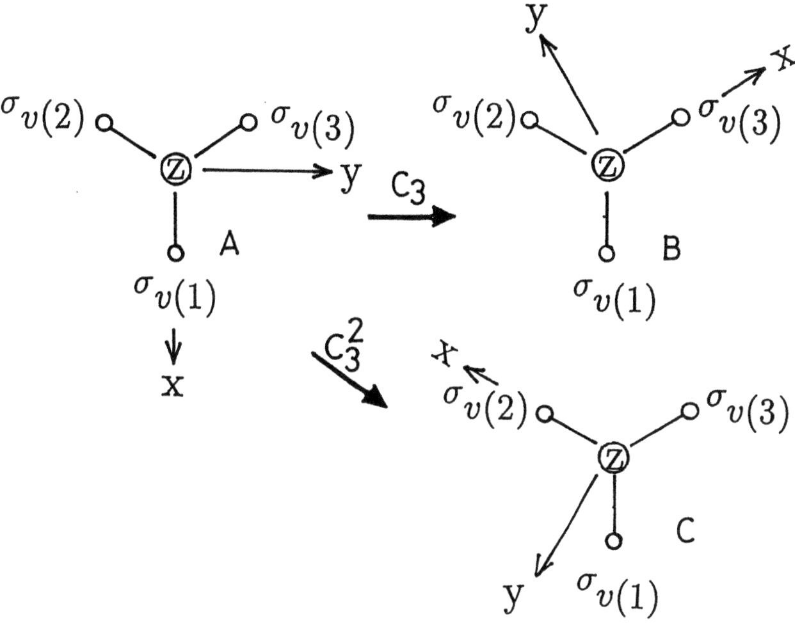

Figure 4.3: A geometrical meaning of a conjugacy class

Example **4.3** Consider a subgroup, $\mathbf{C}_s = \{I, \sigma_{v(1)}\}$, extracted from the $\mathbf{C}_{3v}$ group. We then calculate the corresponding conjugate subgroups, *i.e.*,

$$C_3\mathbf{C}_s C_3^{-1} = \{I, \sigma_{v(3)}\}, C_3^2\mathbf{C}_s C_3^{-2} = \{I, \sigma_{v(2)}\}, \sigma_{v(1)}\mathbf{C}_s\sigma_{v(1)}^{-1} = \{I, \sigma_{v(1)}\},$$

$$\sigma_{v(2)}\mathbf{C}_s\sigma_{v(2)}^{-1} = \{I, \sigma_{v(3)}\}, \quad \text{and} \quad \sigma_{v(1)}\mathbf{C}_s\sigma_{v(3)}^{-1} = \{I, \sigma_{v(2)}\},$$

by means of the multiplication table (Table 4.2). These equations indicate that the resulting three subgroups, $\mathbf{C}_s = \{I, \sigma_{v(1)}\}$, $\mathbf{C}'_s = \{I, \sigma_{v(2)}\}$, and $\mathbf{C}''_s = \{I, \sigma_{v(3)}\}$, are conjugate to each other.

Definition 4.5 (Normal subgroups) *A subgroup* **H** *($\leq$ **G**) is called a* normal *subgroup, if $s\mathbf{H}s^{-1} =$**H** for $\forall s \in$ **G***. *In other words, a normal subgroup is conjugate to itself.*

For exmaple, $\mathbf{C}_3 = \{I, C_3, C_3^2\}$ is a normal subgroup of the $\mathbf{C}_{3v}$ group. This definition results in $s\mathbf{H} = \mathbf{H}s$; that is to say, the right coset by **H** is identical with the left coset.

4.7 Subgroup Lattices

A finite group **G** has a set of subgroups. Suppose we consider a representative for every set of conjugate subgroups. Thereby, we are able to obtain a non-redundant set of such representatives, *i.e.*,

$$SSG = \{\mathbf{G}_1, \mathbf{G}_2, \ldots, \mathbf{G}_s\},$$

where $\mathbf{G}_1$ is an identity group, and $\mathbf{G}_s$ is equal to **G**. For example, the $\mathbf{C}_{3v}$ group yields a non-redundant set of subgroups, $\{\mathbf{C}_1, \mathbf{C}_s, \mathbf{C}_3, \mathbf{C}_{3v}\}$, since a representative is selected from the three conjugate $\mathbf{C}_s$ subgroups (Example 4.3).

If we recognize group-subgroup relationships among the members of such a non-redundant set, we can construct a subgroup lattice for the group **G**. In the case of the $\mathbf{C}_{3v}$ group, finding the corresponding subgroup lattice is a rather easy task. When we manipulate a more complex group, some careful examination is necessary. For example, the subgroup lattice of $\mathbf{T}_d$ group is shown in Fig. 4.4. Such subgroup lattices reveal the structures of respective groups, which will provide us with many versatile concepts to understand molecular symmetry.

4.8 Cyclic Groups

Section 4.7 describes a subgroup lattice of a given group **G**. The determination of all the subgroups is a difficult problem generally. However, there are groups whose subgroups are predictable, *e.g.*, cyclic groups.

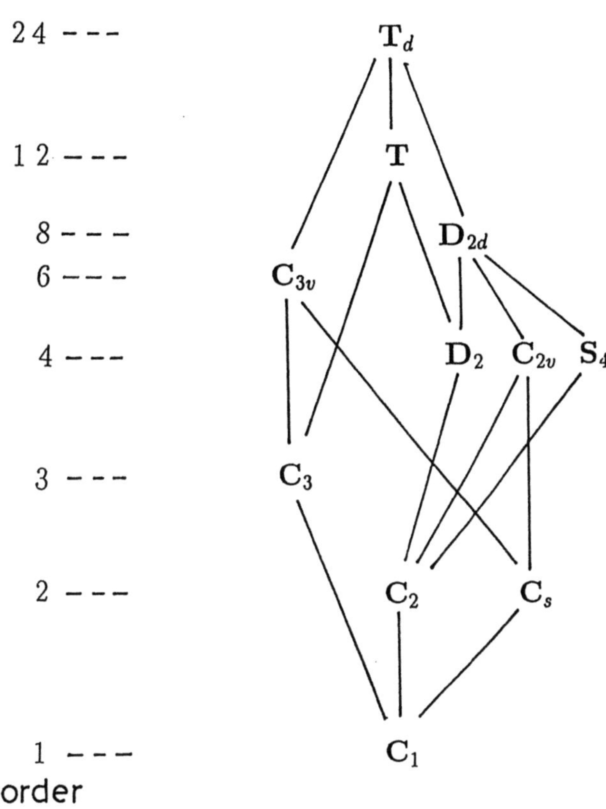

Figure 4.4: The subgroup lattice of $\mathbf{T}_d$ group

Definition 4.6 (Cyclic Groups) *A cyclic group is defined as a group ($\langle a \rangle$) generated from one (a) of its elements. The element (a) is called a generator of the cyclic group.*

If n is a minimum integer that satisfies $a^n = I$, a set represented by $\{I, a, a^2, \ldots, a^{n-1}\}$ is a cyclic group. Since all elements are commutable, cyclic groups are a kind of abelian group. For example, $\mathbf{C}_n$ and $\mathbf{S}_n$ groups are cyclic groups, in which C_n and S_n are respective generators.

When n is an arbitrary positive integer, there exists a cyclic group of the order n. Suppose that ζ is the nth root of 1, *i.e.*, $\zeta = \exp(2\pi i/n) = \cos(2\pi/n) + i\sin(2\pi/n)$. Since $\zeta^n = 1$, we can construct a group, $\{1, \zeta, \ldots, \zeta^{n-1}\}$.

We use the following theorem without proving it.

Theorem 4.10 *Every subgroup of a cyclic group is also cyclic.*

Let $\langle a \rangle$ be a cyclic group of order n, *i.e.*,

$$\langle a \rangle = \{a, a^2, \ldots, a^n (= I)\}.$$

If we select an integer r $(1 < r < n)$ which is not prime to n, we can represent $r = r'q$ and $n = n'q$, where q is the greatest common divisor of r and n. These relations easily provide us with an expression:

$$I = (a^n)^{r'} = (a^{n'q})^{r'} = (a^{r'q})^{n'} = (a^r)^{n'}.$$

This equation indicates that a cyclic group generated by a^r is

$$\langle a^r \rangle = \{a^r, a^{2r}, \ldots, a^{n'r}(= I)\},$$

the order of which is n'.[4] Because n' is a divisor of n, the cyclic group ($\langle a^r \rangle$) is a subgroup of $\langle a \rangle$.

On the other hand, we consider the case in which r and n is coprime. Since $I = (a^n)^r = (a^r)^n$, we obtain a cyclic group,

$$\langle a^r \rangle = \{a^r, a^{2r}, \ldots, a^{nr}\}.$$

Suppose that an integer m $(1 \leq m \leq n)$ satisfies $I = (a^r)^m = a^{mr}$. Since $a^n = I$, mr is exactly divisible by n. Because n and r are coprime, m is divisible by n without remainder. This concludes $m = n$. As a result, $\langle a^r \rangle$ is identical with $\langle a \rangle$.

Thus, there exist one or more generators producing a cyclic group $\langle a \rangle$. The generators are obtained by the following theorem.

Theorem 4.11 *The generators of a cyclic group ($\langle a \rangle$) of order n are a^r ($1 \leq r \leq n - 1$) in all, wherein the integer r is prime to n.*

The number of such generators is equal to Euler's function ($\varphi(n)$), which determines the number of positive integers ($\leq n$) that are coprime to n.[4]

---

[4] For the proof, suppose that m $(1 \leq m \leq n')$ satisfies $I = (a^r)^m = a^{mr} = a^{mr'q}$. Since $a^{n'q} = I$, $mr'q$ is exactly divisible by $n'q$. Because n' and r' are coprime, m is divisible by n' without remainder. This concludes $m = n'$.

Bibliography

[1] K. Asano, H. Nagao, *Gun-Ron (Group Theory)*, (Iwanami Zensyo, Vol. 261), Iwanami, Tokyo (1965).

[2] M. Ōshima, *Gun-Ron (Group Theory)*, (Kyōritsu Zensyo, Vol. 88), Kyōritsu, Tokyo (1954).

[3] I. D. Macdonald, *The Theory of Groups*, Oxford Univ. Press, Oxford (1968).

[4] T. Takagi, *Shotō Seisū-Ron Kōgi (Introductory Lecture on Number Theory)*, 2nd Ed., pp. 41–50, Kyōritsu, Tokyo (1971).

Chapter 5

Coset Representations and Orbits

5.1 Coset Representations

Let $\mathbf{H}$ be a subgroup of a group $\mathbf{G}$ of finite order. Then we have a (right) coset decomposition:

$$\mathbf{G} = \mathbf{H}g_1 + \mathbf{H}g_2 + \cdots + \mathbf{H}g_r, \tag{5.1}$$

where $g_1 = I$ (identity). Let us consider a set of the cosets,

$$\mathbf{G/H} = \{\mathbf{H}g_1, \mathbf{H}g_2, \ldots, \mathbf{H}g_r\}. \tag{5.2}$$

For $\forall g \in \mathbf{G}$, we construct a new set,

$$\{\mathbf{H}g_1g, \mathbf{H}g_2g, \ldots, \mathbf{H}g_rg\}. \tag{5.3}$$

This conversion can be regarded as a permutation represented by

$$\pi_g = \begin{pmatrix} \mathbf{H}g_1 & \mathbf{H}g_2 & \ldots & \mathbf{H}g_r \\ \mathbf{H}g_1g & \mathbf{H}g_2g & \ldots & \mathbf{H}g_rg \end{pmatrix}. \tag{5.4}$$

Since this permutation corresponds to g, we have a representation of $\mathbf{G}$ in terms of

$$\mathbf{G}(/\mathbf{H}) = \{\pi_g \mid \forall g \in \mathbf{G}\}, \tag{5.5}$$

which is called a *coset representation (CR)* of $\mathbf{G}$ by $\mathbf{H}$. When $\mathbf{H}$ is an identity group, the corresponding coset representation is specifically called a *regular representation*.

The following theorem is used without a proof.

Theorem 5.1 *A coset representation is transitive.*

Such a coset representation is originally concerned with a set of cosets. If we number the cosets sequentially, the coset representation can be regarded as a permutation representation acting on a set of positive integers. Hereafter, we do not distinguish between them so long as there emerge no confusions. For example, we examine $\mathbf{D}_{2d}$ group $(\{I, C_{2(1)}, C_{2(2)}, C_{2(3)}, \sigma_{d(1)}, S_4^3, S_4, \sigma_{d(2)}\})$.[1] Since the $\mathbf{D}_{2d}$ group has $\mathbf{C}_s = \{I, \sigma_{d(1)}\}$ as a subgroup, we obtain a coset decomposition:

$$\mathbf{D}_{2d} = \mathbf{C}_s + \mathbf{C}_s C_{2(1)} + \mathbf{C}_s C_{2(2)} + \mathbf{C}_s C_{2(3)}, \tag{5.6}$$

where $\mathbf{C}_s = \{I, \sigma_{d(1)}\}$, $\mathbf{C}_s C_{2(1)} = \{C_{2(1)}, S_4^3\}$, $\mathbf{C}_s C_{2(2)} = \{C_{2(2)}, S_4\}$, and $\mathbf{C}_s C_{2(3)} = \{C_{2(3)}, \sigma_{d(2)}\}$. We number the cosets ($\mathbf{C}_s$, $\mathbf{C}_s C_{2(1)}$, $\mathbf{C}_s C_{2(2)}$, and $\mathbf{C}_s C_{2(3)}$) sequentially to afford the following list.

$\mathbf{D}_{2d}$	$=$	$\mathbf{C}_s$ $+$	$\mathbf{C}_s C_{2(1)}$ $+$	$\mathbf{C}_s C_{2(2)}$ $+$	$\mathbf{C}_s C_{2(3)}$	$\mathbf{D}_{2d}(/\mathbf{C}_s)$
element		1	2	3	4	
I	$\sim$	1	2	3	4	$(1)(2)(3)(4)$
$C_{2(1)}$	$\sim$	2	1	4	3	$(1\ 2)(3\ 4)$
$C_{2(2)}$	$\sim$	3	4	1	2	$(1\ 3)(2\ 4)$
$C_{2(3)}$	$\sim$	4	3	2	1	$(1\ 4)(2\ 3)$
$\sigma_{d(1)}$	$\sim$	1	3	2	4	$(1)(4)(2\ 3)$
S_4	$\sim$	2	4	1	3	$(1\ 2\ 4\ 3)$
S_4^3	$\sim$	3	1	4	2	$(1\ 3\ 4\ 2)$
$\sigma_{d(2)}$	$\sim$	4	2	3	1	$(1\ 4)(2)(3)$

In this list, for example, the set of numbers (2 1 4 3) in the $C_{2(1)}$ row means that

$$
\begin{aligned}
C_{2(1)} &\sim \begin{pmatrix} \mathbf{C}_s & \mathbf{C}_s C_{2(1)} & \mathbf{C}_s C_{2(2)} & \mathbf{C}_s C_{2(3)} \\ \mathbf{C}_s C_{2(1)} & \mathbf{C}_s C_{2(1)} C_{2(1)} & \mathbf{C}_s C_{2(2)} C_{2(1)} & \mathbf{C}_s C_{2(3)} C_{2(1)} \end{pmatrix} \\
&= \begin{pmatrix} \mathbf{C}_s & \mathbf{C}_s C_{2(1)} & \mathbf{C}_s C_{2(2)} & \mathbf{C}_s C_{2(3)} \\ \mathbf{C}_s C_{2(1)} & \mathbf{C}_s & \mathbf{C}_s C_{2(3)} & \mathbf{C}_s C_{2(2)} \end{pmatrix} \\
&= \begin{pmatrix} 1 & 2 & 3 & 4 \\ 2 & 1 & 4 & 3 \end{pmatrix} = (1\ 2)(3\ 4).
\end{aligned}
$$

The coset representation $\mathbf{D}_{2d}(/\mathbf{C}_s)$ is collected in the rightmost column. Since each element of $\mathbf{D}_{2d}$ corresponds to each distinct permutation of the $\mathbf{D}_{2d}(/\mathbf{C}_s)$ representation, this coset representation is a *faithful* representation.

Consider the permutation group: $\mathbf{P} = \{(1)(2)(3)(4), (1\ 2)(3\ 4), (1\ 3)(2\ 4), (1\ 4)(2\ 3), (1)(4)(2\ 3), (1\ 2\ 4\ 3), (1\ 3\ 4\ 2), (1\ 4)(2)(3)\}$. Then, $\mathbf{D}_{2d}$ and $\mathbf{P}$ are isomorphic. The CR $\mathbf{D}_{2d}(/\mathbf{C}_s)$ is regarded as a representation associated with the isomorphic mapping of $\mathbf{D}_{2d}$ onto $\mathbf{P}$.

Since the $\mathbf{D}_{2d}$ group has $\mathbf{D}_2$ as a subgroup, we obtain another coset decomposition:

$$\mathbf{D}_{2d} = \mathbf{D}_2 + \mathbf{D}_2 \sigma_{d(1)}, \tag{5.7}$$

in which the two cosets are represented by $\mathbf{D}_2 = \{I, C_{2(1)}, C_{2(2)}, C_{2(3)}\}$ and $\mathbf{D}_2 \sigma_{d(1)} = \{\sigma_{d(1)}, S_4^3, S_4, \sigma_{d(2)}\}$. If we number the cosets sequetially, we have the coset representation $(\mathbf{D}_{2d}(/\mathbf{D}_2))$ as follows.

$$I, C_{2(1)}, C_{2(2)}, C_{2(3)} \quad \sim \quad \begin{pmatrix} \mathbf{D}_2 & \mathbf{D}_2\sigma_{d(1)} \\ \mathbf{D}_2 & \mathbf{D}_2\sigma_{d(1)} \end{pmatrix} \quad \sim \quad \begin{pmatrix} 1 & 2 \\ 1 & 2 \end{pmatrix} = (1)(2)$$

$$\sigma_{d(1)}, S_4, S_4^3, \sigma_{d(2)} \quad \sim \quad \begin{pmatrix} \mathbf{D}_2 & \mathbf{D}_2\sigma_{d(1)} \\ \mathbf{D}_2\sigma_{d(1)} & \mathbf{D}_2 \end{pmatrix} \quad \sim \quad \begin{pmatrix} 1 & 2 \\ 2 & 1 \end{pmatrix} = (1\ 2).$$

This representation is not faithful.

Consider the permutation group, $\mathbf{P} = \{(1)(2), (1\ 2)\}$. Then $\mathbf{D}_{2d}$ is homomorphic onto $\mathbf{P}$, since the four rotations of the $\mathbf{D}_{2d}$ correspond to the element $(1)(2)$ of $\mathbf{P}$; the four rotoreflections correspond to the $(1\ 2)$ element of $\mathbf{P}$. Thus, the CR $\mathbf{D}_{2d}(/\mathbf{D})$ is a representation associated with the homomorphic mapping of the $\mathbf{D}_{2d}$ group onto the $\mathbf{P}$ group.

Let us next examine the $\mathbf{C}_{3v}$ group $(\{I, C_3, C_3^2, \sigma_{v(1)}, \sigma_{v(2)}, \sigma_{v(3)}\})$. Since the $\mathbf{C}_{3v}$ group has $\mathbf{C}_s = \{I, \sigma_{v(1)}\}$ as a subgroup, we obtain a coset decomposition:

$\mathbf{C}_{3v}$	=	$\mathbf{C}_s$ +	$\mathbf{C}_sC_3$ +	$\mathbf{C}_sC_3^2$	$\mathbf{C}_{3v}(/\mathbf{C}_s)$
element		1	2	3	
I	$\sim$	1	2	3	$(1)(2)(3)$
C_3	$\sim$	2	3	1	$(1\ 2\ 3)$
C_3^2	$\sim$	3	1	2	$(1\ 3\ 2)$
$\sigma_{v(1)}$	$\sim$	1	3	2	$(1)(2\ 3)$
$\sigma_{v(2)}$	$\sim$	2	1	3	$(1\ 2)(3)$
$\sigma_{v(3)}$	$\sim$	3	2	1	$(1\ 3)(2)$

The resulting representation is identical with a symmetric group of degree 3 ($\mathbf{S}^{[3]}$) discussed in Chapter 3.

The $\mathbf{C}_s$ subgroup of the $\mathbf{C}_{3v}$ group has two conjugate subgroups, *i.e.*,

$$(C_3^2)^{-1}\mathbf{C}_sC_3^2 = \{C_3IC_3^2, C_3\sigma_{v(1)}C_3^2\} = \{I, \sigma_{v(2)}\} = \mathbf{C}_s'.$$
$$C_3^{-1}\mathbf{C}_sC_3 = \{C_3^2IC_3, C_3^2\sigma_{v(1)}C_3\} = \{I, \sigma_{v(3)}\} = \mathbf{C}_s''.$$

When we repeat the above procedure, we obtain the corresponding coset representations (CRs), $\mathbf{C}_{3v}(/\mathbf{C}_s')$ and $\mathbf{C}_{3v}(/\mathbf{C}_s'')$, which are equivalent to the $\mathbf{C}_{3v}(/\mathbf{C}_s)$. This fact is proved in general.

Theorem 5.2 *Let* $\mathbf{G}$ *be a finite group and* $\mathbf{H}$ *be its subgroup. The corresponding coset representation* $\mathbf{G}(/\mathbf{H})$ *is equivalent to* $\mathbf{G}(/t\mathbf{H}t^{-1})$ *that is produced by a conjugate subgroup* $t\mathbf{H}t^{-1}$ *(*$t \in \mathbf{G}$*).*

Proof. Consider a set of cosets, $\{t\mathbf{H}t^{-1}t_1, t\mathbf{H}t^{-1}t_2, \cdots, t\mathbf{H}t^{-1}t_r\}$. Then, we have a permutation π_g corresponding to $\forall g \in \mathbf{G}$.

$$\pi_g = \begin{pmatrix} t\mathbf{H}t^{-1}t_1 & t\mathbf{H}t^{-1}t_2 & \cdots & t\mathbf{H}t^{-1}t_r \\ t\mathbf{H}t^{-1}t_1g & t\mathbf{H}t^{-1}t_2g & \cdots & t\mathbf{H}t^{-1}t_rg \end{pmatrix}. \tag{5.8}$$

An appropriate similarity transformation on π_g produces an equivalent permutation (π_g').

$$\begin{pmatrix} tHt^{-1}t_1 & tHt^{-1}t_2 & \cdots & tHt^{-1}t_r \\ Ht^{-1}t_1 & Ht^{-1}t_2 & \cdots & Ht^{-1}t_r \end{pmatrix} \pi_g \begin{pmatrix} Ht^{-1}t_1 & Ht^{-1}t_2 & \cdots & Ht^{-1}t_r \\ tHt^{-1}t_1 & tHt^{-1}t_2 & \cdots & tHt^{-1}t_r \end{pmatrix}$$

$$= \begin{pmatrix} Ht^{-1}t_1 & Ht^{-1}t_2 & \cdots & Ht^{-1}t_r \\ Ht^{-1}t_1g & Ht^{-1}t_2g & \cdots & Ht^{-1}t_rg \end{pmatrix} = \pi_g'$$

A further similarity transformation affords

$$\begin{pmatrix} Ht^{-1}t_1 & Ht^{-1}t_2 & \cdots & Ht^{-1}t_r \\ Ht_1 & Ht_2 & \cdots & Ht_r \end{pmatrix} \pi_g' \begin{pmatrix} Ht_1 & Ht_2 & \cdots & Ht_r \\ Ht^{-1}t_1 & Ht^{-1}t_2 & \cdots & Ht^{-1}t_r \end{pmatrix}$$

$$= \begin{pmatrix} Ht_1 & Ht_2 & \cdots & Ht_r \\ Ht_1g & Ht_2g & \cdots & Ht_rg \end{pmatrix} = \pi_g''.$$

When we combine these transformations, we have a similarity transformation that converts the permutation π_g into π_g'' for all g ($\in \mathbf{G}$). This means that $\mathbf{G}(/tHt^{-1})$ is equivalent to $\mathbf{G}(/\mathbf{H})$.

Theorems 5.1 and 5.2 afford the following important theorem.

Theorem 5.3 *Suppose that a group $\mathbf{G}$ of finite order has a non-redundant set of subgroups,*

$$SSG_{\mathbf{G}} = \{\mathbf{G}_1, \mathbf{G}_2, \ldots, \mathbf{G}_s\}, \tag{5.9}$$

in which the elements are aligned in an ascending order of their orders, i.e., $\mid \mathbf{G}_1 \mid \leq \mid \mathbf{G}_2 \mid \leq \cdots \leq \mid \mathbf{G}_s \mid$; and $\mathbf{G}_1 = identity$ and $\mathbf{G}_s = \mathbf{G}$. The set of CRs,

$$SCR_{\mathbf{G}} = \{\mathbf{G}(/\mathbf{G}_1), \mathbf{G}(/\mathbf{G}_2), \ldots, \mathbf{G}(/\mathbf{G}_s)\}, \tag{5.10}$$

is a complete set of coset representations of $\mathbf{G}$.

Since such coset representations are obtained algebraically by means of coset decompositions, thier properties can be pre-examined. The concrete forms of CRs for the $\mathbf{D}_{2d}$ group are found in Table 5.1.

5.2   Transitive Permutation Representations

A permutation representation (PR) is a homomorphism of a finite group $\mathbf{G}$ onto a permutation group $\mathbf{P}$ on $\Delta = \{\delta_1, \delta_2, \ldots, \delta_n\}$. The PR ($\mathbf{P_G}$) is a set of such permutations ($p_g \in \mathbf{P}$) that satisfy

$$p_g p_{g'} = p_{gg'} \text{ for any } g, g' \in \mathbf{G}. \tag{5.11}$$

When the permutation group $\mathbf{P}$ is transitive, the PR is also transitive. The goal of this section is to prove that any transitive PR is equivalent to a coset representation.

Table 5.1: Coset representations for D_{2d}

Symmetry operation	Coset representation							
	$D_{2d}(/C_1)$	$D_{2d}(/C_2)$	$D_{2d}(/C_2')$	$D_{2d}(/C_s)$	$D_{2d}(/S_4)$	$D_{2d}(/C_{2v})$	$D_{2d}(/D_2)$	$D_{2d}(/D_{2d})$
I	(1)(2)(3)(4)(5)(6)(7)(8)	(1)(2)(3)(4)	(1)(2)(3)(4)	(1)(2)(3)(4)	(1)(2)	(1)(2)	(1)(2)	(1)
$C_{2(1)}$	(1 2)(3 4)(5 6)(7 8)	(1 2)(3 4)	(1)(2)(3 4)	(1 2)(3 4)	(1 2)	(1 2)	(1)(2)	(1)
$C_{2(2)}$	(1 3)(2 4)(5 7)(6 8)	(1 2)(3 4)	(1 2)(3)(4)	(1 3)(2 4)	(1 2)	(1 2)	(1)(2)	(1)
$C_{2(3)}$	(1 4)(2 3)(5 8)(6 7)	(1)(2)(3)(4)	(1 2)(3 4)	(1 4)(2 3)	(1)(2)	(1)(2)	(1)(2)	(1)
$\sigma_{d(1)}$	(1 5)(2 7)(3 6)(4 8)	(1 3)(2 4)	(1 3)(2 4)	(1)(2 3)(4)	(1 2)	(1)(2)	(1 2)	(1)
S_4^3	(1 6 4 7)(2 8 3 5)	(1 4)(2 3)	(1 4 2 3)	(1 3 4 2)	(1)(2)	(1 2)	(1 2)	(1)
S_4	(1 7 4 6)(2 5 3 8)	(1 4)(2 3)	(1 3 2 4)	(1 2 4 3)	(1)(2)	(1 2)	(1 2)	(1)
$\sigma_{d(2)}$	(1 8)(2 6)(3 7)(4 5)	(1 3)(2 4)	(1 4)(2 3)	(1 4)(2)(3)	(1 2)	(1)(2)	(1 2)	(1)
Orbit (O_A)	O_8	–	O_4	O_{4d}	–	O_2		O_1

For illustrating a transitive PR, we examine the symmetry of allene (see also Chapter 2). Figure 5.1 depicts a top view of an allene molecule and its symmetry elements. Suppose we label four hydrogen atoms sequentially. As we conduct each

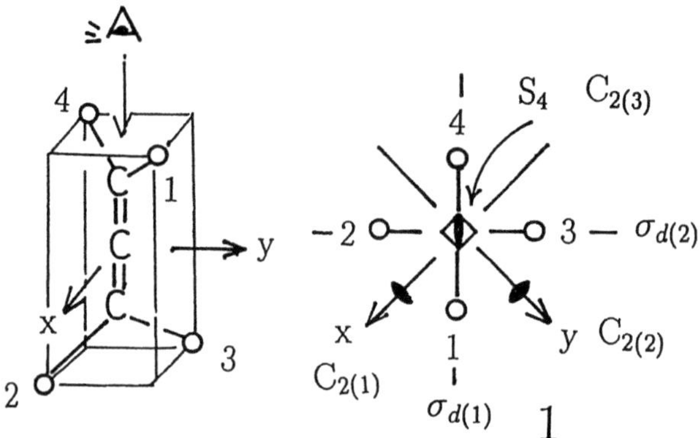

Figure 5.1: Top view of an allene molecule

of the symmetry operations, the sequential numbers are permuted. Thus, each symmetry operation is associated with a permutation of this labeling. When we conduct the symmetry operations, we obtain the following permutations.

$$I \quad \sim \quad p_1 = \begin{pmatrix} 1 & 2 & 3 & 4 \\ 1 & 2 & 3 & 4 \end{pmatrix} = (1)(2)(3)(4)$$

$$C_{2(1)} \quad \sim \quad p_2 = \begin{pmatrix} 1 & 2 & 3 & 4 \\ 2 & 1 & 4 & 3 \end{pmatrix} = (1\ 2)(3\ 4)$$

$$C_{2(2)} \quad \sim \quad p_3 = \begin{pmatrix} 1 & 2 & 3 & 4 \\ 3 & 4 & 1 & 2 \end{pmatrix} = (1\ 3)(2\ 4)$$

$$C_{2(3)} \quad \sim \quad p_4 = \begin{pmatrix} 1 & 2 & 3 & 4 \\ 4 & 3 & 2 & 1 \end{pmatrix} = (1\ 4)(2\ 3)$$

$$\sigma_{d(1)} \quad \sim \quad p_5 = \begin{pmatrix} 1 & 2 & 3 & 4 \\ 1 & 3 & 2 & 4 \end{pmatrix} = (1)(2\ 3)(4)$$

$$S_4^3 \quad \sim \quad p_6 = \begin{pmatrix} 1 & 2 & 3 & 4 \\ 3 & 1 & 4 & 2 \end{pmatrix} = (1\ 3\ 4\ 2)$$

$$S_4 \quad \sim \quad p_7 = \begin{pmatrix} 1 & 2 & 3 & 4 \\ 2 & 4 & 1 & 3 \end{pmatrix} = (1\ 2\ 4\ 3)$$

$$\sigma_{d(2)} \quad \sim \quad p_8 = \begin{pmatrix} 1 & 2 & 3 & 4 \\ 4 & 2 & 3 & 1 \end{pmatrix} = (1\ 4)(2)(3)$$

Obviously, $\mathbf{P_G} = \{p_1, p_2, p_3, p_4, p_5, p_6, p_7, p_8\}$ is a repsesentation of the $\mathbf{D}_{2d}$ group. Since these elements are permutations concerning a domain $\Delta = \{1, 2, 3, 4\}$, this

representation is called a *permutation representation*. It should be noted that the
$\mathbf{P_G}$ representation is equivalent to the coset representation $(\mathbf{D}_{2d}(/\mathbf{C}_s))$.

Chemically speaking, the Δ domain denotes a set of four hydrogen atoms
of allene, which are equivalent to each other. Such a set is called an *orbit* by
applying the terminology of permutation-group theory. Hence, we characterize
the orbit (the set of equivalent atoms) by means of a coset representation.

In the above discussion, we have implicitly used the concept *actions of groups*.
We here give a definition to this concept.

Definition 5.1 (Actions of groups) *A group* $\mathbf{G}$ *is defined as acting on a set* Δ,
when a mapping:

$$\delta \to g(\delta) \quad (\delta \in \Delta, g \in \mathbf{G})$$

is determined for any $g(\in \mathbf{G})$ *and satisfies the following conditions.*

   a) $g_2(g_1(\delta)) = g_2 g_1(\delta) \quad (\delta \in \Delta, g \in \mathbf{G})$
   b) $I(\delta) = \delta \quad$ *(the element* (I) *is an identity of* $\mathbf{G}$, *and* $\delta \in \Delta$*).*

According to this definition, the $\mathbf{P_G}$ can be considered to represent an action of
$\mathbf{G}$ on Δ. Hereafter, we also use this terminology. In the above example in which
a $\mathbf{D}_{2d}$ point group keeps an allene molecule invariant, the $\mathbf{D}_{2d}$ is considered to act
on $\Delta =\{1,2,3,4\}$ corresponding to the four positions of allene. This action creates
permutations of Δ, which construct the permutation representation $\mathbf{P_G}$. The $\mathbf{P_G}$
is associated with a permutation group $\mathbf{P}$. In this case, we have an isomorphism
$\mathbf{P_G} \cong \mathbf{P}$, which indicates that the $\mathbf{P_G}$ is a faithful representation.

Let $\mathbf{P}$ be a permutation group on $\Delta = \{\delta_1, \delta_2, \ldots, \delta_n\}$. For simplicity's sake,
we take account only of the subsrcipts, considering $\mathbf{P}$ on $\Delta = \{1, 2 \ldots, n\}$ without
losing generality.

Lemma 5.1 *Let* p $(\in \mathbf{P})$ *be a permutaion that fixes a point* i $(\in \Delta)$. *The set of
all of the permutations* p's *constructs a group,*

$$\mathbf{P}_i = \{p \mid p \in \mathbf{P}, i^p = i\},$$

which is called the stablizer *of the point* (i).

Proof. For $\forall p$, $\forall q \in \mathbf{P}_i$, we have $(i^q)^p = i^p = i$. Hence, $pq \in \mathbf{P}_i$. On the other
hand, if p is an element of $\mathbf{P}_i$, we have $i = i^p$. Then, $i^{p^{-1}} = (i^p)^{p^{-1}} = i^{p^{-1}p} = i$.
Hence, we have $p^{-1} \in \mathbf{P}_i$. It follows that $\mathbf{P}_i$ is a group.

Lemma 5.2 *Let* $\mathbf{P}$ *be a transitive permutation group on* $\Delta = \{1, 2 \ldots, n\}$ *to be
associated with* $\mathbf{P_G}$. *Let* $p_i(\in \mathbf{P})$ *be a permutation converting* i *into* $1 \in \Delta$, *where
$p_1 = I$. Then, we have a coset decomposition,*

$$\mathbf{P} = \mathbf{P}_1 p_1 + \mathbf{P}_1 p_2 + \cdots + \mathbf{P}_1 p_n, \tag{5.12}$$

where $\mathbf{P}_1$ *is the stabilizer of the integer* (1).

Proof. (1) Suppose that $i^{p_i} = 1$ and $j^{p_j} = 1$ $(i \neq j)$. Then, we have $p_i \neq p_j$ and $\mathbf{P}_1 p_i \neq \mathbf{P}_1 p_j$. We assume that the two cosets, $\mathbf{P}_1 p_i$ and $\mathbf{P}_1 p_j$, have a common element (p), *i.e.*, $p \in \mathbf{P}_1 p_i$ and $p \in \mathbf{P}_1 p_j$. Then, there exist appropriate elements $(q_i, q_j \in \mathbf{P}_1)$ that satisfy $p = q_i p_i$ and $p = q_j p_j$. It follows that $q_i p_i = q_j p_j \Longrightarrow p_i p_j^{-1} = q_i^{-1} q_j$. Since $q_i^{-1} q_j \in \mathbf{P}_1$, We can select $q \equiv q_i^{-1} q_j \in \mathbf{P}_1$. Hence, $p_i p_j^{-1} = q$ $\Longrightarrow p_i = q p_j \in \mathbf{P}_1 p_j$. This is against the original assumption. Therefore, we obtain $\mathbf{P}_1 p_i \cap \mathbf{P}_1 p_j = \emptyset$.

(2) Let $\mathbf{P}'$ be the sum of such cosets, *i.e.*,

$$\mathbf{P}' = \mathbf{P}_1 p_1 + \mathbf{P}_1 p_2 + \cdots + \mathbf{P}_1 p_n. \tag{5.13}$$

Obviously, $\mathbf{P} \supseteq \mathbf{P}'$. Suppose that

$$\mathbf{P} = \mathbf{P}' + \mathbf{P}_1 p_z. \tag{5.14}$$

Then we have $1^{p_z^{-1}} = k$ for $\exists k \in \Delta$, because $p_z^{-1} \in \mathbf{P}$. Hence, $k^{p_z} = 1$. This means that p_z is identical with p_k. Therefore, we have $\mathbf{P} = \mathbf{P}'$.

By means of Lemma 5.2, we have the following equation.

$$\mid \mathbf{P} \mid = \mid \mathbf{P}_1 \mid \mid \Delta \mid,$$

Lemma 5.2 holds for any element i $(\in \Delta)$; the corresponding equation is obtained for any $\mathbf{P}_i$ that is the stabilizer of the point i $(\in \Delta)$; thus, we have the following lemma.

Lemma 5.3 *If* $\mathbf{P}$ *is a transitive permutation group, we have*

$$\mid \mathbf{P} \mid = \mid \mathbf{P}_i \mid \mid \Delta \mid, \; for \; i = 1, 2, \ldots, \mid \Delta \mid, \tag{5.15}$$

where $\mid \mathbf{P} \mid$ *and* $\mid \mathbf{P}_i \mid$ *denote the orders of the respective groups and the symbol* $\mid \Delta \mid$ *represents the length (or size) of* Δ.

Lemma 5.3 will be used to prove so-called Burnside's lemma (see Chapter 13).

Now we arrive at the goal to prove that such a transitive PR is equivalent to a coset representation.

Theorem 5.4 *Let* $\mathbf{G}$ *be a finite group and* $\mathbf{P_G}$ *be a transitive permutation representation. Then, the* $\mathbf{P_G}$ *is equivalent to an appropriate coset representation* $\mathbf{G}(/\mathbf{H})$, *where* $\mathbf{H}$ *is a subgroup of* $\mathbf{G}$.

Proof. Let $\mathbf{P}$ be a transitive permutation group on $\Delta = \{1, 2 \ldots, n\}$ to be associated with $\mathbf{P_G}$. Let $\mathbf{P}_1$ be the stabilizer of the letter 1 $(\in \Delta)$. Since there exists a homomorphic mapping of $\mathbf{G}$ onto $\mathbf{P}$ concerning $\mathbf{P_G}$, we can select a subgroup $\mathbf{H}$ $(\leq \mathbf{G})$ that corresponds to $\mathbf{P}_1$ $(\leq \mathbf{P})$.

Let $p_i (\in \mathbf{P})$ be a permutation converting i to $1 \in \Delta$. Then, we have the following coset decomposition by means of Lemma 5.2.

$$\mathbf{P} = \mathbf{P}_1 p_1 + \mathbf{P}_1 p_2 + \cdots + \mathbf{P}_1 p_n, \tag{5.16}$$

where $p_1 = I$.

Since we are able to select g_i $(\in \mathbf{G})$ that corresponds to each p_i, we have a coset decomposition,

$$\mathbf{G} = \mathbf{H}g_1 + \mathbf{H}g_2 + \cdots + \mathbf{H}g_n, \tag{5.17}$$

where $g_1 = I$. Equation 5.17 affords a permutation,

$$\pi_g = \begin{pmatrix} \mathbf{H}g_1 & \mathbf{H}g_2 & \dots & \mathbf{H}g_n \\ \mathbf{H}g_1 g & \mathbf{H}g_2 g & \dots & \mathbf{H}g_n g \end{pmatrix}, \tag{5.18}$$

which constructs a coset representation $\mathbf{G}(/\mathbf{H})$ when g runs over $\mathbf{G}$.

Let p_g $(\in \mathbf{P})$ correspond to $g \in \mathbf{G}$, i.e.,

$$p_g = \begin{pmatrix} 1 & 2 & \dots & n \\ 1^{p_g} & 2^{p_g} & \dots & n^{p_g} \end{pmatrix}. \tag{5.19}$$

Comparison between the coset decompositions (eqs. 5.16 and 5.17) reveals the one-to-one correspondence ($i \Longleftrightarrow \mathbf{H}g_i$) between $\Delta = \{1, 2 \dots, n\}$ and $\mathbf{G}/\mathbf{H} = \{\mathbf{H}g_1, \mathbf{H}g_2, \dots, \mathbf{H}g_n\}$. Then, we obtain $i^{p_g} \Longleftrightarrow \mathbf{H}g_i g$. That is to say, $p_g \cong \pi_g$. ∎

In the light of Theorem 5.4, Theorem 5.3 is restated as follows.

Theorem 5.5 *A complete set of different transitive permutation representations of* $\mathbf{G}$ *is identical with the complete set of coset representations of* $\mathbf{G}$.

Such a coset representation is originally concerned with a set of cosets. However, the above discussions indicate that any set of objectives (in addition to the set of cosets) can be regarded as *being subject to* or as *being governed by* an appropriate coset representation (CR).

5.3 Mark Tables

In the preceding section, any transitive permutation representation is proved to be equivalent to an appropiate CR. Such CRs can be precalculated by using coset decompositions.

Definition 5.2 *Let* $\mathbf{G}(/\mathbf{G}_i)$ $(i = 1, 2, \dots, s)$ *be coset representations of* $\mathbf{G}$ *by* $\mathbf{G}_i$, *where* $|\mathbf{G}_1| \le |\mathbf{G}_2| \le \cdots \le |\mathbf{G}_s|$; *and* $\mathbf{G}_1$ *is an identity group and* $\mathbf{G}_s$ *is equal to* $\mathbf{G}$. *A mark* (m_{ij}) *of* $\mathbf{G}_j$ $(\le \mathbf{G})$ *in* $\mathbf{G}(/\mathbf{G}_i)$ *is defined as the number of fixed points (or cosets) in* $\mathbf{G}(/\mathbf{G}_i)$ *on the action of* $\mathbf{G}_j$.

Table 5.2: Mark table for $\mathbf{G}$ group

$j \; i$	$\mathbf{G}_1$	$\mathbf{G}_2$	$\cdots$	$\mathbf{G}_j$	$\cdots$	$\mathbf{G}_s$
$\mathbf{G}(/\mathbf{G}_1)$	m_{11}	0	$\cdots$	0	$\cdots$	0
$\mathbf{G}(/\mathbf{G}_2)$	m_{21}	m_{22}	$\cdots$	0	$\cdots$	0
$\vdots$	$\vdots$	$\vdots$		$\vdots$		$\vdots$
$\mathbf{G}(/\mathbf{G}_i)$	m_{i1}	m_{i2}	$\cdots$	m_{ij}	$\cdots$	0
$\vdots$	$\vdots$	$\vdots$		$\vdots$		$\vdots$
$\mathbf{G}(/\mathbf{G}_s)$	1	1	$\cdots$	1	$\cdots$	1

Obviously, $m_{ij} = 0$ for $i > j$; $m_{ij} = 1$ for $j = s$. A table listing all m_{ij}'s is called a *table of marks* or a *mark table* (Table 5.2). In the light of this table, we find that any two CRs of $\mathbf{G}(/\mathbf{G}_i)$ $(1, 2, \ldots, s)$ are different from each other in their marks. The marks of conjugate subgroups are equal.

In order to illustrate a procedure of constructing a mark table, we calculate marks for the CR $(\mathbf{D}_{2d}(/\mathbf{C}_s))$ using the data collected in Table 5.1. We count fixed points (*i.e.* 1-cycles) over the elements marked with $\surd$ for each of the subgroups. For example, when we gather the checked elements of $\mathbf{C}_s$ (*i.e.*, $(1)(2)(3)(4)$ and $(1)(2\ 3)(4)$), we find (1) and (4) to be common 1-cycles. Hence, we have 2 as the number of fixed points. When we repeat this procedure over all of the subgroups, we obtain the $\mathbf{D}_{2d}(/\mathbf{C}_s)$ row of marks as shown at the bottom of the following marksheet.

symmetry operation		CR $\mathbf{D}_{2d}(/\mathbf{C}_s)$	$\mathbf{C}_1$	$\mathbf{C}_2$	$\mathbf{C}_2'$	$\mathbf{C}_s$	$\mathbf{S}_4$	$\mathbf{C}_{2v}$	$\mathbf{D}_2$	$\mathbf{D}_{2d}$
I	$\sim$	$(1)(2)(3)(4)$	$\surd$	$\surd$	$\surd$	$\surd$	$\surd$	$\surd$	$\surd$	$\surd$
$C_{2(1)}$	$\sim$	$(1\ 2)(3\ 4)$			$\surd$				$\surd$	$\surd$
$C_{2(2)}$	$\sim$	$(1\ 3)(2\ 4)$							$\surd$	$\surd$
$C_{2(3)}$	$\sim$	$(1\ 4)(2\ 3)$		$\surd$			$\surd$	$\surd$	$\surd$	$\surd$
$\sigma_{d(1)}$	$\sim$	$(1)(2\ 3)(4)$				$\surd$		$\surd$		$\surd$
S_4^3	$\sim$	$(1\ 3\ 4\ 2)$					$\surd$			$\surd$
S_4	$\sim$	$(1\ 2\ 4\ 3)$					$\surd$			$\surd$
$\sigma_{d(2)}$	$\sim$	$(1\ 4)(2)(3)$				$\surd$				$\surd$
		mark	4	0	0	2	0	0	0	0

Marks for the other CRs are calculated in a similar way; the collection of these values gives the mark table shown in Table 5.3. The corresponding inverse matrix is also calculated (Table 5.4).

The method of constructing the mark table indicates how to assign a CR to a given orbit (*e.g.*, a set of equivalent atoms). First, we find a set of equivalent atoms of a given molecule. We then count fixed atoms (points) for every subgroup,

Table 5.3: Mark table of D_{2d}

	C_1	C_2	C_2'	C_s	S_4	C_{2v}	D_2	D_{2d}
$D_{2d}(/C_1)$	8	0	0	0	0	0	0	0
$D_{2d}(/C_2)$	4	4	0	0	0	0	0	0
$D_{2d}(/C_2')$	4	0	2	0	0	0	0	0
$D_{2d}(/C_s)$	4	0	0	2	0	0	0	0
$D_{2d}(/S_4)$	2	2	0	0	2	0	0	0
$D_{2d}(/C_{2v})$	2	2	0	2	0	2	0	0
$D_{2d}(/D_2)$	2	2	2	0	0	0	2	0
$D_{2d}(/D_{2d})$	1	1	1	1	1	1	1	1

Table 5.4: The inverse of the mark table for D_{2d}

	D_{2d} $(/C_1)$	D_{2d} $(/C_2)$	D_{2d} $(/C_2')$	D_{2d} $(/C_s)$	D_{2d} $(/S_4)$	D_{2d} $(/C_{2v})$	D_{2d} $(/D_2)$	D_{2d} $(/D_{2d})$	sum[a]
C_1	1/8	0	0	0	0	0	0	0	1/8
C_2	-1/8	1/4	0	0	0	0	0	0	1/8
C_2'	-1/4	0	1/2	0	0	0	0	0	1/4
C_s	-1/4	0	0	1/2	0	0	0	0	1/4
S_4	0	-1/4	0	0	1/2	0	0	0	1/4
C_{2v}	1/4	-1/4	0	-1/2	0	1/2	0	0	0
D_2	1/4	-1/4	-1/2	0	0	0	1/2	0	0
D_{2d}	0	1/2	0	0	-1/2	-1/2	-1/2	1	0

[a] sum $= \sum_{i=1}^{s} \overline{m}_{ji}$

where all of its symmetry operations are applied. The resulting values are listed to generate a row vector, which is referred to as a *fixed-point vector (FPV)*. Finally, we compare the vector with the rows of Table 5.3. The identical row is a CR to be assigned.

We have clarified the correspondence between the orbit of four equivalent hydrogens in allene and the CR ($D_{2d}(/C_s)$) in the light of direct comparison of the permutaion representation with the CR. When we manipulate more complex cases, however, such direct comparison is not always an easy task, because both the explicit form of the permutation representation and that of the CR vary with the modes of numbering. This task is conveniently accomplished by means of a mark table (*e.g.*, Table 5.3), since an FPV is independent of such modes of numbering. When we examine the top view of the allene molecule (Fig. 5.1), the orbit of the four hydrogens gives an FPV, (4 0 0 2 0 0 0 0), the elements of which are aligned in the order of the SSG, $\{C_1, C_2, C_2', C_s, S_4, C_{2v}, D_2, D_{2d}\}$. This vector is equal to the $D_{2d}(/C_s)$ row of Table 5.3. Thereby, we conclude that the orbit is subject to the CR ($D_{2d}(/C_s)$).

Construction of such mark tables for point groups to be examined is a rewarding task. Several tables are found in Appendix A. The inverse matrices of the mark tables are calculated and collected in Appendix B.

5.4　Permutation Representations and Orbits

Let P be a permutation group acting on the set represented by $\Delta = \{\ \delta_1, \delta_2, \ldots, \delta_{|\Delta|}\ \}$. When there exists a $p \in P$ that satifies $\delta_j = \delta_i^p$ for two elements (δ_i and δ_j), we define a binary relation as being $\delta_i \sim \delta_j$. This relation can be easily proved to be an equivalence relation. According to this equivalence relation, the Δ is partitioned into an appropriate number of equivalence classes: $\Delta_1, \Delta_2, \ldots, \Delta_r$. Each Δ_i is called an *orbit* of P. The number of members of each orbit ($|\ \Delta_i\ |$) is called the *length* of the orbit (Δ_i). Obviously, a transitive permutation group has a single orbit; an intransitive permutation group corresponds to two or more orbits.

A homomorphism of a finite group G onto the permutation group P produces a permutation representation (PR) P_G. This PR, transitive or intransitive, is also associated with such orbit(s). For illustrating an intransitive permutation representation, let us examine adamantane-2,6-dione (2) of D_{2d} symmetry (Fig. 5.2). This compound contains 12 hydrogen atoms to be considered. If we number them sequentially in an appropriate fashion shown in this figure, we have the following permutations.

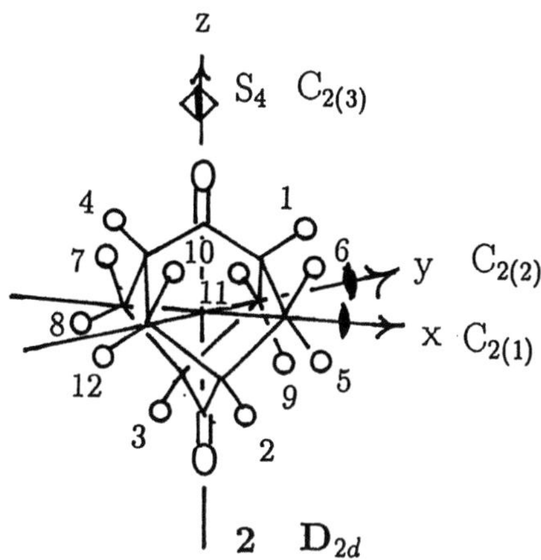

Figure 5.2: Symmetry of adamantane-2,6-dione

$$
\begin{array}{lll}
I & \sim & q_1 = (1)(2)(3)(4) \mid (5)(6)(7)(8)(9)(10)(11)(12) \\
C_{2(1)} & \sim & q_2 = (1\ 2)(3\ 4) \mid (5\ 6)(7\ 8)(9\ 10)(11\ 12) \\
C_{2(2)} & \sim & q_3 = (1\ 3)(2\ 4) \mid (5\ 7)(6\ 8)(9\ 11)(10\ 12) \\
C_{2(3)} & \sim & q_4 = (1\ 4)(2\ 3) \mid (5\ 8)(6\ 7)(9\ 12)(10\ 11) \\
\sigma_{d(1)} & \sim & q_5 = (1)(2\ 3)(4) \mid (5\ 9)(6\ 11)(7\ 10)(8\ 12) \\
S_4^3 & \sim & q_6 = (1\ 3\ 4\ 2) \mid (5\ 11\ 8\ 10)(6\ 9\ 7\ 12) \\
S_4 & \sim & q_7 = (1\ 2\ 4\ 3) \mid (5\ 10\ 8\ 11)(6\ 12\ 7\ 9) \\
\sigma_{d(2)} & \sim & q_8 = (1\ 4)(2)(3) \mid (5\ 12)(6\ 10)(7\ 11)(8\ 9)
\end{array}
$$

Since these permutations correspond to the symmetry operations of the $\mathbf{D}_{2d}$ point groups, the resulting set $\mathbf{Q} = \{q_1, q_2, q_3, q_4, q_5, q_6, q_7, q_8\}$ constructs a permutation group on $\Delta' = \{1, 2, \ldots, 12\}$. When we examine the concrete forms of these permutations carefully, we find that the twelve integers are partitioned into two sets, *i.e.*, $\{1, 2, 3, 4\}$ and $\{5, 6, \ldots, 12\}$. Hence, the permutation representation ($\mathbf{Q}$) is intransitive; the original set (Δ') is divided into the two orbits. The permutations concerning the orbit $\{1, 2, 3, 4\}$ construct a representation that is easily shown to be equivalent to the CR ($\mathbf{D}_{2d}(/\mathbf{C}_s)$) listed in Table 5.1. On the other hand, a representation concerning $\{5, 6, \ldots, 12\}$ is concluded to be equivalent to the CR ($\mathbf{D}_{2d}(/\mathbf{C}_1)$), if we renumber the members of the orbit using integers 1 to 8.

The intransitivity can be correlated to the non-equivalency of the 12 hydrogen atoms in **2**. Obviously, the four methine hydrogens at the bridgehead positions are equivalent; they are different from the remaining 8 methylene hydrogens at the bridge positions.

The above procedure can be summarized by a scheme: a set of atoms $\Longrightarrow$ an intransitive permutation representation $\Longrightarrow$ a set of transitive permutation representations $\Longleftrightarrow$ a set of orbits. If we first find such orbits by inspection, we can modify this scheme as follows: a set of atoms $\Longrightarrow$ a set of orbits $\Longleftrightarrow$ a set of transitive permutation representations. Moreover, the last step can be simplified by means of a mark table since a transitive permutation representation is equivalent to a coset representation.

Let us illustrate the modified procedure.

Example **5.1** By inspection of Fig. 5.2, we can easily find two orbits of hydrogen atoms; four methine hydrogens at the bridgehead positions and 8 methylene hydrogens at the bridge positions. As a result, we have two FPVs, (4 0 0 2 0 0 0 0) and (8 0 0 0 0 0 0 0). These vectors are equal to the $\mathbf{D}_{2d}(/\mathbf{C}_s)$ and $\mathbf{D}_{2d}(/\mathbf{C}_1)$ rows of Table 5.3. Thereby, we conclude that the orbits are subject to the CRs ($\mathbf{D}_{2d}(/\mathbf{C}_s)$ and $\mathbf{D}_{2d}(/\mathbf{C}_1)$). The concrete forms of these CRs are in turn found in Table 5.1.

Since Lemmas 5.1 and 5.2 are concerned with a transitive permutation group, they are also applicable to the present transitive permutation group acting on each orbit. Hence, Theorem 5.4 can also be rewritten so as to hold for this case.

Theorem 5.6 *Suppose that* $\Delta = \{\ \delta_1, \delta_2, \ldots, \delta_{|\Delta|}\ \}$ *is partitioned into orbits* Δ_1, Δ_2, ..., Δ_r *by the action of a group* $\mathbf{G}$ *of a finite order. Let* $\mathbf{Q}_{\mathbf{G}}^{(\alpha)}$ *be a transitive permutation representation that associates with the orbit* Δ_α *(* $\alpha = 1, 2, \ldots, r$ *). Then,* $\mathbf{Q}_{\mathbf{G}}^{(\alpha)}$ *is equivalent to an appropriate coset represetation* $\mathbf{G}(/\mathbf{H})$, *where* $\mathbf{H}$ *is a subgroup of* $\mathbf{G}$.

Since Theorem 5.3 indicates that such CRs are represented by $\mathbf{G}(/\mathbf{G}_s)$ ($i = 1, 2, \ldots, s$), Theorem 5.6 permits us to consider that a PR $\mathbf{P_G}$ consists of α_i coset representations $\mathbf{G}(/\mathbf{G}_i)$ ($i = 1, 2, \ldots, s$), where each α_i is a multiplicity of the CR and i covers over 1 to s. Hence, we end up with the following theorem.

Theorem 5.7 *Any permutation representation* $\mathbf{P_G}$ *of a finite group* $\mathbf{G}$ *acting on* Δ *can be represented by a sum of CRs.*

$$\mathbf{P_G} = \sum_{i=1}^{s} \alpha_i \mathbf{G}(/\mathbf{G}_i), \tag{5.20}$$

wherein the multiplicities α_i *are non-negative integers. The multiplicities are obtained by using a mark table (Table 5.2):*

$$\mu_j = \sum_{i=1}^{s} \alpha_i m_{ij} \ (j = 1, 2, \ldots, s), \tag{5.21}$$

where μ_j *is the mark (the number of fixed points) of* $\mathbf{G}_j$ *in* $\mathbf{P_G}$.

Proof. Equation 5.20 is obtained easily from Theorem 5.6. Equation 5.21 is obtained from the mark table (Table 5.2). Since the mark m_{ij} is associated with each $G(/G_i)$, the mark μ_j is obtained by the summation of the marks over i, where the multiplicities (α_i) are taken into account. It follows that $\mu_j = \sum_{i=1}^{s} \alpha_i m_{ij}$ for each G_j.

When we have the marks μ_j calculated by an appropriate method, we can obtain α_i by solving eq. 5.21. Note that the determinant of the mark table is not equal to zero. Hence, we have

$$\alpha_i = \sum_{j=1}^{s} \mu_j \overline{m}_{ji} \ (i = 1, 2, \ldots, s), \tag{5.22}$$

where $\overline{m}_{ji}$ denotes the ji-element of the inverse of the mark table.

These equations can be expressed by matrices,

$$F = EM \tag{5.23}$$

or

$$E = FM^{-1}, \tag{5.24}$$

wherein

$$E = (\alpha_1 \ \alpha_2 \ \cdots \ \alpha_s) \tag{5.25}$$

$$F = (\mu_1 \ \mu_2 \ \cdots \ \mu_s) \tag{5.26}$$

$$M = \begin{pmatrix} m_{11} & m_{12} & \cdots & m_{1s} \\ m_{21} & m_{22} & \cdots & m_{2s} \\ \vdots & \vdots & & \vdots \\ m_{s1} & m_{s2} & \cdots & m_{ss} \end{pmatrix} \tag{5.27}$$

and

$$M^{-1} = \begin{pmatrix} \overline{m}_{11} & \overline{m}_{12} & \cdots & \overline{m}_{1s} \\ \overline{m}_{21} & \overline{m}_{22} & \cdots & \overline{m}_{2s} \\ \vdots & \vdots & & \vdots \\ \overline{m}_{s1} & \overline{m}_{s2} & \cdots & \overline{m}_{ss} \end{pmatrix}. \tag{5.28}$$

The matrix M is a mark table and M^{-1} is the inverse. We refer to E as a *muliplicity vector (MV)*. The row vector (F) is a fixed-point vector (FPV). The latter has previously been introduced for a single orbit, but it is now redefined for manipulating two or more orbits.

Theorem 5.7 provides a partition of Δ into α_i orbits,

$$\Delta_{i1}, \Delta_{i2}, \ldots, \Delta_{i\alpha_i} \tag{5.29}$$

for each $G(/G_i)$ $(i = 1, 2, \ldots, s)$. Chemically speaking, each orbit $(\Delta_{i\alpha})$ represents a set of equivalent atoms (or other objects).

If we pay our attention to the lengths of the orbits, we have

$$\mid \Delta_{i\alpha} \mid = \mid G \mid / \mid G_i \mid . \tag{5.30}$$

Hence, eq. 5.20 affords the following corollary.

Corollary 5.1

$$\mid \Delta \mid = \sum_{i=1}^{s} \alpha_i \mid G \mid / \mid G_i \mid . \tag{5.31}$$

The total number (r) of orbits can be calculated in terms of eq. 5.22.

Corollary 5.2

$$r = \sum_{i=1}^{s} \alpha_i = \sum_{i=1}^{s} \sum_{j=1}^{s} \mu_j \overline{m}_{ji.} = \sum_{j=1}^{s} \mu_j (\sum_{i=1}^{s} \overline{m}_{ji}). \tag{5.32}$$

The last summations $\sum_{i=1}^{s} \overline{m}_{ji}$ $(j = 1, 2, \ldots, s)$ are obtained by summing up each row of the inverse of a mark table. As for D_{2d}, these values are listed in the rightmost column of Table 5.4. This corollary is an alternative expression of so-called Burnside's lemma (the Cauchy-Frobenius lemma), which will be derived later.

Theorem 5.7 affords an alternative method of assigning CRs to orbits. For exemplifying this theorem, we re-examine the problem described in Example 5.1

Example **5.2** The 12 hydrogen atoms of **2** (Fig. 5.2) in a lump are taken into consideration. By inspection, we have an FPV, (12 0 0 2 0 0 0 0), which is multiplied by the inverse of the mark table for D_{2d} (Table 5.4). Thereby, we obtain a row vector, (1 0 0 1 0 0 0 0), which means that there appear one $D_{2d}(/C_1)$ and one $D_{2d}(/C_s)$. This result is identical with that of Example 5.1 and summarized by a formal expression,

$$\mathbf{P_{D_{2d}}} = \mathbf{D}_{2d}(/C_1) + \mathbf{D}_{2d}(/C_s). \tag{5.33}$$

The number of such orbits is given by applying eq. 5.32 to the FPV. Thus, we obtain

$$r = 12 \times \frac{1}{8} + 2 \times \frac{1}{4} = 2.$$

Table 5.1 contains $\mathbf{O_A}$ notations for orbits in the bottom row. The absence of some of the notations will be discussed later (Chapter 7).

Bibliography

[1] W. Burnside, *Theory of Groups of Finite Order*, 2nd ed., Cambridge Univ. Press, Cambridge (1911).

[2] M. Ōshima, *Gun-Ron (Group Theory)*, (Kyōritsu-Zensyo, Vol. 88), Kyōritsu, Tokyo (1954).

[3] B. Baumslag, B. Chandler, *Theory and Problems of Group Theory*, McGraw-Hill, New York (1968).

Chapter 6

Systematic Classification of Molecular Symmetries [1]

Point groups are frequently incomplete to afford full symmetry information about molecules; various molecules belong to the same point group. For example, a list of C_{2v} molecules contains water, hydrogen sulfide, formaldehyde, phosgene, chlorobenzene, fluorobenzene, 1,2- and 1,3-difluorobezenes, 1,2-benzoquinone, 1,4-dichloronaphthalene, 1,4-naphthoquinone, pyridine, 4-chloropyridine, 2,6-dichloro-pyridine, pyridine N-oxide, furan, thiophene, cyclopropanone, cyclobutanone, cyclopentanone, tetrahydrofuran, tetrahydrothiophene, dichloromethane, difluoro-methane, oxirane, phenanthrene, bicyclo[2.2.1]heptane, basketane, adamantanone, noradamantane and so on. Such a list should be classified into several categories in the light of a rational criterion. For this purpose, Pople[1] has proposed the concept of "framework group". By this method, difluoromethane is designated as $C_{2v}[C_2(C), \sigma_v(F_2), \sigma'_v(H_2)]$. Flurry[2] pointed out that the framework group is related to local (site) symmetries and proposed his notation based on the local symmetries. Thus, the difluoromethane is designated as $C_{2v}[C_{2v}(C), C_s(F_2), C'_s(H_2)]$. Although Flurry's method has a potential applicability, it is not so easy to determine such local symmetries especially in the cases of complex molecules. We here propose the SCR (set-of-coset-representation) notation for classifying molecular symmetry.

6.1 Assignment of Coset Representations to Orbits

Atoms in a molecule can be classified into several equivalence classes in the light of molecular symmetry (a point group). In order to obtain a more versatile method of classification beyond a point group, we should develop an effective tool of characterizing such classes. In the preceding chapter, we have referred to them as

---

[1]Reprinted in part with permission from S. Fujita, *Bull. Chem. Soc. Jpn.*, **63**, 315–327 (1990). ©(1990) The Chemical Society of Japan.

orbits; and furthermore, we have clarified one-to-one correspondence between an orbit and a coset representaiton (CR). This correspondence provides us with a characterization tool.

Assignment of a CR to an orbit is accomplished by means of two methods. Since we have discussed the principles of these methods in Chapter 5, we will here describe recipes for carrying out the assignment.

Method I. The 1st method consists of the following steps:

1. Find a point group (**G**) to describe a given molecule.

2. Partition atoms of the molecule into orbits (sets of equivalent atoms).

3. Count fixed atoms to produce an fixed-point vector (FPV) with respect to each of the orbits.

4. Compare the FPV with each of the rows of a mark table for the **G** group; an identical row is a CR to be assigned.

These steps should be repeated over all of the orbits that have been found in the 2nd step. Step 4 requires a mark table for the **G** point group. Such mark tables are found in Appendix A. Example 5.1 (Chapter 5) has exemplified Method I. We examine an additional example as follows.

Example **6.1** Let us examine an oxirane molecule (**1**). This molecule belongs to C_{2v} point group, which has SSG = $\{C_1, C_2, C_s, C'_s, C_{2v}\}$.

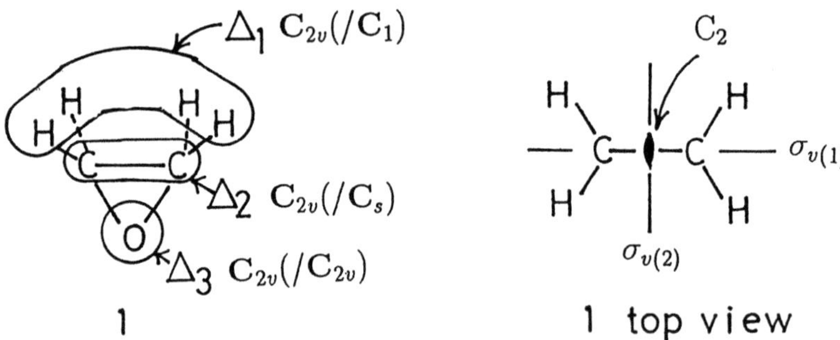

By inspection, we find three orbits, *i.e.*, a four-membered orbit Δ_1 (H_4), a two-membered orbit Δ_2 (C_2) and a one-membered orbit Δ_3 (O). When we apply all the operations of each subgroup of the C_{2v} point group, we obtain the following FPVs:

Table 6.1: Mark table of C_{2v}

coset	subgroup				
reprentation	C_1	C_2	C_s	C_s'	C_{2v}
$C_{2v}(/C_1)$	4	0	0	0	0
$C_{2v}(/C_2)$	2	2	0	0	0
$C_{2v}(/C_s)$	2	0	2	0	0
$C_{2v}(/C_s')$	2	0	0	2	0
$C_{2v}(/C_{2v})$	1	1	1	1	1

$$(4\ 0\ 0\ 0\ 0) \quad \text{for the orbit } (\Delta_1) \text{ of } H_4,$$
$$(2\ 0\ 2\ 0\ 0) \quad \text{for the orbit } (\Delta_2) \text{ of } C_2, \text{ and}$$
$$(1\ 1\ 1\ 1\ 1) \quad \text{for the orbit } (\Delta_3) \text{ of } O.$$

By the comparison of these FPVs with the mark table of C_{2v} (Table 6.1), we have $C_{2v}(/C_1)$ for the Δ_1 orbit, $C_{2v}(/C_s)$ for the Δ_2 orbit, and $C_{2v}(/C_{2v})$ for the Δ_3 orbit. Note $|\ C_{2v}\ |/|\ C_1\ | = 4/1 = 4$, $|\ C_{2v}\ |/|\ C_s\ | = 4/2 = 2$, and $|\ C_{2v}\ |/|\ C_{2v}\ | = 4/4 = 1$, which are equal to the lengths of the respective orbits.

Method II. On the other hand, the 2nd method contains the following steps:

1. Find a point group (G) to describe a given molecule.

2. Count fixed atoms to produce an fixed-point vector (FPV) with respect to all of the atoms to be examined.

3. Multiply the FPV by the inverse of the mark table for the G group.

4. Refer the the resulting vector to SSG $= \{G_1, G_2, \ldots, G_s\}$ and find the multiplicity of the CR $G(/G_i)$ to be assigned.

Method II requires the inverses of mark tables for point groups. They are found in Appendix B. Example 5.2 (Chapter 5) explains Method II by using D_{2d} group. The following example is concerned with C_{2v} group.

Example **6.2** For illustrating the 2nd method, let us examine the same oxirane molecule (**1**). When we apply all the operations of each subgroup of the C_{2v} point group to the seven atoms, we obtain an FPV, $(7\ 1\ 3\ 1\ 1)$. This vector is multiplied by the inverse of the mark table, *i.e.*,

$$(7\ 1\ 3\ 1\ 1)\begin{pmatrix} 1/4 & 0 & 0 & 0 & 0 \\ -1/4 & 1/2 & 0 & 0 & 0 \\ -1/4 & 0 & 1/2 & 0 & 0 \\ -1/4 & 0 & 0 & 1/2 & 0 \\ 1/2 & -1/2 & -1/2 & -1/2 & 1 \end{pmatrix} = (1\ 0\ 1\ 0\ 1), \qquad (6.1)$$

wherein the second 5×5 matrix is the inverse. The resulting vector contains respective multiplicities of coset representations (CRs). The multiplicities are aligned in the order of a set of CRs, *i.e.*, SCR = $\{C_{2v}(/C_1), C_{2v}(/C_2), C_{2v}(/C_s), C_{2v}(/C'_s), C_{2v}(/C_{2v})\}$. Hence, we have a formal expression,

$$P_{C_{2v}} = C_{2v}(/C_1) + C_{2v}(/C_s) + C_{2v}(/C_{2v}). \tag{6.2}$$

We can easily assign these CRs to the sets of equivalent atoms, *i.e.*, $C_{2v}(/C_1)$ to the Δ_1 orbit (H_4), $C_{2v}(/C_s)$ to the Δ_2 orbit (C_2), and $C_{2v}(/C_{2v})$ to the Δ_3 orbit (O).

It should be noted that there is arbitrariness in selecting C_s or C'_s. Accordingly, an alternative assignment of $C_{2v}(/C'_s)$ to the Δ_2 orbit (C_2) is possible. However, this arbitrariness is not essential in discussing symmetrical properties.

6.2   SCR Notation

The discussions in the preceding section provide a basis for characterizing the symmetry of a given molecule. Atoms contained in the molecule are partitioned into several orbits or sets of equivalent atoms, which correspond to an appropriate sum of CRs, $\sum_{i=1}^{s} \alpha_i G(/G_i)$, where G is the point group of the molecule; G_i is a subgroup of G; and the symbol α_i represents the multiplicity of each CR (Theorem 5.7 in Chapter 5). Let the symbol $\Delta_{i\alpha}$ denote an α-th orbit that is subject to the CR $G(/G_i)$, where $\alpha = 1, 2, \ldots, \alpha_i$ for each i. Suppose that the orbit ($\Delta_{i\alpha}$) is occupied by the atoms ($A^{(i\alpha)}$) of the same kind and that the number of the atoms is equal to $r = |\Delta_{i\alpha}| = |G|/|G_i|$. Then, the molecule is represented by

$$G[\cdots; /G_i(\underbrace{A_r^{(i1)}, A_r^{(i2)}, A_r^{(i\alpha_i)}}_{\alpha_i}); \cdots], \tag{6.3}$$

which contains each CR after a slush (/) and its members in parentheses. We call this symmol an *SCR (set-of-CR) notation*. Since the orbits of the oxirane (**1**) is represented by eq. 6.2, the SCR notation of **1** is obtained as $C_{2v}[/C_1(H_4); /C_s(C_2); /C_{2v}(O)]$. This SCR notation means that four hydrogen atoms of **1** construct an orbit subject to $C_{2v}(/C_1)$; that two carbon atoms construct an orbit subject to $C_{2v}(/C_s)$; and that one oxygen atom constructs an orbit subject to $C_{2v}(/C_{2v})$.

Any molecules can be specified in terms of eq. 6.3. Table 6.2 lists the SCR notations of molecules of C_{2v} symmetry, where several structures are shown.

Table 6.2: SCR Notations of C_{2v}-molecules

molecule	SCR notation
water	$C_{2v}[/C_s(H_2); /C_{2v}(O)]$
hydrogen sulfide	$C_{2v}[/C_s(H_2); /C_{2v}(S)]$
formaldehyde	$C_{2v}[/C_s(H_2); /C_{2v}(C,O)]$
phosgene	$C_{2v}[/C_s(Cl_2); /C_{2v}(C,O)]$
chlorobenzene	$C_{2v}[/C_s(2H_2,2C_2); /C_{2v}(H,2C,Cl)]$
fluorobenzene	$C_{2v}[/C_s(2H_2,2C_2); /C_{2v}(H,2C,F)]$
1,2-difluorobezene	$C_{2v}[/C_s(2H_2,3C_2, F_2)]$
1,3-difluorobezene	$C_{2v}[/C_s(H_2,2C_2, F_2); /C_{2v}(2H,2C)]$
1,2-benzoquinone	$C_{2v}[/C_s(2H_2,3C_2, O_2)]$
1,4-dichloronaphthalene	$C_{2v}[/C_s(3H_2,5C_2, Cl_2)]$
1,4-naphthoquinone	$C_{2v}[/C_s(3H_2,5C_2, O_2)]$
pyridine	$C_{2v}[/C_s(2H_2,2C_2); /C_{2v}(H,C,N)]$
4-chloropyridine	$C_{2v}[/C_s(2H_2,2C_2); /C_{2v}(C,N,Cl)]$
2,6-dichloropyridine	$C_{2v}[/C_s(H_2,2C_2,Cl_2); /C_{2v}(H,C,N)]$
pyridine N-oxide	$C_{2v}[/C_s(2H_2,2C_2); /C_{2v}(H,C,N,O)]$
furan	$C_{2v}[/C_s(2H_2,2C_2); /C_{2v}(O)]$
thiophene	$C_{2v}[/C_s(2H_2,2C_2); /C_{2v}(S)]$
cyclopropanone	$C_{2v}[/C_1(H_4); /C_s(C_2); /C_{2v}(C,O)]$
cyclobutanone	$C_{2v}[/C_1(H_4); /C_s(C_2); /C_s'(H_2); /C_{2v}(2C,O)]$
cyclopentanone	$C_{2v}[/C_1(2H_4); /C_s(2C_2); /C_{2v}(C,O)]$
tetrahydrofuran	$C_{2v}[/C_1(2H_4); /C_s(2C_2); /C_{2v}(O)]$
tetrahydrothiophene	$C_{2v}[/C_1(2H_4); /C_s(2C_2); /C_{2v}(S)]$
dichloromethane	$C_{2v}[/C_s(H_2); /C_s'(Cl_2); /C_{2v}(C)]$
difluoromethane	$C_{2v}[/C_s(H_2); /C_s'(F_2); /C_{2v}(C)]$
oxirane (**1**)	$C_{2v}[/C_1(H_4); /C_s(Cl_2); /C_{2v}(O)]$
phenanthrene (**2**)	$C_{2v}[/C_s(7C_2, 5H_2)]$
iceanedione (**3**)	$C_{2v}[/C_1(2C_4,3H_4); /C_s(2C_2,H_2,O_2)]$
noradamantane (**4**)	$C_{2v}[/C_1(C_4,2H_4); /C_s(C_2, 2H_2); /C_s'(C_2, H_2); /C_{2v}(C)]$
bicyclo[2.2.1]heptane (**5**)	$C_{2v}[/C_1(C_4,2H_4); /C_s(C_2, H_2); /C_s'(H_2); /C_{2v}(C)]$
basketane (**6**)	$C_{2v}[/C_1(C_4,2H_4); /C_s(3C_2, 2H_2)]$
adamantanone (**7**)	$C_{2v}[/C_1(C_4,2H_4); /C_s(C_2, 2H_2); /C_s'(C_2,H_2); /C_{2v}(2C,O)]$
1,1-difluorocyclopropane (**8**)	$C_{2v}[/C_1(H_4); /C_s(C_2); /C_s'(F_2); /C_{2v}(C)]$
1,1-difluoroallene (**9**)	$C_{2v}[/C_s(H_2); /C_s'(F_2); /C_{2v}(3C)]$

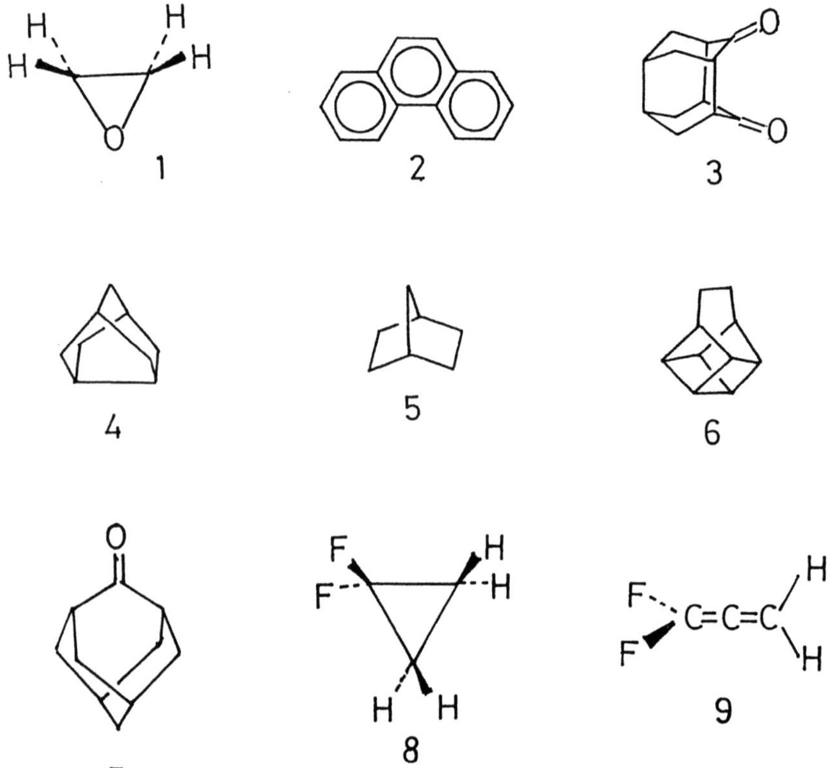

Example **6.3** Let us now examine molecules of $\mathbf{D}_{3h}$ symmetry. For the $\mathbf{D}_{3h}$ point group, we obtain SSG $= \{\mathbf{C}_1, \mathbf{C}_2, \mathbf{C}_s, \mathbf{C}_s', \mathbf{C}_3, \mathbf{C}_{2v}, \mathbf{C}_{3v}, \mathbf{C}_{3h}, \mathbf{D}_3, \mathbf{D}_{3h}\}$.

Let us work out an X_3Y_2 derivative (**10**) of phosphorane. When we count fixed points for every subgroup of the SSG, we have an FPV $= (6 \;\; 2 \;\; 4 \;\; 4 \;\; 3 \;\; 2 \;\; 3 \;\; 1 \;\; 1 \;\; 1)$. The FPV is multiplied by the inverse of a mark table for $\mathbf{D}_{3h}$ to afford

$(6\;2\;4\;4\;3\;2\;3\;1\;1\;1) \times$

$$\begin{pmatrix}
1/12 & 0 & 0 & 0 & 0 & 0 & 0 & 0 & 0 & 0 \\
-1/4 & 1/2 & 0 & 0 & 0 & 0 & 0 & 0 & 0 & 0 \\
-1/4 & 0 & 1/2 & 0 & 0 & 0 & 0 & 0 & 0 & 0 \\
-1/12 & 0 & 0 & 1/6 & 0 & 0 & 0 & 0 & 0 & 0 \\
-1/12 & 0 & 0 & 0 & 1/4 & 0 & 0 & 0 & 0 & 0 \\
1/2 & -1/2 & -1/2 & -1/2 & 0 & 1 & 0 & 0 & 0 & 0 \\
1/4 & 0 & -1/2 & 0 & -1/4 & 0 & 1/2 & 0 & 0 & 0 \\
1/12 & 0 & 0 & -1/6 & -1/4 & 0 & 0 & 1/2 & 0 & 0 \\
1/4 & -1/2 & 0 & 0 & -1/4 & 0 & 0 & 0 & 1/2 & 0 \\
-1/2 & 1/2 & 1/2 & 1/2 & 1/2 & -1 & -1/2 & -1/2 & -1/2 & 1
\end{pmatrix}$$

$$= (0\,0\,0\,0\,0\,1\,1\,0\,0\,1).$$

Note that the second 10×10 matrix is the inverse. The resulting vector indicates

$$\mathbf{P_{D_{3h}}} = \mathbf{D}_{3h}(/\mathbf{C}_{2v}) + \mathbf{D}_{3h}(/\mathbf{C}_{3v}) + \mathbf{D}_{3h}(/\mathbf{D}_{3h}), \tag{6.4}$$

since a set of CRs is represented by SCR = $\{\mathbf{D}_{3h}(/\mathbf{C}_1)$, $\mathbf{D}_{3h}(/\mathbf{C}_2)$, $\mathbf{D}_{3h}(/\mathbf{C}_s)$, $\mathbf{D}_{3h}(/\mathbf{C}'_s)$, $\mathbf{D}_{3h}(/\mathbf{C}_3)$, $\mathbf{D}_{3h}(/\mathbf{C}_{2v})$, $\mathbf{D}_{3h}(/\mathbf{C}_{3v})$, $\mathbf{D}_{3h}(/\mathbf{C}_{3h})$, $\mathbf{D}_{3h}(/\mathbf{D}_3)$, $\mathbf{D}_{3h}(/\mathbf{D}_{3h})\}$.

We can easily assign these CRs to the sets of equivalent atoms, *i.e.*, $\mathbf{D}_{3h}(/\mathbf{C}_{2v})$ to the Δ_1 orbit (X_3), $\mathbf{D}_{3h}(/\mathbf{C}_{3v})$ to the Δ_2 orbit (Y_2), and $\mathbf{D}_{3h}(/\mathbf{D}_{3h})$ to the Δ_3 orbit (P). This result is summarized by the SCR notation, $\mathbf{D}_{3h}[/\mathbf{C}_{2v}(X_3);$ $/\mathbf{C}_{3v}(Y_2);/\mathbf{D}_{3h}(P)]$.

Example 6.3 is based on Method II. Of course, Method I is applicable to this case, where the mark table listed in Appendix A is used. There are various molecules belonging to $\mathbf{D}_{3h}$ symmetry. Table 6.3 lists SCR notations for representatives of such $\mathbf{D}_{3h}$-molecules.

Example **6.4** Let us now examine methane (**16**) that is simple but belongs to complicated $\mathbf{T}_d$ symmetry. According to SSG = $\{\mathbf{C}_1$, $\mathbf{C}_2$, $\mathbf{C}_s$, $\mathbf{C}_3$, $\mathbf{S}_4$, $\mathbf{D}_2$, $\mathbf{C}_{2v}$, $\mathbf{C}_{3v}$, $\mathbf{D}_{2d}$, $\mathbf{T}$, $\mathbf{T}_d\}$, we count fixed points among 5 atoms. Then, we have FPV = $(5\ 1\ 3\ 2\ 1\ 1\ 1\ 2\ 1\ 1\ 1)$. This FPV is multiplied by the inverse of a mark table for $\mathbf{T}_d$ to afford

$$(5\ 1\ 3\ 2\ 1\ 1\ 1\ 2\ 1\ 1\ 1)\times$$

Table 6.3: SCR Notations of D_{3h}-molecules

molecule	SCR notation
10	$D_{3h}[/C_{2v}(X_3); /C_{3v}(Y_2);/D_{3h}(P)]$
11	$D_{3h}[/C_s(H_6); /C_{2v}(C_3)]$
12	$D_{3h}[/C_s(C_6,H_6)]$
13	$D_{3h}[/C_s(3C_6,2H_6); /C_{3v}(C_2,H_2)]$
14	$D_{3h}[/C_1(H_{12}); /C'_s(2C_6)]$
15	$D_{3h}[/C_{2v}(F_3); /D_{3h}(B)]$

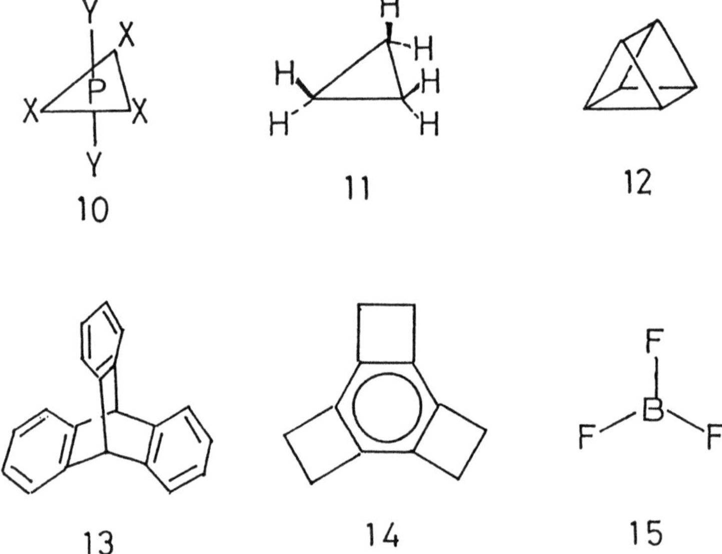

$$\begin{pmatrix}
1/24 & 0 & 0 & 0 & 0 & 0 & 0 & 0 & 0 & 0 & 0 \\
-1/8 & 1/4 & 0 & 0 & 0 & 0 & 0 & 0 & 0 & 0 & 0 \\
-1/4 & 0 & 1/2 & 0 & 0 & 0 & 0 & 0 & 0 & 0 & 0 \\
-1/6 & 0 & 0 & 1/2 & 0 & 0 & 0 & 0 & 0 & 0 & 0 \\
0 & -1/4 & 0 & 0 & 1/2 & 0 & 0 & 0 & 0 & 0 & 0 \\
1/12 & -1/4 & 0 & 0 & 0 & 1/6 & 0 & 0 & 0 & 0 & 0 \\
1/4 & -1/4 & -1/2 & 0 & 0 & 0 & 1/2 & 0 & 0 & 0 & 0 \\
1/2 & 0 & -1 & -1/2 & 0 & 0 & 0 & 1 & 0 & 0 & 0 \\
0 & 1/2 & 0 & 0 & -1/2 & -1/2 & -1/2 & 0 & 1 & 0 & 0 \\
1/6 & 0 & 0 & -1/2 & 0 & -1/6 & 0 & 0 & 0 & 1/2 & 0 \\
-1/2 & 0 & 1 & 1/2 & 0 & 1/2 & 0 & -1 & -1 & -1/2 & 1
\end{pmatrix}$$

$$= (0\ 0\ 0\ 0\ 0\ 0\ 0\ 1\ 0\ 0\ 1),$$

wherein the second 11×11 matrix is the inverse matrix. The resulting vector indicates

$$\mathbf{P_{T_d}} = \mathbf{T}_d(/\mathbf{C}_{3v}) + \mathbf{T}_d(/\mathbf{T}_d), \tag{6.5}$$

because a set of CRs is SCR = { $\mathbf{T}_d(/\mathbf{C}_1)$, $\mathbf{T}_d(/\mathbf{C}_2)$, $\mathbf{T}_d(/\mathbf{C}_s)$, $\mathbf{T}_d(/\mathbf{C}_3)$, $\mathbf{T}_d(/\mathbf{S}_4)$, $\mathbf{T}_d(/\mathbf{D}_2)$, $\mathbf{T}_d(/\mathbf{C}_{2v})$, $\mathbf{T}_d(/\mathbf{C}_{3v})$, $\mathbf{T}_d(/\mathbf{D}_{2d})$, $\mathbf{T}_d(/\mathbf{T})$, $\mathbf{T}_d(/\mathbf{T}_d)$} in this case. We can easily assign these CRs to the sets of equivalent atoms, *i.e.*, $\mathbf{T}_d(/\mathbf{C}_{3v})$ to the Δ_1 orbit (H_4) and $\mathbf{T}_d(/\mathbf{T}_d)$ to the Δ_2 orbit (C). This result is summarized by the SCR notation, $\mathbf{T}_d[/\mathbf{C}_{3v}(H_4); /\mathbf{T}_d(C)]$.

Table 6.4 collects SCR notations for several $\mathbf{T}_d$-molecules. The compound (**17**) is named tetrahedrane after its geometrical form. The compound (**18**) is called adamantane because of its diamond structure (Greek: adamas). The geometrical relationship between **17** and **18** will be discussed in Chapter 17. The compound (**19**) is derived by substituting nitrogen atoms for four methines of cubane. The substitution reduces the original $\mathbf{O}_h$ symmetry of cubane into $\mathbf{T}_d$.

The original version[3] of the SCR notation consists of type I and II notations. The type II notation is based on the subduction of coset representations, characterizing the symmetry of a molecule as well as that of a parent skeleton. The type I notation is an abbreviation of the type I notation, where information about the skeleton is omitted. The present version essentially succeeds to the type I notation, whereas the distinction between molecules and skeletons are not taken into consideration.

In Chapter 7, we will show that any point of a $\mathbf{G}(/\mathbf{G}_i)$ orbit belongs to the local symmatry $\mathbf{G}_i$. The $\mathbf{G}(/\mathbf{G}_i)$ determines the permutational properties as well as the local symmetry of the orbit. The symbol $(\mathbf{G}(/\mathbf{G}_i))$ is convenient to indicate such inherent nature of the orbit.

Table 6.4: SCR Notations of T_d-molecules

molecule	SCR notation
16	$T_d[/C_{3v}(H_4);\ /T_d(C)]$
17	$T_d[/C_{3v}(C_4,\ H_4)]$
18	$T_d[/C_s(H_{12});\ /C_{2v}(C_6);\ /C_{3v}(C_4,\ H_4)]$
19	$T_d[/C_{3v}(C_4,H_4,N_4)]$

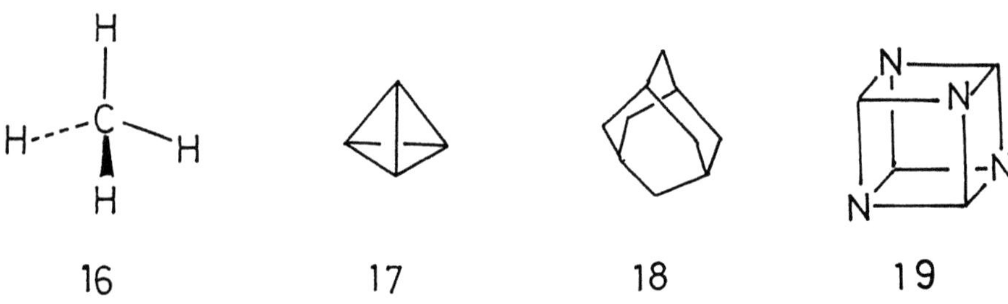

16   17   18   19

Bibliography

[1] J. A. Pople, *J. Am. Chem. Soc.*, **102**, 4615 (1980).

[2] R. L. Flurry, Jr. *J. Am. Chem. Soc.*, **103**, 2901 (1981).

[3] S. Fujita, *Bull. Chem. Soc. Jpn.*, **63**, 315 (1990).

Chapter 7

Local Symmetries and Forbidden Coset Representations [1]

In the preceding chapter, we have discussed SCR notations, which are based on the correspondence between coset representations (CRs) and orbits (sets of equivalent atoms). Without taking account of this correspondence, enumeration of such orbits has been done by Brester[1], Jahn-Teller[2], Boyle[3] and Fowler-Quinn[4] for most point groups. By these conventional methods, each orbit (O_A) is usually characterized by the site symmetry group (H_A) that stabilizes (or fixes) one site (A) of the orbit. This characterization is accomplished by using tables presented by Fowler and Quinn.[4] The site symmetry group (H_A) for a non-center atom is shown to be one of C_1, C_s, C_n, or C_{nv} ($n \geq 2$);[3] several subsymmetries of G are incapable of being site symmetry groups.

This incapability becomes clearer if we consider the correspondence between CRs and orbits. For example, Table 5.1 (Chapter 5) indicates such correspondence between the CRs of the D_{2d} group and O_A-notations[4] for orbits. By the inspection of this table, there naturally emerges a question: Why are the orbits corresponding to $D_{2d}(/C_2)$ *etc.* absent? We find other examples if we examine the SCR notations described in Chapter 6. For example, the CR $C_{2v}(/C_2)$ has not emerged in such SCR notations, whereas the other CRs are present.

There has been, however, no mathematical rationalization for this type of absence. The present chapter aims at providing such a selection that is based on a more logical framework in terms of coset representations. For this purpose, we here clarify chemical and geometrical meanings of CRs and their relationship to orbits in such a manner that is different from the treatment in the preceding chapters. This analysis reveals the inherent identities of Brester's method[1], the so-called k-values in framework groups,[5], the σ-character technique[4], and the site symmetry method[6] in the light of CRs.

---

[1]Reprinted in part from S. Fujita, *Theor. Chim. Acta*, **78**, 45–63 (1990). ©(1990) Springer-Verlag.

7.1 Blocks and Local Symmetries

Local symmetries in a regular body. First, we explain the geometrical meaning of a coset representation (CR) by introducing the concept of a *regular body*.

Let $\mathbf{G}$ be a finite point group of order $|\,\mathbf{G}\,|$ and act on a set of $|\,\mathbf{G}\,|$ points in a 3D-space:

$$\Delta^R = \{\delta_1, \delta_2, \ldots, \delta_{|\mathbf{G}|}\}. \tag{7.1}$$

Suppose that there is only one stabilizer of the point (δ_1) which is an identity group, *i.e.*, $\mathbf{G}_{\delta_1} = \mathbf{G}_1 = \{\mathrm{I}\}$. If g_i $(\in \mathbf{G})$ transforms the point δ_i to δ_1, we obtain a coset decomposition represented by

$$\begin{aligned}
\mathbf{G} &= \mathbf{G}_{\delta_1} g_1 + \mathbf{G}_{\delta_1} g_2 + \ldots + \mathbf{G}_{\delta_1} g_{|\mathbf{G}|} \\
&= \mathbf{G}_1 g_1 + \mathbf{G}_1 g_2 + \ldots + \mathbf{G}_1 g_{|\mathbf{G}|}.
\end{aligned} \tag{7.2}$$

Hence, each point δ_i corresponds to the coset $\mathbf{G}_1 g_i$ (*i.e.*, g_i itself) in one-to-one fashion. Since $\mathbf{G}(/\mathbf{G}_1)$ is the regular representation (RR), Δ^R defines the regular orbit about δ_1. We call this 3D-object (Δ^R) a *regular body*.

Let us consider a subset of the regular body (Δ^R), which belongs to a subgroup $\mathbf{G}_i$ ($\leq \mathbf{G}$). We call the subset *a block of* $\mathbf{G}_i$ *symmetry* or a $\mathbf{G}_i$-*block*. This selection is realized by considering a subduction of the RR,

$$\mathbf{G}(/\mathbf{G}_1) \downarrow \mathbf{G}_i = \frac{|\,\mathbf{G}\,|}{|\,\mathbf{G}_i\,|} \mathbf{G}_i(/\mathbf{G}_{i1}), \tag{7.3}$$

where $\mathbf{G}_i(/\mathbf{G}_{i1})$ is a regular representation of the subgroup $\mathbf{G}_i$. The subduction of RR is a special case of subductions of coset representations, which will be discussed later.

Equation 7.3 indicates that the regular body (Δ^R) is divided into $|\,\mathbf{G}\,|/|\,\mathbf{G}_i\,|$ blocks of $\mathbf{G}_i$-symmetry, the sizes of which are equal to $|\,\mathbf{G}_i\,|$. We represent these blocks by the symbols,

$$\omega_1, \omega_2, \ldots, \omega_r,$$

where $r = |\,\mathbf{G}\,| / |\,\mathbf{G}_i\,|$. Since we can select an arbitrary block from these blocks, we examine ω_1 as a representative case. Let us consider a coset decomposition of $\mathbf{G}$ by $\mathbf{G}_i$,

$$\mathbf{G} = \mathbf{G}_i g_1 + \mathbf{G}_i g_2 + \ldots + \mathbf{G}_i g_r, \tag{7.4}$$

where $g_1 = I$. If we operate the representatives, $\{g_1, g_2, \ldots, g_r\}$, onto the block ω_1, we obtain r blocks,

$$\Omega_1 = g_1 \omega_1 = \omega_1, \ \Omega_2 = g_2 \omega_1, \ \ldots, \ \Omega_r = g_r \omega_1.$$

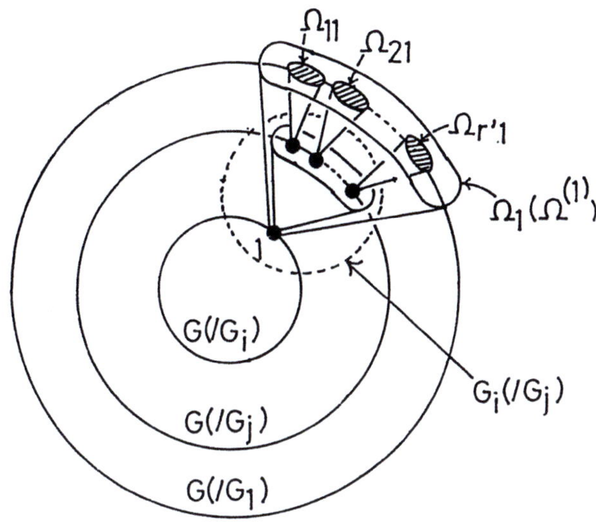

Figure 7.1: Blocks in a regular body

Since each Ω_k corresponds to the coset $G_i g_k$ through g_k in one-to-one fashion, the coset representation $G(/G_i)$ based on eq. 7.4 governs a set of blocks,

$$\Omega = \{\Omega_1, \Omega_2, \ldots, \Omega_r\}. \tag{7.5}$$

Figure 7.1 illustrates these blocks $(\Omega_k; k = 1, 2, \ldots, r)$, where we pay our attention to the relationship between $G(/G_i)$ and $G(/G_1)$. Since each of the blocks (Ω_k) belongs to $G_i(/G_{i1})$, we arrive at the following lemma.

Lemma 7.1 (Local symmetry in a regular body).
Suppose that a regular body of G *symmetry is subject to* $G(/G_1)$. *Then, the coset representation* $G(/G_i)$ *governs a* G_i *block that has a* $G_i(/G_{i1})$ *orbit.*

Note that the $G_i(/G_{i1})$ representation is an RR of G_i. This lemma reveals the meaning implied by the conventional term "site symmetry". We refer to the G_i group in the parentheses in the CR $G(/G_i)$ as a *local symmetry*, while we refer to G as a *global symmetry*. The term "site symmetry" is originally used to designate the symmetry of a site or an atom. On the other hand, the term "local symmetry" defined here is regarded as an attribute of the $G(/G_i)$ orbit. In addition, Lemma 7.1 indicates that the term "local symmetry" is applicable to such a site (a member of the orbit) as well as to a block attaching to the site.

The above formulation is purely mathematical and rather tedious to follow; however, its chemical meaning is quite simple. The following examples give intuitive explanation that is helpful to understand the mathematical formulation.

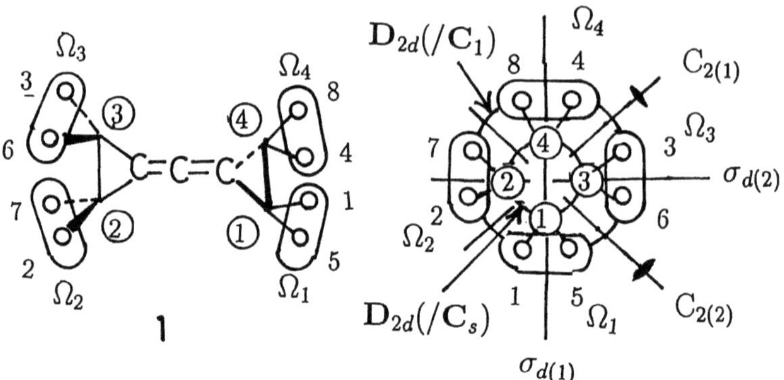

Figure 7.2: Orbits and local symmetry in a $\mathbf{D}_{2d}$-molecule

Example **7.1** Let us first work out an allene derivative (**1**) depicted in Figure 7.2. The 8 hydrogen atoms of this molecule construct an orbit governed by $\mathbf{D}_{2d}(/\mathbf{C}_1)$. The four methylene carbons in the two cyclopropane rings belong to a $\mathbf{D}_{2d}(/\mathbf{C}_s)$ orbit. The central carbon of the allene group is subject to the CR ($\mathbf{D}_{2d}(/\mathbf{D}_{2d})$); two terminal allene carbons belong to the $\mathbf{D}_{2d}(/\mathbf{C}_{2v})$ orbit. The assignment of the CRs to the respective orbits is accomplished by referring to the table of marks (Chapter 5).

We focus our attention on the $\mathbf{D}_{2d}(/\mathbf{C}_1)$ orbit (8 hydrogens), which is subject to the regular representation. Thus, the orbit can be considered to be a regular body. Consider a subduction by $\mathbf{C}_s$ group. Then, we have the following blocks: $\omega_1 = \{\mathrm{H}^1, \mathrm{H}^5\}$, $\omega_2 = \{\mathrm{H}^2, \mathrm{H}^6\}$, $\omega_3 = \{\mathrm{H}^3, \mathrm{H}^7\}$, and $\omega_4 = \{\mathrm{H}^4, \mathrm{H}^8\}$, all of which belong to the $\mathbf{C}_s$ local symmetry. Among them, we select the first block, *i.e.*, $\Omega_1 = \omega_1 = \{\mathrm{H}^1, \mathrm{H}^5\}$. When every symmetry operation of $\mathbf{D}_{2d}$ is operated on Ω_1, there emerge $\Omega_2 = \{\mathrm{H}^2, \mathrm{H}^7\}$, $\Omega_3 = \{\mathrm{H}^3, \mathrm{H}^6\}$, and $\Omega_4 = \{\mathrm{H}^4, \mathrm{H}^8\}$. As we can see easily, each of the blocks belongs to $\mathbf{C}_s$ symmetry. More precisely, Ω_1 and Ω_2 belong to the $\mathbf{C}_s$ ($\leq \mathbf{D}_{2d}$); on the other hand, Ω_3 and Ω_4 belong to the $\mathbf{C}'_s$ ($\leq \mathbf{D}_{2d}$) that is conjugate to the $\mathbf{C}_s$.

We now consider a set of the blocks, $\Omega = \{\Omega_1, \Omega_2, \Omega_3, \Omega_4\}$ as a $\mathbf{D}_{2d}$-set. Because the blocks Ω_i ($i = 1, 2, 3, 4$) correspond to cyclopropane carbons $\mathrm{C}^{(i)}$ in one-to-one fashion and because the $\mathrm{C}^{(i)}$'s belong to $\mathbf{D}_{2d}(/\mathbf{C}_s)$, the orbit Ω is concluded to be subject to $\mathbf{D}_{2d}(/\mathbf{C}_s)$. This intuitive explanation indicates that the blocks Ω_i ($i = 1, 2, 3, 4$) having $\mathbf{C}_s$ symmetry constitute a $\mathbf{D}_{2d}(/\mathbf{C}_s)$ orbit. Thus, the CR $\mathbf{D}_{2d}(/\mathbf{C}_s)$ is found to correspond to the local symmetry $\mathbf{C}_s$ in one-to-one fashion. An alternative selection of $\mathbf{C}_s$ blocks is possible *i.e.*, $\Omega_1 = \{\mathrm{H}^2, \mathrm{H}^6\}$, $\Omega_2 = \{\mathrm{H}^1, \mathrm{H}^8\}$, $\Omega_3 = \{\mathrm{H}^4, \mathrm{H}^5\}$, and $\Omega_4 = \{\mathrm{H}^3, \mathrm{H}^7\}$. This selection is essentially

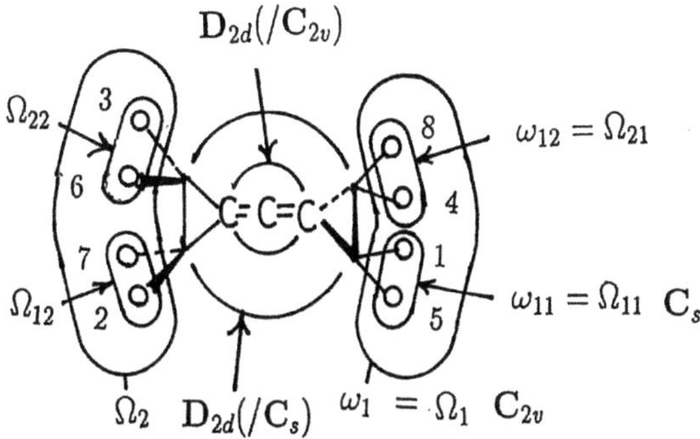

Figure 7.3: Orbits and local symmetry in a D_{2d}-molecule

equivalent to the former selection in a geometric sense.

Example **7.2** We re-examine the molecule (1) described in Example 7.1. Consider a subduction of the $D_{2d}(/C_1)$ orbit (8 hydrogens) by C_{2v} group (Fig. 7.3). Then, we have the following blocks: $\omega_1 = \{H^1, H^5, H^4, H^8\}$ and $\omega_2 = \{H^2, H^7, H^3, H^6\}$. which belong to the C_{2v} local symmetry. Among them, we select the first block, *i.e.*, $\Omega_1 = \omega_1 = \{H^1, H^5, H^4, H^8\}$. If every symmetry operation of D_{2d} is operated on Ω_1, there emerges $\Omega_2 = \{H^2, H^7, H^3, H^6\}$, which is identical with ω_2.

The set of such blocks, $\Omega = \{\Omega_1, \Omega_2\}$, is now regarded as a D_{2d}-set. Because the blocks Ω_i $(i = 1, 2)$ correspond to the allene carbons in one-to-one fashion and because the allene terminals belong to $D_{2d}(/C_{2v})$, the orbit Ω is concluded to be subject to $D_{2d}(/C_{2v})$. This intuitive explanation indicates that the blocks Ω_i $(i = 1, 2)$ having C_{2v} symmetry constitute a $D_{2d}(/C_{2v})$ orbit. Thus, the CR $D_{2d}(/C_{2v})$ is found to correspond to the local symmetry C_{2v} in one-to-one fashion. This treatment is understandable if we pay our attention to two cyclopropane rings.

Regular body for a subgroup. Let us now consider a regular body which corresponds to $G_i(/G_{i1})$. If an appropriate group G_j satisfies $G_j \leq G_i \leq G$, we can consider a subduction of $G_i(/G_{i1})$ in terms of

$$G_i(/G_{i1}) \downarrow G_j = \frac{|G_i|}{|G_j|} G_j(/G_{j1}). \tag{7.6}$$

According to this equation, the orbit (ω_p) governed by the RR $(G_i(/G_{i1}))$ is divided into r' blocks,

$$\omega_{1p}, \omega_{2p}, \ldots, \omega_{r'p},$$

where $r' =\mid G_i \mid / \mid G_j \mid$ and $p = 1, 2, \ldots, r$. Let us consider a coset decomposition,

$$G_i = G_j h_1 + G_j h_2 + \ldots + G_j h_{r'}, \tag{7.7}$$

where $h_1 = I$. Then the same discussion as above holds for this case. If we consider ω_{11} only (*i.e*, $p = 1$), the set of blocks

$$\Omega_1 = \{\Omega_{11}, \Omega_{21}, \ldots, \Omega_{r'1}\} \tag{7.8}$$

is subject to a CR $G_i(/G_j)$, where

$$\Omega_{11} = h_1 \omega_{11} = \omega_{11}, \Omega_{21} = h_2 \omega_{21}, \ldots, \Omega_{r'1} = h_{r'} \omega_{r'1}.$$

Note here that all of these blocks belong to G_j symmetry. Figure 7.1 depicts the relationship between these blocks. Because the block encircled with a broken line is associated with Ω_1 on the $G(/G_1)$ orbit and because Ω_1 is subject to $G_i(/G_j)$, the encircled block is concluded to be subject to $G_i(/G_j)$. In other words, the coset representation $G(/G_i)$ governs a G_i-block that has a $G_i(/G_j)$ suborbit if $G_j \leq G_i \leq G$.

When we introduce eq. 7.7 into 7.4, we obtain

$$\begin{aligned}
G = \ & G_j h_{11} + G_j h_{12} + \ldots + G_j h_{1r} + \\
& G_j h_{21} + G_j h_{22} + \ldots + G_j h_{2r} + \\
& \vdots \quad + \quad \vdots \quad + \quad + \quad \vdots \quad + \\
& G_j h_{r'1} + G_j h_{r'2} + \ldots + G_j h_{r'r},
\end{aligned} \tag{7.9}$$

where $h_q g_p = h_{qp}$ ($p = 1, 2, \ldots, r; q = 1, 2, \ldots, r'$). This is a coset decomposition of G by G_j. If we transform ω_{11} by means of h_{qp} ($p = 1, 2, \ldots, r; q = 1, 2, \ldots, r'$) to produce $r'r$ blocks,

$$\widehat{\Omega} = \left.\begin{cases} \begin{array}{cccc} \Omega_{11} & \Omega_{12} & \cdots & \Omega_{1r} \\ \Omega_{21} & \Omega_{22} & \cdots & \Omega_{2r} \\ \vdots & \vdots & & \vdots \\ \Omega_{r'1} & \Omega_{r'2} & \cdots & \Omega_{r'r} \end{array} \end{cases}\right\} \cdots G(/G_j)$$
$$\begin{array}{cccc} \vdots & \vdots & & \vdots \end{array} \tag{7.10}$$
$$\Omega = \{\Omega_1, \quad \Omega_2, \quad \cdots, \quad \Omega_r\} \quad \cdots G(/G_i)$$

where $\Omega_{qp} = h_{qp} \omega_{11}$ ($p = 1, 2, \ldots, r; q = 1, 2, \ldots, r'$). This equation indicates the one-to-one correspondence between Ω_{pq} and the coset $(G_j h_{qp})$ *via* h_{qp}. Hence, the set of blocks $\widehat{\Omega}$ is concluded to be subject to the CR $(G(/G_j))$. Note that Ω_1 is identical with the first column of $\widehat{\Omega}$. Since $h_{qp} \omega_1 = h_q g_p \omega_1 = h_q \Omega_p$, the remaining p-th column of $\widehat{\Omega}$ is associated with Ω_p ($p = 1, 2, \ldots, r$). These discussions are summarized as follows.

Lemma 7.2 (Local symmetry of a $G(/G_i)$ orbit having one suborbit).
*a) If $G_j \leq G_i \leq G$, the coset representation $G(/G_i)$ governs a G_i-block that has
a $G_i(/G_j)$ suborbit.*
*b) The orbit produced by all of such blocks that are equivalent to the G_i-block is
subject to $G(/G_j)$.*

The second proposition is formally represented by an induction,

$$G_i(/G_j) \uparrow G(/G_i) = G(/G_j). \tag{7.11}$$

This lemma describes the mode of substitution in which a G_i-block having a
$G_i(/G_j)$ suborbit is introduced onto every point of a $G(/G_i)$ orbit. Obviously, a
similar discussion can generalize this lemma so as to be applicable to the case in
which the G_i-block has a given set of suborbits. Hence, we obtain a theorem.

Theorem 7.1 (Local symmetry of a $G(/G_i)$ orbit having several suborbits).
*a) If $G_j \leq G_i \leq G$, the coset representation $G(/G_i)$ governs a G_i-block that has
a set of suborbits represented by*

$$\sum_j \beta_j G_i(/G_j), \tag{7.12}$$

*where the summation is over a given set of subgroups satisfying the above condition;
and β_j's are non-negative integers.*
*b) The orbits produced by all of such blocks that are equivalent to the G_i-block are
subject to the CRs represented by*

$$\sum_j \beta_j G(/G_j). \tag{7.13}$$

The first proposition of this theorem means that an orbit (a set of equivalent atoms
or ligands) appearing in a given molecule corresponds to a coset representation,
$G(/G_i)$, in one-to-one fashion, and that each member of the orbit belongs to a local
symmetry of G_i. The second proposition is formally represented by an induction,

$$\sum_j \beta_j G_i(/G_j) \uparrow G(/G_i) = \sum_j \beta_j G(/G_j). \tag{7.14}$$

Theorem 7.1 can be summarized in general: *When a G molecule has a $G(/G_i)$
orbit, each member of the orbit belongs to the local symmetry of G_i.* This theorem
affords mathematical foundations to the SCR (set-of-coset-representation) notation (Chapter 6)[7] and to the concept of chirality fittingness (Chapter 8).[8]

It should be noted here that the concept of "block" is related to but conceptually different from the concept of "ligand". The former is a mathematical concept
and the latter is a chemical one. A block has a chemical meaning if it satisfies
the selection rules described in the next section. We refer to an allowed block as
a *ligand*.

Example **7.3** This is a continuation of Example 7.2 (Fig. 7.3) . Let us examine the four hydrogen atoms that are contained in one of the cyclopropane rings (ω_1 = {H^1, H^5,H^4, H^8}) This is a regular body of the C_{2v} symmetry. Hence, the orbit is divided into ω_{11} = {H^1, H^5} and ω_{12} = {H^4, H^8} by the subduction into C_s. If we select ω_{11} as Ω_{11}, equation 7.8 gives

$$\Omega_1 = \{\Omega_{11}, \Omega_{21}\},$$

where $\Omega_{11} = \omega_{11}$ = {H^1, H^5} and $\Omega_{21} = \omega_{12}$ = {H^4, H^8}. Lemma 7.1 indicates that the Ω_1 is subject to $C_{2v}(/C_s)$.

If we select ω_{11} as Ω_{11}, eq. 7.10 affords

$$\widehat{\Omega} = \left\{ \begin{matrix} \Omega_{11} & \Omega_{12} \\ \Omega_{21} & \Omega_{22} \end{matrix} \right\} \quad \cdots \quad \mathbf{D}_{2d}(/\mathbf{C}_s)$$

$$\begin{matrix} \vdots & \vdots \end{matrix}$$

$$\Omega = \{\Omega_1, \quad \Omega_2\} \quad \cdots \quad \mathbf{D}_{2d}(/\mathbf{C}_{2v})$$

where $\Omega_{12} \doteq$ {H^2, H^7} and Ω_{22} = {H^3, H^6}. We now consider a set of the blocks, $\widehat{\Omega}$. The four members are subject to $\mathbf{D}_{2d}(/\mathbf{C}_s)$. On the other hand, $\Omega = \{\Omega_1, \Omega_2\}$ is subject to $\mathbf{D}_{2d}(/\mathbf{C}_{2v})$.

Figure 7.4 illustrates orbits and CRs for $\mathbf{T}_d$ group, in which a set of atoms marked with open circles in each molecule are the members of each orbit. The orbit (**2**) contains 24 hydrogen atoms, which construct a regular body of the $\mathbf{T}_d$ symmetry. This orbit is governed by RR ($\mathbf{T}_d(/\mathbf{C}_1)$). The other orbits (**3–6**) can be determined to correspond to the respective CRs. The assignment of each orbit to a coset representation is conducted easily as follows. Consider orbit **4** of Figure 7.4, the six-membered orbit of $\mathbf{T}_d$ found in adamantane. When we use all the symmetry operations concerning every subsymmetry of $\mathbf{T}_d$ on **4**, we obtain an FPV: F = (6 2 2 0 0 0 2 0 0 0 0), the elements of which are aligned in the order of $SSG_{\mathbf{T}_d}$ = {$\mathbf{C}_1, \mathbf{C}_2, \mathbf{C}_s, \mathbf{C}_3, \mathbf{S}_4, \mathbf{D}_2, \mathbf{C}_{2v}, \mathbf{C}_{3v}, \mathbf{D}_{2d}, \mathbf{T}, \mathbf{T}_d$}. Since this vector is identical with the $\mathbf{T}_d(/\mathbf{C}_{2v})$ row of a mark table (Appendix A), the orbit (**4**) is subject to the CR $\mathbf{T}_d(/\mathbf{C}_{2v})$.

Figure 7.4 also exemplifies the interrelation of these orbits, which has been shown generally in Fig. 7.1. The twelve methylene carbons of **2**, each of which is attached by a pair of hydrogens selected from the 24 hydrogen atoms of the $\mathbf{T}_d(/\mathbf{C}_1)$ orbit, construct a $\mathbf{T}_d(/\mathbf{C}_s)$ orbit. Note that this orbit is equivalent to the orbit (**3**) in a geometrical sense. This fact is an example of Lemma 7.1, since each pair of hydrogens in the methylene group is considered to be a $\mathbf{C}_s$-block governed by $\mathbf{C}_s(/\mathbf{C}_1)$.

These methylene carbons (belonging to the 12-membered $\mathbf{T}_d(/\mathbf{C}_s)$ orbit) are, in turn, divided into six pairs ($\mathbf{C}_{2v}$-blocks) if every cyclopropane ring is taken

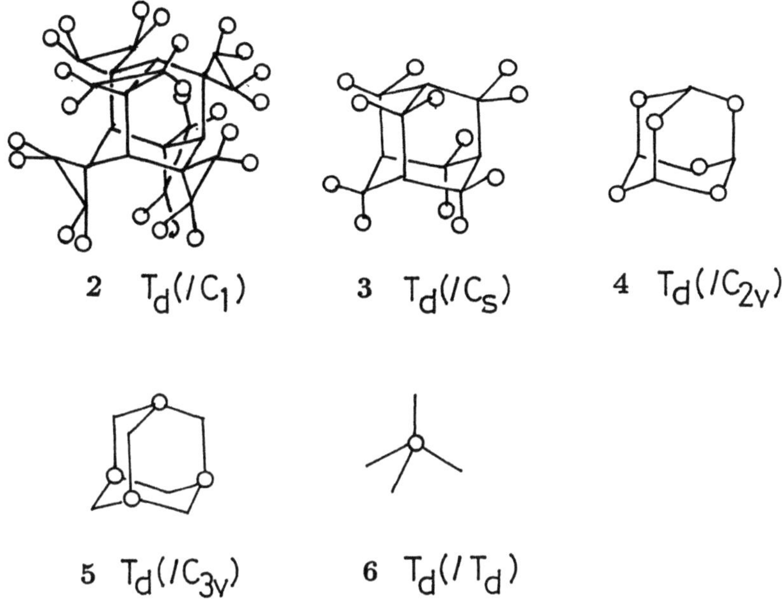

Figure 7.4: Orbits of the T_d point group

into consideration. Each pair in the cyclopropane ring is considered to link to the remaining cyclopropane carbon at each spiro-position. Then, the latter six carbon atoms at the spiro-positions are concluded to belong to a $T_d(/C_{2v})$ orbit, which is equivalent to the orbit (**4**). This fact exemplifies Lemma 7.2, since the pair in such a cyclopropane ring is regarded as a C_{2v}-block having a $C_{2v}(/C_s)$ orbit.

From an alternative point of view, each of the six cyclopropane rings as a whole can be considered to be a C_{2v}-block that contains four hydrogen atoms ($C_{2v}(/C_1)$), two carbon atoms ($C_{2v}(/C_s)$) and one carbon atom ($C_{2v}(/C_{2v})$). Six of such C_{2v} blocks construct a $T_d(/C_{2v})$ orbit. This fact verifies Theorem 7.1. The relationship between 24 hydrogens ($T_d(/C_1)$) and 4 brigdehead carbons ($T_d(/C_{3v})$) can be also recognized by inspection.

It is worthwhile mentioning the differences between our approach and the Fowler-Quinn method.[4] Fowler and Quinn have considered that each orbit (O_A) is characterized by a site-symmetry group, H_A, a subgroup of G which describes the symmetry of a given molecule, and that the reducible representation of G generated by a σ function on each member of the orbit is a permutation representation. Thus, their concepts of orbit, local symmetry, and permutation representation are presented rather separately.

On the other hand, our approach indicates that an orbit is governed by a coset representation $G(/H)$ which comes from a coset decomposition of G by a subgroup H. The assignment of the CR to the orbit can be algebraically accomplished by means of a mark table (Chapters 5 and 6). Moreover, each member of the orbit belongs to the H local symmetry that appears in the symbol $G(/H)$, as shown in this section. Hence, our concepts of "orbit", "local symmetry", and "global symmetry" are closely unified in terms of such a key concept as CR.

7.2   Forbidden Coset Representations

Selection rules for coset representations. Lemmas 7.1 and 7.2 clarify whether a coset representation is allowed or forbidden to govern a set of points as a 3D-object.

Theorem 7.2 (Forbidden CRs).
The coset representation $G(/G_j)$ is forbidden,
a) if G_j ($< G$) has an S_n ($n \geq 2$) as an symmetry operation,
b) if G_j ($< G$) is one of polyhedral groups (T, O and I) or of dihedral groups (D_n),
c) if G_j ($< G_i \leq G$) is C_n ($n \geq 2$) and G_i ($\leq G$) is C_{nv}, or
d) if G_j ($< G_i \leq G$) is $C_{n(v)}$ ($n \geq 2$) and G_i ($\leq G$) is $C_{m(v)}$.

Proof. Consider a regular body of $\mathbf{G}$ symmetry. Lemma 7.1 indicates that the CR $\mathbf{G}(/\mathbf{G}_j)$ governs $\Omega = \{\Omega_1, \Omega_2, \ldots, \Omega_r\}$ each element of which has a $\mathbf{G}_j(/\mathbf{G}_{j1})$ orbit. The problem may be restated: Can every element correspond to a different point at a general position or not?

The condition described in a) requires that the center of each $\mathbf{G}_j$ object (Ω_p; $p = 1, 2, \ldots$, or r) is identical with the origin of the $\mathbf{G}$-body. This means that none of the $\mathbf{G}_j$ objects corresponds to points at general positions. Hence, $\mathbf{G}(/\mathbf{G}_j)$ is forbidden. If $\mathbf{G}_j$ is equal to $\mathbf{G}$, the $\mathbf{G}(/\mathbf{G}_j)$ (*i.e.* $\mathbf{G}(/\mathbf{G})$) is allowed to govern the origin as an orbit. Obviously, The same restriction gives rise to the rule b).

If c) is true, Lemma 7.2 indicates that $\Omega_1 = \{\Omega_{11}, \Omega_{21}\}$ is subject to $\mathbf{C}_{nv}(/\mathbf{C}_n)$. The symmetries of Ω_{11} and Ω_{21} are both $\mathbf{C}_n$. In addition, they are antipodal with each other. This fact means that the centers of the two blocks are identical with each other. Since the total effect is represented by the CR $\mathbf{G}(/\mathbf{C}_n)$ according to Lemma 7.2, the CR $\mathbf{G}(/\mathbf{C}_n)$ is forbidden.

If d) holds, Lemma 7.2 indicates the relationship among $\mathbf{G}(/\mathbf{C}_{mv})$, $\mathbf{C}_{mv}(/\mathbf{C}_{nv})$ and $\mathbf{G}(/\mathbf{C}_{nv})$. Note that the $\mathbf{G}(/\mathbf{C}_{mv})$ governs Ω, the $\mathbf{C}_{mv}(/\mathbf{C}_{nv})$ governs Ω_1, and the $\mathbf{G}(/\mathbf{C}_{nv})$ governs $\widehat{\Omega}$. As a result, the symmetry of Ω_1 ($\in \Omega$) is $\mathbf{C}_{mv}$, while that of subblocks (Ω_{11} *etc.* $\in \Omega_1$) is $\mathbf{C}_{nv}$. This means that the center of the former block is identical with those of the latter subblocks. Hence, the CR $\mathbf{G}(/\mathbf{C}_{nv})$ is forbidden. The same situation holds for the case of $\mathbf{C}_n$ and $\mathbf{C}_m$.

In order to illustrate Theorem 7.2, we consider a set of coset representations for $\mathbf{T}_d$. The rule a) of Theorem 7.1 forbids $\mathbf{T}_d(/\mathbf{S}_4)$ and $\mathbf{T}_d(/\mathbf{D}_{2d})$. The rule b) indicates the absence of $\mathbf{T}_d(/\mathbf{D}_2)$ and $\mathbf{T}_d(/\mathbf{T})$. The rule c) does not allow $\mathbf{T}_d(/\mathbf{C}_2)$ and $\mathbf{T}_d(/\mathbf{C}_3)$. There are no cases avoided by the rule d). As a result, five CRs remain to be allowed, *i.e.*, $\mathbf{T}_d(/\mathbf{C}_1)$, $\mathbf{T}_d(/\mathbf{C}_s)$, $\mathbf{T}_d(/\mathbf{C}_{2v})$, $\mathbf{T}_d(/\mathbf{C}_{3v})$, and $\mathbf{T}_d(/\mathbf{T}_d)$. Figure 7.4 depicts the allowed CRs for the $\mathbf{T}_d$ group.

Application of Theorem 7.2 to $\mathbf{D}_{2d}$ gives $\mathbf{D}_{2d}(/\mathbf{C}_2)$ (rule c), $\mathbf{D}_{2d}(/\mathbf{S}_4)$ (rule a), and $\mathbf{D}_{2d}(/\mathbf{D}_2)$ (rule b) as forbidden CRs. The remaining CRs are allowed: $\mathbf{D}_{2d}(/\mathbf{C}_1)$, $\mathbf{D}_{2d}(/\mathbf{C}_2')$, $\mathbf{D}_{2d}(/\mathbf{C}_s)$, $\mathbf{D}_{2d}(/\mathbf{C}_{2v})$, and $\mathbf{D}_{2d}(/\mathbf{D}_{2d})$. This result verifies the data described in Table 5.1 (Chapter 5).

In the light of Theorem 7.2, we are able to determine the allowance of CRs for every point group. Several results are shown in Appendix C, where a forbidden CR is marked with an asterisk. Note that this theorem holds only for an orbit containing atoms or ligands (not for an orbit containing bonds or faces).

Derivation of Brester's equations and of the k-value equations of framework groups. In Chapter 5, we have already obtained the following equation.

$$| \Delta | = \sum_{i=1}^{s} \alpha_i \, | \, \mathbf{G} \, | \, / \, | \, \mathbf{G}_i \, | \, . \tag{7.15}$$

Theorem 7.2 indicates that α_i is equal to 0 for a forbidden $\mathbf{G}_i$. Hence, eq. 7.15 contains the Brester equation[1] and an equivalent framework-group equation[5] as special cases, although they are different in their explicit forms.

In order to illustrate this equivalency, we here refer to the $\mathbf{T}_d$ symmetry as an example. Since $\mid \mathbf{T}_d \mid / \mid \mathbf{C}_1 \mid = 24$, $\mid \mathbf{T}_d \mid / \mid \mathbf{C}_2 \mid = 12$, and so on, eq. 7.15 is converted into

$$\mid \Delta \mid = 24\alpha_{\mathbf{C}_1} + 12\alpha_{\mathbf{C}_2} + 12\alpha_{\mathbf{C}_s} + 8\alpha_{\mathbf{C}_3} + 6\alpha_{\mathbf{S}_4} + 6\alpha_{\mathbf{D}_2} + 6\alpha_{\mathbf{C}_{2v}} + 4\alpha_{\mathbf{C}_{3v}} + 3\alpha_{\mathbf{D}_{2d}} + 2\alpha_{\mathbf{T}} + \alpha_{\mathbf{T}_d}. \tag{7.16}$$

The above discussion on the allowability of CRs for the $\mathbf{T}_d$ group indicates that

$$\alpha_{\mathbf{C}_2} = 0, \ \alpha_{\mathbf{C}_3} = 0, \ \alpha_{\mathbf{S}_4} = 0,$$
$$\alpha_{\mathbf{D}_2} = 0, \ \alpha_{\mathbf{D}_{2d}} = 0, \ \text{and} \ \alpha_{\mathbf{T}} = 0,$$

for the forbidden CRs. With respect to the allowed CRs, we denote the corresponding terms as

$$\alpha_{\mathbf{C}_1} = m = k_1, \ \alpha_{\mathbf{C}_s} = m_d = k_2, \ \alpha_{\mathbf{C}_{2v}} = m_2 = k_3,$$
$$\alpha_{\mathbf{C}_{3v}} = m_3 = k_4, \ \text{and} \ \alpha_{\mathbf{T}_d} = m_0 = k_5.$$

Thereby, eq. 7.16 is converted into

$$N = 24m + 12m_d + 6m_2 + 4m_3 + m_0, \tag{7.17}$$

where $N = \mid \Delta \mid$, and into

$$\Sigma = 24k_1 + 12k_2 + 6k_3 + 4k_4 + k_5, \tag{7.18}$$

where $\Sigma = \mid \Delta \mid$. These equations are identical with those derived alternatively by Brester's method[1] and the framework-group method.[5]

For practical purposes, however, it is unnecessary to consider such allowability of CRs in an explicit fashion. Forbidden CRs are spontaneously excluded by the procedure of assigning CRs to orbits that has been discussed in detail in Chapters 5 and 6.

Bibliography

[1] G. Herzberg, *Infrared and Raman Spectra of Polyatomic Molecules*, Chapter 2, Van Nostrand, New York (1945).

[2] H. A. Jahn, E. Teller, *Proc. Roy. Soc. (A)*, **161**, 220 (1937).

[3] L. L. Boyle, *Spectrochim. Acta*, **28A**, 1347 (1972).

[4] P. W. Fowler, C. M. Quinn, *Theor. Chim. Acta*, **70**, 333 (1986).

[5] J. A. Pople, *J. Am. Chem. Soc.*, **102**, 4615 (1980).

[6] R. L. Flurry, Jr., *Theor. Chim. Acta*, **31**, 221 (1973).

[7] S. Fujita, *Bull. Chem. Soc. Jpn.*, **63**, 315 (1990).

[8] S. Fujita, *J. Am. Chem. Soc.*, **112**, 3390 (1990).

Chapter 8

Chirality Fittingness of an Orbit [1]

In the preceding chapter, we have introduced the concept of "local symmetry" that is associated with a coset representation. This concept strictly controls symmetrical properties of any three-dimensional (3D) objects such as molecular models. The concept should be extended into a slightly different form, because most problems in stereochemistry are concerned with chirality/achirality dichotomy.

8.1 Ligands

Theorem 7.1 (Chapter 7) can be summarized in general: *When a* $\mathbf{G}$ *molecule has a* $\mathbf{G}(/\mathbf{G}_i)$ *orbit, each member of the orbit belongs to the local symmetry of* $\mathbf{G}_i$. This theorem indicates that 3D objects along with atoms are allowed to be such members. Chemically speaking, the 3D objects are ligands.[2] Such ligands have been formulated as "point ligands" in conventional stereochemical discussions; in other words, the ligands have been treated as being atoms. This treatment should be re-examined from a more logical standpoint. The present formulation gives a mathematical foundation to the conventional treatments; and we are able to discuss stereochemistry in a stricter fashion.

In the light of the discussions in Chapter 7, we have

$$
\begin{aligned}
\widehat{\Omega} &= \left\{
\begin{array}{cccc}
\Omega_{11} & \Omega_{12} & \cdots & \Omega_{1r} \\
\Omega_{21} & \Omega_{22} & \cdots & \Omega_{2r} \\
\vdots & \vdots & & \vdots \\
\Omega_{r'1} & \Omega_{r'2} & \cdots & \Omega_{r'r}
\end{array}
\right\} \cdots \mathbf{G}(/\mathbf{G}_j) \\
& \qquad\qquad \vdots \qquad\quad \vdots \qquad\qquad \vdots \\
\Omega &= \{\Omega_1, \quad \Omega_2, \quad \cdots, \quad \Omega_r\} \quad \cdots \mathbf{G}(/\mathbf{G}_i).
\end{aligned}
\tag{8.1}
$$

---

[1] This chapter is based on the articles published in S. Fujita, *J. Math. Chem.*, **5**, 121–156 (1990). See also S. Fujita, *J. Am. Chem. Soc.*, **112**, 3390–3397 (1990).

[2] We use the term "ligand" in place of the chemical term "(functional) group". The term "group" is used only in mathematical meaning.

Each block Ω_k corresponds to the coset $\mathbf{G}_i t_k$ that is given by

$$\mathbf{G} = \mathbf{G}_i t_1 + \mathbf{G}_i t_2 + \ldots + \mathbf{G}_i t_r. \tag{8.2}$$

The $\mathbf{G}(/\mathbf{G}_i)$ CR is a coset representation on the following set of cosets.

$$\mathbf{G}/\mathbf{G}_i = \{\mathbf{G}_i t_1, \mathbf{G}_i t_2, \ldots, \mathbf{G}_i t_r\}, \tag{8.3}$$

where each coset $(\mathbf{G}_i t_k)$ corresponds to Ω_k *via* t_k. Equation 8.2 implies that $\mathbf{G}(/\mathbf{G}_i)$ acts on Ω, the members of which can be regarded as ligands (or atoms or other 3D objects). This correspondence is schematically represented by

$$\mathbf{G}_i t_k \longleftrightarrow \Omega_k \equiv \text{ a ligand.} \tag{8.4}$$

In this case, each Ω_k contains one suborbit. Since these results can be extended to the case that contains a set of suborbits, we take account only of a case in which the Ω_k contains one suborbit without losing generality.

Figure 8.1 depicts such blocks that appear in tricyclo[4.2.1.0^{4,10}]decane (**1**) belonging to $\mathbf{C}_{3v}$ symmetry. Note that $\mathbf{C}_{3v}(/\mathbf{C}_3)$ is forbidden, since the center of gravity of Ω_1 is identical with that of Ω_2. For each coset representation, we obtain the following correspondence:

$\mathbf{C}_{3v}(/\mathbf{C}_1)$	$\mathbf{C}_{3v}(/\mathbf{C}_s)$	$\mathbf{C}_{3v}(/\mathbf{C}_3)$	$\mathbf{C}_{3v}(/\mathbf{C}_{3v})$
$\Omega_1 \longleftrightarrow I$	$\Omega_1 \longleftrightarrow \mathbf{C}_s$	$\Omega_1 \longleftrightarrow C_3$	$\Omega_1 \longleftrightarrow \mathbf{C}_{3v}$
$\Omega_2 \longleftrightarrow C_3$	$\Omega_2 \longleftrightarrow \mathbf{C}_s C_3$	$\Omega_2 \longleftrightarrow C_3 \mathbf{C}_s$	
$\Omega_3 \longleftrightarrow C_3^2$	$\Omega_3 \longleftrightarrow \mathbf{C}_s C_3^2$		
$\Omega_4 \longleftrightarrow \sigma_{v(1)}$			
$\Omega_5 \longleftrightarrow \sigma_{v(3)}$			
$\Omega_6 \longleftrightarrow \sigma_{v(2)}$			

8.2   Behavior of Cosets on the Action of a CR

Since an orbit is subject to a coset representation (CR), the examination of the CR produces an essential understanding on stereochemical equivalency. This task is accomplished by scrutinizing the mode of action of the CR $(\mathbf{G}(/\mathbf{G}_i))$ on the set of blocks (Ω) introduced in the preceding paragaraph. In particular, the correspondence represented by eq. 8.4 plays an important role.

First, we examine $\mathbf{G}/\mathbf{G}_i$ that is subject to the CR $(\mathbf{G}(/\mathbf{G}_i))$. The CR is constructed by the coset decomposition of a point group $\mathbf{G}$ by $\mathbf{G}_i$ (eq. 8.2). There are three cases in such combinations of $\mathbf{G}$ and $\mathbf{G}_i$, *i.e.* achiral-achiral, aciral-chiral, and chiral-chiral. Thereby, we differentiate such $\mathbf{G}/\mathbf{G}_i$'s (eq. 8.3) into three cases:

a) an *achiral part* in which $\mathbf{G}$ is an achiral group and $\mathbf{G}_i$ is an achiral subgroup,

b) a *neutral part* in which both $\mathbf{G}$ and $\mathbf{G}_i$ ($\leq \mathbf{G}$) are chiral groups, and

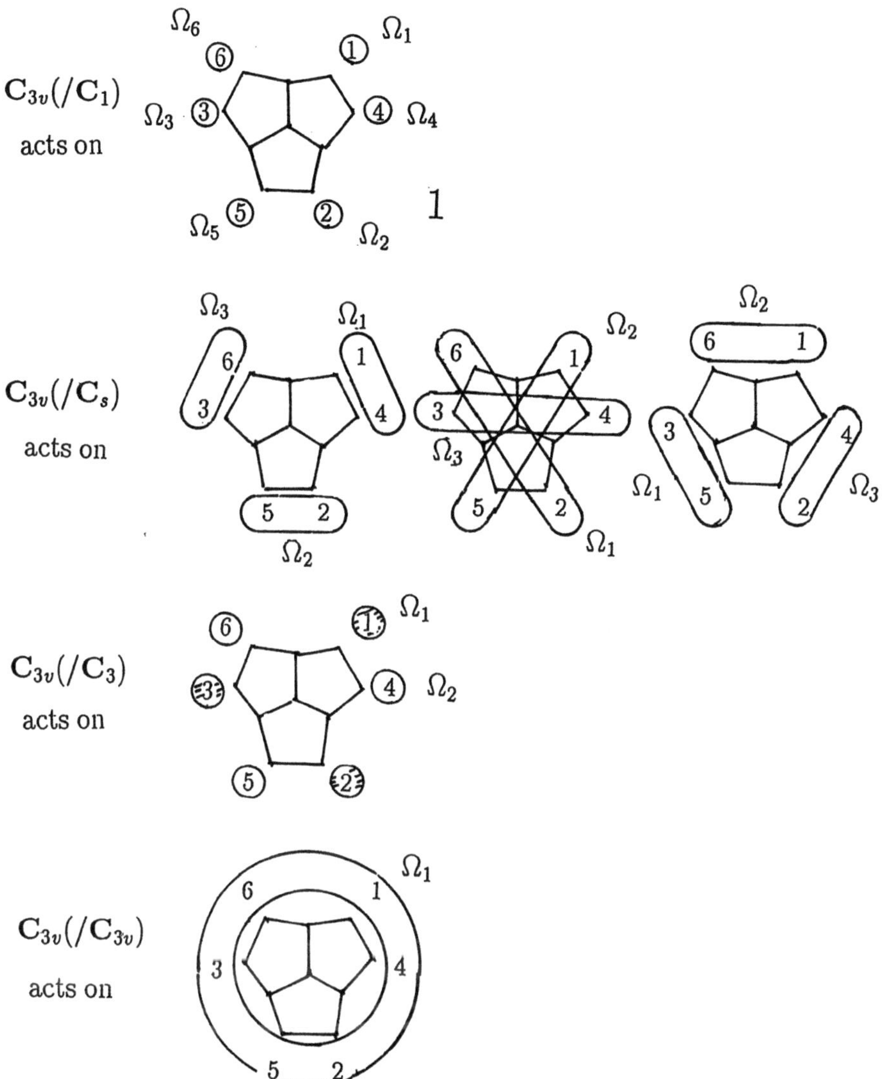

Figure 8.1: The relationship between coset representations and blocks

c) a *prochiral part* in which $\mathbf{G}$ is an achiral group and $\mathbf{G}_i$ is a chiral subgroup.

The term *chiral point group* is used for designating a point group that has proper rotations only; and the term *achiral point group* denotes a point group that contains improper rotations as well as proper rotations.

Achiral parts. Since $\mathbf{G}$ and $\mathbf{G}_i$ are achiral groups in Case (a), each coset in the $\mathbf{G}/\mathbf{G}_i$ contains proper and improper rotations in equal numbers. In this meaning, we can state that each coset in an achiral part is *achiral*. All of the elements of the corresponding transversal $\{t_1, t_2, \ldots, t_r\}$ can be selected to be proper rotations. If we select proper rotations only, eq. 8.2 can be converted into

$$\mathbf{G}^{max} = \mathbf{G}_i^{max} t_1 + \mathbf{G}_i^{max} t_2 + \ldots + \mathbf{G}_i^{max} t_r, \qquad (8.5)$$

where $\mathbf{G}^{max}$ is a maximum subgroup of $\mathbf{G}$ having proper rotations only; and $\mathbf{G}_i^{max}$ is a maximum subgroup of $\mathbf{G}_i$ having proper rotations only. Note that $\mid \mathbf{G} \mid/2 = \mid \mathbf{G}^{max} \mid$ and $\mid \mathbf{G}_i \mid/2 = \mid \mathbf{G}_i^{max} \mid$. Comparison between eqs. 8.2 and 8.5 indicates that each $\mathbf{G}_i t_k$ corresponds to $\mathbf{G}_i^{max} t_k$ in one-to-one fashion ($k = 1, 2, \ldots, r$). For all of the proper rotations (*i.e.* $\forall g \in \mathbf{G}^{max}$), the cosets ($\mathbf{G}_i^{max} t_k g$'s) run over all of the cosets determined by eq. 8.5. At the same time, the cosets ($\mathbf{G}_i t_k g$'s) run over all of the cosets determined by eq. 8.2 for all of the proper rotations (*i.e.* $\forall g \in \mathbf{G}^{max}$), because of the correspondence above. In other words, $\mathbf{G}/\mathbf{G}_i$ is transitive even if only the proper rotations are taken into consideraton. We depict this fact schematically as follows.

$$\boxed{\mathbf{G}_i t_1 \mid \mathbf{G}_i t_2 \mid \ldots \mid \mathbf{G}_i t_r} \quad \longleftrightarrow \quad \boxed{\Omega_1 \mid \Omega_2 \mid \ldots \mid \Omega_r} \qquad (8.6)$$
$$\underbrace{\qquad\qquad\qquad}_{r} \qquad\qquad\qquad \underbrace{\qquad\qquad\qquad}_{r}$$

where r is equal to $\mid \mathbf{G} \mid/\mid \mathbf{G}_i \mid$; and $\mathbf{G}_i$ is an achiral subgroup.

Example **8.1** For illustrating an achiral part, let us examine the coset representation, $\mathbf{C}_{3v}(/\mathbf{C}_s)$. The 2nd row of Figure 8.1 depicts blocks ($\sim$ ligands) to be examined. We have already shown the correspondence: $\mathbf{C}_s \longleftrightarrow \Omega_1$, $\mathbf{C}_s C_3 \longleftrightarrow \Omega_2$, and $\mathbf{C}_s C_3^2 \longleftrightarrow \Omega_3$. Since $\mathbf{C}_{3v}/\mathbf{C}_s = \{\mathbf{C}_s, \mathbf{C}_s C_3, \mathbf{C}_s C_3^2\}$ is an achiral part, we have a scheme,

$$\boxed{\mathbf{C}_s \mid \mathbf{C}_s C_3 \mid \mathbf{C}_s C_3^2} \quad \longleftrightarrow \quad \boxed{\Omega_1 \mid \Omega_2 \mid \Omega_3} \qquad (8.7)$$
$$\underbrace{\qquad\qquad\qquad}_{3} \qquad\qquad\qquad \underbrace{\qquad\qquad}_{3}$$

When we effect each symmetry operation, we obtain

$$\begin{array}{lccc}
I & \boxed{\mathbf{C}_s \mid \mathbf{C}_s C_3 \mid \mathbf{C}_s C_3^2} & \longleftrightarrow & \boxed{\Omega_1 \mid \Omega_2 \mid \Omega_3} \\
C_3 & \boxed{\mathbf{C}_s C_3 \mid \mathbf{C}_s C_3^2 \mid \mathbf{C}_s} & \longleftrightarrow & \boxed{\Omega_2 \mid \Omega_3 \mid \Omega_1} \\
C_3^2 & \boxed{\mathbf{C}_s C_3^2 \mid \mathbf{C}_s \mid \mathbf{C}_s C_3} & \longleftrightarrow & \boxed{\Omega_3 \mid \Omega_1 \mid \Omega_2} \\
\sigma_{v(1)} & \boxed{\mathbf{C}_s \mid \mathbf{C}_s C_3^2 \mid \mathbf{C}_s C_3} & \longleftrightarrow & \boxed{\Omega_1 \mid \Omega_3 \mid \Omega_2} \\
\sigma_{v(2)} & \boxed{\mathbf{C}_s C_3 \mid \mathbf{C}_s \mid \mathbf{C}_s C_3^2} & \longleftrightarrow & \boxed{\Omega_2 \mid \Omega_1 \mid \Omega_3} \\
\sigma_{v(3)} & \boxed{\mathbf{C}_s C_3^2 \mid \mathbf{C}_s C_3 \mid \mathbf{C}_s} & \longleftrightarrow & \boxed{\Omega_3 \mid \Omega_2 \mid \Omega_1}
\end{array} \tag{8.8}$$

Neutral parts. Case (b) is trivial. Since $\mathbf{G} = \mathbf{G}^{max}$ in this case, eqs.8.2 and 8.5 are identical. We have

$$\underbrace{\boxed{\mathbf{G}_i t_1 \mid \mathbf{G}_i t_2 \mid \dots \mid \mathbf{G}_i t_r}}_{r} \longleftrightarrow \underbrace{\boxed{\Omega_1 \mid \Omega_2 \mid \dots \mid \Omega_r}}_{r} \tag{8.9}$$

where $\mathbf{G}_i$ is a chiral subgroup. It should be noted that each coset in a neutral part is *chiral.*

Prochiral parts. Finally, we examine Case (c). Because $\mathbf{G}_i$ contains only proper rotations, eq. 8.2 indicates that the transversal, $\{t_1, t_2, \dots, t_r\}$, consists of $r/2$ improper rotations and $r/2$ proper ones. Note that $\mathbf{G}$ posesses $\mid \mathbf{G} \mid /2$ improper rotations and the same number of proper rotations. Suppose that the elements (t_1 to $t_{r/2}$) are proper rotations. Then, we have the following equation in place of eq. 8.5.

$$\mathbf{G}^{max} = \mathbf{G}_i t_1 + \mathbf{G}_i t_2 + \dots + \mathbf{G}_i t_{r/2}, \tag{8.10}$$

where $\mathbf{G}_i$ can be easily proven to be a subgroup of $\mathbf{G}^{max}$ as well as of $\mathbf{G}$. Comparison between eqs. 8.2 and 8.10 indicates that each $\mathbf{G}_i t_k$ ($k = 1, 2, \dots, r/2$) corresponds to $\mathbf{G}_i^{max} t_k$ in a one-to-one fashion. Let $t_{r/2+1}$ be an appropriate improper rotation. Then, we can take $t_{r/2+k} = t_k t_{r/2+1}$. This equation means that each $\mathbf{G}_i t_{r/2+k}$ ($k = 1, 2, \dots, r/2$) also corresponds to $\mathbf{G}_i^{max} t_k$ in a one-to-one fashion. It follows that

$$\begin{array}{c} \overbrace{\hphantom{\mathbf{G}_i t_{r/2+1} \mid \mathbf{G}_i t_{r/2+2} \mid \dots \mid \mathbf{G}_i t_r}}^{r/2} \\ \begin{array}{|c|c|c|c|} \hline \mathbf{G}_i t_1 & \mathbf{G}_i t_2 & \dots & \mathbf{G}_i t_{r/2} \\ \hline \mathbf{G}_i t_{r/2+1} & \mathbf{G}_i t_{r/2+2} & \dots & \mathbf{G}_i t_r \\ \hline \end{array} \\ \underbrace{\hphantom{\mathbf{G}_i t_{r/2+1} \mid \mathbf{G}_i t_{r/2+2} \mid \dots \mid \mathbf{G}_i t_r}}_{r/2} \end{array} \longleftrightarrow \begin{array}{c} \overbrace{\hphantom{\Omega_{r/2+1} \mid \Omega_{r/2+2} \mid \dots \mid \Omega_r}}^{r/2} \\ \begin{array}{|c|c|c|c|} \hline \Omega_1 & \Omega_2 & \dots & \Omega_{r/2} \\ \hline \Omega_{r/2+1} & \Omega_{r/2+2} & \dots & \Omega_r \\ \hline \end{array} \\ \underbrace{\hphantom{\Omega_{r/2+1} \mid \Omega_{r/2+2} \mid \dots \mid \Omega_r}}_{r/2} \end{array} \tag{8.11}$$

where the first row is concerned with the cosets containing proper rotations only; and the 2nd row is a set of the cosets containing improper rotations only. Hence, Scheme 8.11 shows that, if $g \in \mathbf{G}$ is any proper rotation, $\mathbf{G}_i t_1 g$'s ($g \in \mathbf{G}^{max}$) run over the half of the cosets determined by eq. 8.2 and $\mathbf{G}_i t_{r/2+1} g$ runs over the remaining half. Moreover, the action of every improper rotation results in the interchange of the two rows.

Example **8.2** The following example deals with $\mathbf{C}_{3v}(/\mathbf{C}_1)$ that is a prochiral part. Since this CR is a regular representation, each coset consists of a symmetry operation itself. We differentiate proper rotations from improper ones. Then, we obtain the following correspondence (Fig. 8.1):

$$
\begin{array}{|c|c|c|}
\hline
I & C_3 & C_3^2 \\
\hline
\sigma_{v(1)} & \sigma_{v(2)} & \sigma_{v(3)} \\
\hline
\end{array}
\longleftrightarrow
\begin{array}{|c|c|c|}
\hline
\Omega_1 & \Omega_2 & \Omega_3 \\
\hline
\Omega_4 & \Omega_5 & \Omega_6 \\
\hline
\end{array}
\tag{8.12}
$$

When each symmetry operation acts on $\mathbf{C}_{3v}/\mathbf{C}_1$ and Ω, we have

$$
\begin{aligned}
I \quad &
\begin{array}{|c|c|c|}
\hline
I & C_3 & C_3^2 \\
\hline
\sigma_{v(1)} & \sigma_{v(2)} & \sigma_{v(3)} \\
\hline
\end{array}
\longleftrightarrow
\begin{array}{|c|c|c|}
\hline
\Omega_1 & \Omega_2 & \Omega_3 \\
\hline
\Omega_4 & \Omega_5 & \Omega_6 \\
\hline
\end{array} \\[4pt]
C_3 \quad &
\begin{array}{|c|c|c|}
\hline
C_3 & C_3^2 & I \\
\hline
\sigma_{v(2)} & \sigma_{v(3)} & \sigma_{v(1)} \\
\hline
\end{array}
\longleftrightarrow
\begin{array}{|c|c|c|}
\hline
\Omega_2 & \Omega_3 & \Omega_1 \\
\hline
\Omega_5 & \Omega_6 & \Omega_4 \\
\hline
\end{array} \\[4pt]
C_3^2 \quad &
\begin{array}{|c|c|c|}
\hline
C_3^2 & I & C_3 \\
\hline
\sigma_{v(3)} & \sigma_{v(1)} & \sigma_{v(2)} \\
\hline
\end{array}
\longleftrightarrow
\begin{array}{|c|c|c|}
\hline
\Omega_3 & \Omega_1 & \Omega_2 \\
\hline
\Omega_6 & \Omega_4 & \Omega_5 \\
\hline
\end{array} \\[4pt]
\sigma_{v(1)} \quad &
\begin{array}{|c|c|c|}
\hline
\sigma_{v(1)} & \sigma_{v(3)} & \sigma_{v(2)} \\
\hline
I & C_3^2 & C_3 \\
\hline
\end{array}
\longleftrightarrow
\begin{array}{|c|c|c|}
\hline
\Omega_4 & \Omega_6 & \Omega_5 \\
\hline
\Omega_1 & \Omega_3 & \Omega_2 \\
\hline
\end{array} \\[4pt]
\sigma_{v(2)} \quad &
\begin{array}{|c|c|c|}
\hline
\sigma_{v(2)} & \sigma_{v(1)} & \sigma_{v(3)} \\
\hline
C_3 & I & C_3^2 \\
\hline
\end{array}
\longleftrightarrow
\begin{array}{|c|c|c|}
\hline
\Omega_5 & \Omega_4 & \Omega_6 \\
\hline
\Omega_2 & \Omega_1 & \Omega_3 \\
\hline
\end{array} \\[4pt]
\sigma_{v(2)} \quad &
\begin{array}{|c|c|c|}
\hline
\sigma_{v(3)} & \sigma_{v(2)} & \sigma_{v(1)} \\
\hline
C_3^2 & C_3 & I \\
\hline
\end{array}
\longleftrightarrow
\begin{array}{|c|c|c|}
\hline
\Omega_6 & \Omega_5 & \Omega_4 \\
\hline
\Omega_3 & \Omega_2 & \Omega_1 \\
\hline
\end{array}
\end{aligned}
\tag{8.13}
$$

8.3   Chirality Fittingness of an Orbit

The above discussion reveals that the three cases are different from each other in their behavior towards $\mathbf{G}$ and $\mathbf{G}^{max}$. The hehavior is also maintained in the action of the CR on Ω because of eq. 8.4. For simplicity of discussion, we equalize Ω and a set of ligands. Then, let the symbol A denote an achiral ligand selected from Ω; and let Q and Q' ($\in \Omega$) denote chiral ligands that are antipodal with each other.

1. **Homospheric orbits.** We first examine Case (a). The corresponding orbit is called a *homospheric orbit*. Scheme 8.6 allows an achiral ligand (A) to occupy each cell in place of each coset. This occupation is schematically represented by

$$\underbrace{\boxed{A}\,\boxed{A}\,\ldots\,\boxed{A}}_{r}\,, \qquad (8.14)$$

where r is equal to $\mid G \mid / \mid G_i \mid$. However, any chiral ligand (Q) is proved to be incapable of occupying each cell of Scheme 8.6 as follows. Suppose that a chiral ligand (Q) is assigned to each cell of Scheme 8.6. Since each $G_i t_k$ contains improper rotations in this case, anyone of the improper rotations of G converts this ligand into the corresponding antipode (Q') on an appropriate $G_i t_{k'}$ cell. On the other hand, there exists a proper rotation converting the $G_i t_k$ cell into $G_i t_{k'}$ cell, which keeps the ligand Q invariant. Hence, the chiral ligand (Q) is equal to its antipode (Q'). This means that the ligand should be achiral. Therefore, an achiral part (Case (a)) permits achiral ligands only.

2. **Hemispheric orbits.** In Case (b), the CR ($G(/G_i)$) is not concerned with improper rotations. The corresponding orbit is called a *hemispheric orbit*. Since Scheme 8.9 holds for this case, the neutral part permits any ligands, whether they are chiral or achiral.

$$\underbrace{\boxed{A}\,\boxed{A}\,\ldots\,\boxed{A}}_{r} \qquad (8.15)$$

or

$$\underbrace{\boxed{Q}\,\boxed{Q}\,\ldots\,\boxed{Q}}_{r}\,, \qquad (8.16)$$

3. **Enantiospheric orbits.** For discussing Case (c), we have two possibilities. The corresponding orbit is referred to as an *enantiospheric orbit*.

Suppose that we assign a chiral ligand (Q) to $G_i t_1$ ($1 \leq k \leq r/2$). Since $G_i t_1 g$'s ($g \in G$) run over the first row of Scheme 8.11, the $r/2$ cells are assigned to Q. On the other hand, Since $G_i t_1 g$'s ($g \in G^{max}$) run over the 2nd row of Scheme 8.11, the remaining $r/2$ are assigned to the antipode (Q'). This packing of Ω with Q and Q' is illustrated as follows. We refer to this type as *a compensated chiral packing* of a prochiral part, since this produces an *achiral* molecule of G symmetry.

$$\underbrace{\begin{array}{cccc}\boxed{Q}&\boxed{Q}&\ldots&\boxed{Q}\\\boxed{Q'}&\boxed{Q'}&\ldots&\boxed{Q'}\end{array}}_{r/2}\,, \qquad (8.17)$$

If we exchange the two rows of the packing 8.18, we have another molecule that is diastereomeric (not enantiomeric) to the first packing.

$$\begin{array}{|c|c|c|c|}\hline Q' & Q' & \ldots & Q' \\\hline Q & Q & \ldots & Q \\\hline\end{array}\;.$$

$$\underbrace{\qquad\qquad}_{r/2}$$

(8.18)

Thus, we have another achiral molecule of **G** symmetry. These two molecules are general expressions of *meso*-compounds that are diastereomeric to each other.

An additional assignment is possible in Case (c). Suppose that an achiral ligand (A) is assigned to each cell, producing a packing in which all of r cells in Scheme 8.11 take A for realizing the **G**-symmetry. However, note that the r cells are divided into two halves having $r/2$ cells. Thus, each half composed of $r/2$ cells is superimposable upon itself by proper rotations of **G**; the two halves are exchanged not by proper rotations but by improper ones. This type of packing is referred to as *a paired achiral packing* of a prochiral part.

$$\begin{array}{|c|c|c|c|}\hline A & A & \ldots & A \\\hline A & A & \ldots & A \\\hline\end{array}\;,$$

$$\underbrace{\qquad\qquad}_{r/2}$$

(8.19)

The above discussions afford the following theorem concerning the chirality fittingness of an orbit. The term "chirality fittingness" is defined as its capability of accepting chiral or achiral ligands. Since the CR is concerned with an orbit (or sphere), we propose the terms having a suffix *-spheric* (Greek: globe, sphere).

Theorem 8.1 *Suppose that a coset representation* $\mathbf{G}(/\mathbf{G}_i)$ *governs an orbit, the members of which are ligands (or atoms). The chirality fittingness of the orbit is determined by the CR as follows.*

  a) *An orbit of ligands is defined as being* homospheric, *if associated with an achiral part. The homospheric orbit takes only achiral ligands of the same kind.*

  b) *An orbit of ligands is defined as being* hemispheric, *if associated with a neutral part. The hemispheric orbit takes achiral as well as chiral ligands in the manner that the orbit is filled with such ligands of the same kind.*

  c) *An orbit of ligands is defined as being* enantiospheric, *if associated with a prochiral part. The enantiospheric orbit takes achiral ligands in such a way as a paired achiral packing. On the other hand, chiral ligands occupy the enantiospheric orbit in such a way as a compensated chiral packing.*

Strictly speaking, we use the terms "achiral", "neutral" and "prochiral parts" to classify G/G_i in terms of coset representations. On the other hand, we employ the terms concerning chirality fittingness (homospheric, hemispheric, and enantiospheric) to classify orbits that are composed of any objects to be examined. Since there are one-to-one correspondences:

$$\begin{array}{rcl}
\text{achiral part} & \longleftrightarrow & \text{homospheric} \\
\text{neutral part} & \longleftrightarrow & \text{hemispheric} \\
\text{prochiral part} & \longleftrightarrow & \text{enantiospheric,}
\end{array}$$

we mainly use the latter sphericity terms rather than the former so long as such usage produces no confusions. Table 8.1 summarizes these three possibilities. In terms of Theorem 8.1, we refer to the chirality fittingness of an orbit as *homospheric*, *hemispheric*, or *enantiospheric*. The terms introduced here are based on the relationship between G and G_i constructing coset representations. These three terms provide the concept of stereochemical equivalency with a more general foundation than conventional terminology does. We introduce the terms *homospheric*

Table 8.1: The chirality fittingness of an orbit subject to $G(/G_i)$

G	G_i	G/G_i	chirality fittingness of a $G(/G_i)$ orbit	ligands allowed[a]
achiral	achiral	achiral part	homospheric	achiral
achiral	chiral	prochiral part	enantiospheric	achiral, chiral
chiral	chiral	neutral part	hemispheric	achiral, chiral

[a]The original symmetry of each ligand is restricted within the local symmetry G_i of the orbit. See the text.

and *enantiospheric* in order to differentiate them from the conventional terms (homotopic and enantiotopic). A homospheric orbit of a molecule has its mirror image that can be superimposed on itself by proper rotations. An enantiospheric orbit of a molecule has the two halves that are mirror images to each other and cannot be superimposed by proper rotations.

Table 8.2 collects several molecules of D_{2d} symmetry, where orbits and chirality fittingness are found. By inspection of these molecules, we find geometrical correspondence between orbits that are subject to the same CR. However, the determination of such correspondence should be done carefully in some cases. Let us examine the $D_{2d}(/C_1)$ orbit of **3** (eight hydrogen atoms) and that of **4** (eight hydrogen atoms where the symbol (o) denotes a hydrogen atom). They are somewhat different. This difference can be recognized more clearly if we focus our attention on the four methylene groups of each molecule. The methylene of **3** is

Table 8.2: Chirality fittingness of orbits in D_{2d}-molecules

molecule	member of the orbit	coset representation	chirality fittingness (sphericity)
2	H_4	$D_{2d}(/C_s)$	homospheric
	C_2	$D_{2d}(/C_{2v})$	homospheric
	C	$D_{2d}(/D_{2d})$	homospheric
3	H_8	$D_{2d}(/C_1)$	enantiospheric
	C_4 (methylene)	$D_{2d}(/C_s)$	homospheric
	C_2 (allene)	$D_{2d}(/C_{2v})$	homospheric
	C (center)	$D_{2d}(/D_{2d})$	homospheric
4	H_8 (methylene)	$D_{2d}(/C_1)$	enantiospheric
	C_4 (methylene)	$D_{2d}(/C_2')$	enantiospheric
	C_4 (methine)	$D_{2d}(/C_s)$	homospheric
	H_4 (methine)	$D_{2d}(/C_s)$	homospheric
	C_2 (carbonyl)	$D_{2d}(/C_{2v})$	homospheric
	O_2 (carbonyl)	$D_{2d}(/C_{2v})$	homospheric

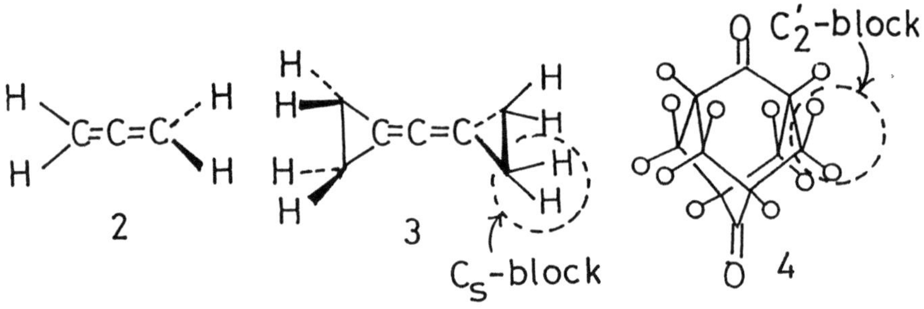

a C_s-ligand while that of **4** is a C_2'-ligand. The two protons of each methylene of **3** are superimposable to each other by any improper rotation of the D_{2d} group but they are not superposed by proper rotations. On the other hand, the two protons of the methylene of **4** are not superposed by the proper rotations but are superimposable by the improper rotations. A selection of a C_2'-block from **3** or a selection of a C_s-block from **4** would be a good exercise to the readers.

The four methylene ligands of **3** are interchangeable by proper and improper rotations. This is a result that is produced by the homosphericity of the methylene carbons of **3**. In the molecule (**4**), the four methylenes are divided into two halves. One of the halves is composed of two methylenes in a pair of diagonal places; the other half consists of the remaining two methylenes in another pair of diagonal places. The two halves are superimposed by improper rotations only. This behavior comes from the fact that the corresponding methylene carbons belong to an enantiospheric orbit.

The term *hemispheric* is introduced to clarify the course of asymmetric syntheses. If a molecule has a hemispheric orbit, the mirror image of the orbit is absent in the molecule but present in the antipode of the molecule. For an illustration, we examine the orbits of twistane (**5**). Since the orbit having four methylene hydrogens (H^b) and that having four methine hydrogens (H^d) are subject to $D_2(/C_1)$, they are both hemispheric. One hydrogen on every methylene carbon of an ethylene bridge (H^a or H^d) constructs an hemispheric orbit that is also subject to $D_2(/C_1)$. Hence, the mirror image of the orbit is absent in the molecule. The 10 carbons of twistane (**5**) are partitioned into three orbits, *i.e.*, four C^α ($D_2(/C_1)$), two C^β ($D_2(/C_2')$), and four C^γ ($D_2(/C_1)$). They are all hemispheric in terms of the above definition.

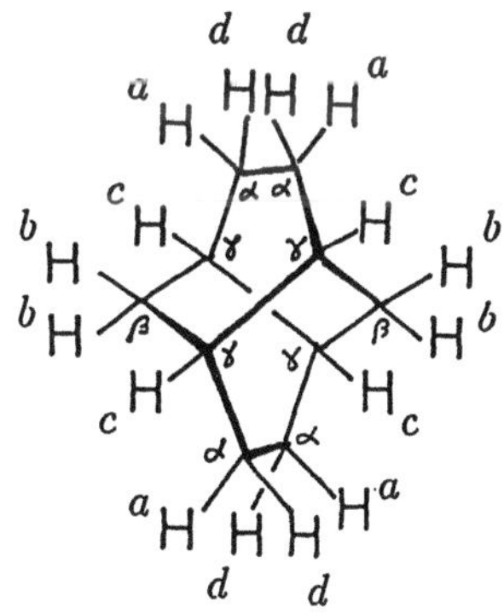

5

In the preceding paragraph, we have introduced the terms (homospheric, hemispheric and enantiospheric), which stem from the chirality fittingnesses of orbits. These terms afford a dintinct characterization of chiral and achiral compounds.

1. A chiral compound posesses hemispheric orbits only.

2. An achiral compound has enantiospheric orbits and/or homospheric orbits.

Achiral compounds are further classified into prochairal and para-achiral compounds; this point will be discussed in detail in Chapter 10.

The terms "enantiomeric" and "diastereomeric" have been used in stereochemistry. These terms are defined clearly in terms of the present terminology.

1. Two isomers are *enantiomeric*, if they have orbits of the same kind and if all hemispheric orbits of the one isomer are the mirror images of those of the other isomer.

2. Two isomers are *diastereomeric*, if they have orbits of the same kind but they are not enantiomeric.

3. Two isomers are *homomeric*, if they have orbits of the same kind and are superimposable by proper rotations.

4. Otherwise, two isomers are *heteromeric.*

By this definition, an achiral compound is alternatively defined as a self-enantiomeric compound. The term *homomeric* is introduced to simplify mathematical treatment. Chemically, homomeric isomers are regarded as the same compound. It should be noted that the term *enantiomeric* is not connected with the term *enantiospheric* in the present definition.

Chapter 9

Subduction of Coset Representations[1]

In this book, we deal with two types of desymmetrization processes, (1) asymmetric syntheses and (2) derivation of a molecule from a high-symmetry parent molecule. As a basis of solving such problems, we introduce *subduction of coset representations*.

9.1 Subduction of Coset Representations

Consider CRs $G(/G_i)$ $(i = 1, 2, \ldots, s)$ derived from coset decompositions of G by G_i (Chapter 5),

$$G = G_i g_1 + G_i g_2 + \cdots + G_i g_r, \tag{9.1}$$

where $g_1 = I$ (an identity). Suppose that we have permutations represented by

$$\pi_g = \begin{pmatrix} G_i g_1 & G_i g_2 & \cdots & G_i g_r \\ G_i g_1 g & G_i g_2 g & \cdots & G_i g_r g \end{pmatrix} \tag{9.2}$$

for all $g \in G$. Then we have a coset representation,

$$G(/G_i) = \{\pi_g \mid \forall g \in G\}. \tag{9.3}$$

Each of these CRs is defined over $\forall g \in G$. If we restrict such elements within those of G_j $(\leq G)$, we have a permutation representation of the subgroup G_j. We call this permutation representation a *subduced representation (SR)* of $G(/G_i)$ by G_j. Let the symbol $G(/G_i) \downarrow G_j$ denote such a subduced representation. Hence, we arrive at a definition of the subduced representation:

$$G(/G_i) \downarrow G_j = \{\pi_g \mid \forall g \in G_j\}. \tag{9.4}$$

Although a coset representation is transitive, such an SR is intransitive in general. Suppose that SSG_{G_j} and SCR_{G_j}[2] for G_j are defined as being

$$SSG_{G_j} = \{G_1^{(j)}, G_2^{(j)}, \ldots, G_{v_j}^{(j)}\}, \tag{9.5}$$

---

[1]This chapter is based on the articles that appeared in S. Fujita, *Theor. Chim. Acta*, **76**, 247-268 (1989). See also S. Fujita, *J. Am. Chem. Soc.*, **112**, 3390-3397 (1990).

[2]For the definition, see Chapters 2 and 5.

Table 9.1: Subduction table of G group

	$\cdots$	$\downarrow \mathbf{G}_j$	$\cdots$
$\mathbf{G}(/\mathbf{G}_i)$	$\cdots$	$\sum_{k=1}^{v_j} \beta_k^{(ij)} \mathbf{G}_j(/\mathbf{G}_k^{(j)})$	$\cdots$

and
$$SCR_{\mathbf{G}_j} = \{\mathbf{G}_j(/\mathbf{G}_1^{(j)}), \mathbf{G}_j(/\mathbf{G}_2^{(j)}), \ldots, \mathbf{G}_j(/\mathbf{G}_{v_j}^{(j)})\}. \tag{9.6}$$

Since the SR is a permutation representation of the $\mathbf{G}_j$ group, Theorem 5.6 (Chapter 5) is also applicable to this case. Hence, we obtain the following theorem.

Theorem 9.1 (Subduction of coset representations).
The SR defined by eq. 9.4 is reduced by the following equation,

$$\mathbf{G}(/\mathbf{G}_i) \downarrow \mathbf{G}_j = \sum_{k=1}^{v_j} \beta_k^{(ij)} \mathbf{G}_j(/\mathbf{G}_k^{(j)}), \tag{9.7}$$
$$for \qquad i = 1, 2, \ldots, s \ and \ j = 1, 2, \ldots, s$$

where each $\beta_k^{(ij)}$ multiplicity is a non-negative integer. Each multiplicity is obtained by solving

$$\mu_\ell^{(j)} = \sum_{k=1}^{v_j} \beta_k^{(ij)} m_{k\ell}^{(j)} \quad (for \ \ell = 1, 2, \ldots, v_j), \tag{9.8}$$

or inversely

$$\beta_k^{(ij)} = \sum_{\ell=1}^{v_j} \mu_\ell^{(j)} \overline{m}_{\ell k}^{(j)} \quad (for \ k = 1, 2, \ldots, v_j), \tag{9.9}$$

where $\mu_\ell^{(j)}$ is the mark of $\mathbf{G}_\ell^{(j)}$ in $\mathbf{G}(/\mathbf{G}_i) \downarrow \mathbf{G}_j$. The matrices, $(m_{k\ell}^{(j)})$ and $(\overline{m}_{\ell k}^{(j)})$, denote the mark table of the $\mathbf{G}_j$ group and its inverse, respectively.

Such a subduction can be precalculated, since the $\mathbf{G}(/\mathbf{G}_i)$ CR and the subgroup $\mathbf{G}_j$ are given. Tabulation of the precalculated subductions is useful to further applications (Table 9.1). We call the $s \times s$ table a *subduction table* of the G group.

For illustrating the procedure of the precalculation, we construct the subduction table of $\mathbf{D}_{2d}$ group.

Example 9.1 Let us examine the CR $\mathbf{D}_{2d}(/\mathbf{C}_s)$, which consists of $I \sim (1)(2)(3)(4)$, $C_{2(1)} \sim (1\ 2)(3\ 4)$, $C_{2(2)} \sim (1\ 3)(2\ 4)$, $C_{2(3)} \sim (1\ 4)(2\ 3)$, $\sigma_{d(1)} \sim (1)(4)(2\ 3)$,

$S_4^3 \sim (1\ 2\ 4\ 3)$, $S_4 \sim (1\ 3\ 4\ 2)$, and $\sigma_{d(2)} \sim (1\ 4)(2)(3)$. If we select the elements of $\mathbf{C}_{2v}$, then we have a subduced representation (SR) derived from $\mathbf{D}_{2d}(/\mathbf{C}_s)$ by $\mathbf{C}_{2v}$:

$$\mathbf{D}_{2d}(/\mathbf{C}_s) \downarrow \mathbf{C}_{2v} = \begin{cases} I \sim & (1)(4) \mid (2)(3) \\ C_{2(3)} \sim & (1\ 4) \mid (2\ 3) \\ \sigma_{d(1)} \sim & (1)(4) \mid (2\ 3) \\ \sigma_{d(2)} \sim & (1\ 4) \mid (2)(3) \end{cases}$$

This SR is an intransitive permutation of the $\mathbf{C}_{2v}$ group. If we inspect the concrete form of each permutation, we can easily find that there emerge two orbits, *i.e.*, $\{1, 4\}$ and $\{2, 3\}$. When we count fixed points for every subgroup of $\mathbf{C}_{2v}$, we obtain an FPV = (4 0 2 2 0). Method II (Chapter 6) affords

$$(4\ 0\ 2\ 2\ 0)\begin{pmatrix} 1/4 & 0 & 0 & 0 & 0 \\ -1/4 & 1/2 & 0 & 0 & 0 \\ -1/4 & 0 & 1/2 & 0 & 0 \\ -1/4 & 0 & 0 & 1/2 & 0 \\ 1/2 & -1/2 & -1/2 & -1/2 & 1 \end{pmatrix} = (0\ 0\ 1\ 1\ 0). \qquad (9.10)$$

This equation means that

$$\mathbf{D}_{2d}(/\mathbf{C}_s) \downarrow \mathbf{C}_{2v} = \mathbf{C}_{2v}(/\mathbf{C}_s) + \mathbf{C}_{2v}(/\mathbf{C}_s'). \qquad (9.11)$$

This procedure is repeated over all i and all j to afford Table 9.2.

9.2 Subduced Mark Table

In the above procedure, the marks $(\mu_\ell^{(j)})$ have been obtained by using the concrete forms of CRs.[1] If we remember the process of constructing a mark table (Chapter 5), we become aware of an alternative and more convenient treatment. Obviously, the task of constructing a mark table has in common with the task of subducing a CR so that they are both concerned with desymmetrization (or restriction) of the CR. For example, the process of obtaining the FPV in Example 9.1 is essentially the same as the process of obtaining marks of subgroups ($\leq \mathbf{C}_{2v}$) in $\mathbf{D}_{2d}(/\mathbf{C}_s)$, except that the $\mathbf{C}_s'$ group is contained in the SSG of the $\mathbf{C}_{2v}$ but not in the SSG of the $\mathbf{D}_{2d}$ group. Note that the $\mathbf{C}_s'$ group is conjugate to the $\mathbf{C}_s$ group in the $\mathbf{D}_{2d}$ group. The 4th integer 2 in the FPV = (4 0 2 2 0) for the $\mathbf{C}_s'$ group can be obtained by duplicating the value for the $\mathbf{C}_s$ (the 3rd value), since the marks of any two conjugate subgroups are equal in general.

We first define a *subduced mark table (SMT)*.

Definition 9.1 (Subduced mark table).
From the mark table of $\mathbf{G}$ *described in Chapter 5, we select the respective column*

Table 9.2: Subduction of coset representations for D_{2d} Point Group

	$\downarrow C_1$	$\downarrow C_2$	$\downarrow C_2'$	$\downarrow C_s$	$\downarrow S_4$	$\downarrow C_{2v}$	$\downarrow D_2$	$\downarrow D_{2d}$
$D_{2d}(/C_1)$	$8C_1(/C_1)$	$4C_2(/C_1)$	$4C_2'(/C_1)$	$4C_s(/C_1)$	$2S_4(/C_1)$	$2C_{2v}(/C_1)$	$2D_2(/C_1)$	$D_{2d}(/C_1)$
$D_{2d}(/C_2)^*$	$4C_1(/C_1)$	$4C_2(/C_2)$	$2C_2'(/C_1)$	$2C_s(/C_1)$	$2S_4(/C_2)$	$2C_{2v}(/C_2)$	$2D_2(/C_2)$	$D_{2d}(/C_2)$
$D_{2d}(/C_2')$	$4C_1(/C_1)$	$2C_2(/C_1)$	$C_2'(/C_1)$ $+2C_2'(/C_2)$	$2C_s(/C_1)$	$S_4(/C_1)$	$C_{2v}(/C_1)$	$D_2(/C_2')$ $+D_2(/C_2'')$	$D_{2d}(/C_2')$
$D_{2d}(/C_s)$	$4C_1(/C_1)$	$2C_2(/C_1)$	$2C_2'(/C_1)$	$C_s(/C_1)$ $+2C_s(/C_s)$	$S_4(/C_1)$	$C_{2v}(/C_s)$ $+C_{2v}(/C_s')$	$D_2(/C_1)$	$D_{2d}(/C_s)$
$D_{2d}(/S_4)^*$	$2C_1(/C_1)$	$2C_2(/C_2)$	$C_2'(/C_1)$	$C_s(/C_1)$	$2S_4(/S_4)$	$C_{2v}(/C_2)$	$D_2(/C_2)$	$D_{2d}(/S_4)$
$D_{2d}(/C_{2v})$	$2C_1(/C_1)$	$2C_2(/C_2)$	$C_2'(/C_1)$	$2C_s(/C_s)$	$S_4(/C_2)$	$2C_{2v}(/C_{2v})$	$D_2(/C_2)$	$D_{2d}(/C_{2v})$
$D_{2d}(/D_2)^*$	$2C_1(/C_1)$	$2C_2(/C_2)$	$2C_2'(/C_2)$	$C_s(/C_1)$	$S_4(/C_2)$	$C_{2v}(/C_2)$	$2D_2(/D_2)$	$D_{2d}(/D_2)$
$D_{2d}(/D_{2d})$	$C_1(/C_1)$	$C_2(/C_2)$	$C_2'(/C_2)$	$C_s(/C_s)$	$S_4(/S_4)$	$C_{2v}(/C_{2v})$	$D_2(/D_2)$	$D_{2d}(/D_{2d})$

* Forbidden CR. See Chapter 7.

that corresponds to each subgroup equal to that of SSG_{G_j} (eq. 9.5). If SSG_G contains no counterpart corresponding to a subgroup in the SSG_{G_j}, we duplicate the column corresponding to the conjugate subgroup of the latter subgroup. The resulting $s \times v_j$ matrix is called a subduced mark table (SMT) *of G by G$_j$.*

Let the symbol $\widehat{M}^{(j)}$ denote the SMT; $\widehat{M}^{(j)} = (\widehat{M}^{(j)}_{i\ell})$, where j is tentatively fixed, i runs over 1 to s, and ℓ runs over 1 to v_j.

The $\mu^{(j)}_\ell$ mark, which is originally defined as the mark of $G^{(j)}_\ell$ in $G(/G_i) \downarrow G_j$ (eq. 9.9), is equal to the mark of $G^{(j)}_\ell$ and equal to the mark of its conjugate subgroup in $G(/G_i)$. Hence, we have the following equation.

$$\mu^{(j)}_\ell = \widehat{M}^{(j)}_{i\ell} \text{ (for } i = 1, 2, \ldots, s; \ell = 1, 2, \ldots, v_j), \tag{9.12}$$

where j is tentatively fixed. Equation 9.12 is introduced into eq. 9.9 to afford the following lemma.

Lemma 9.1 *The multiplicities $\beta^{(ij)}_k$'s are calculated by*

$$\beta^{(ij)}_k = \sum_{\ell=1}^{v_j} \widehat{M}^{(j)}_{i\ell} \overline{m}^{(j)}_{\ell k} \tag{9.13}$$
$$\text{for} \quad k = 1, 2, \ldots, v_j; i = 1, 2, \ldots, s; j = 1, 2, \ldots, s.$$

In Chapter 5, we have already otained the mark table of D_{2d} point group (See also Appendix A). From the data of this table, we can easily obtain an SMT for the D_{2d}. Thereby, we are able to re-examine Example 9.1 in the light of Lemma 9.1.

Example 9.2 Suppose that G is equal to D_{2d} and G_i's are selected from $SSR_{D_{2d}}$. Thus, G_i ($i = 1, 2, \ldots, 8$) are $G_1 = C_1$, $G_2 = C_2$, $G_3 = C'_2$, $G_4 = C_s$, $G_5 = S_4$, $G_6 = C_{2v}$, $G_7 = D_2$, and $G_8 = D_{2d}$. Consider that G_j is equal to C_{2v}. For the C_{2v} group, we select subgroups, $G^{(j)}_k$, as being $G^{(j)}_1 = C_1$, $G^{(j)}_2 = C_2$, $G^{(j)}_3 = C_s$, $G^{(j)}_4 = C'_s$, and $G^{(j)}_5 = C_{2v}$. The $\beta^{(ij)}_k$'s for $D_{2d}(/G_i) \downarrow C_{2v}$ ($G_i \leq D_{2d}$) are calculated by means of eq. 9.13. From the mark table of the D_{2d} group, we adopt the C_1, C_2, C_s, and C_{2v} columns to produce the corresponding SMT. The C'_s column of the SMT is a duplication of the C_s column. Thus, eq. 9.13 holds for this case in the following form.

$$
\begin{array}{ccccc}
\mathbf{C}_1 & \mathbf{C}_2 & \mathbf{C}_s & \mathbf{C}_s' & \mathbf{C}_{2v}
\end{array}
$$

$$
\begin{pmatrix}
8 & 0 & 0 & 0 & 0 \\
4 & 4 & 0 & 0 & 0 \\
4 & 0 & 0 & 0 & 0 \\
4 & 0 & 2 & 2 & 0 \\
2 & 2 & 0 & 0 & 0 \\
2 & 2 & 2 & 2 & 2 \\
2 & 2 & 0 & 0 & 0 \\
1 & 1 & 1 & 1 & 1
\end{pmatrix}
\begin{pmatrix}
1/4 & 0 & 0 & 0 & 0 \\
-1/4 & 1/2 & 0 & 0 & 0 \\
-1/4 & 0 & 1/2 & 0 & 0 \\
-1/4 & 0 & 0 & 1/2 & 0 \\
1/2 & -1/2 & -1/2 & -1/2 & 1
\end{pmatrix}
$$

$$
\begin{array}{ccccc}
\mathbf{C}_1 & \mathbf{C}_2 & \mathbf{C}_s & \mathbf{C}_s' & \mathbf{C}_{2v}
\end{array}
$$

$$
= \begin{pmatrix}
2 & 0 & 0 & 0 & 0 \\
0 & 2 & 0 & 0 & 0 \\
1 & 0 & 0 & 0 & 0 \\
0 & 0 & 1 & 1 & 0 \\
0 & 1 & 0 & 0 & 0 \\
0 & 0 & 0 & 0 & 2 \\
0 & 1 & 0 & 0 & 0 \\
0 & 0 & 0 & 0 & 1
\end{pmatrix}
\begin{array}{l}
\cdots\ \mathbf{D}_{2d}(/\mathbf{C}_1) \downarrow \mathbf{C}_{2v} \\
\cdots\ \mathbf{D}_{2d}(/\mathbf{C}_2) \downarrow \mathbf{C}_{2v} \\
\cdots\ \mathbf{D}_{2d}(/\mathbf{C}_2') \downarrow \mathbf{C}_{2v} \\
\cdots\ \mathbf{D}_{2d}(/\mathbf{C}_s) \downarrow \mathbf{C}_{2v} \\
\cdots\ \mathbf{D}_{2d}(/\mathbf{S}_4) \downarrow \mathbf{C}_{2v} \\
\cdots\ \mathbf{D}_{2d}(/\mathbf{C}_{2v}) \downarrow \mathbf{C}_{2v} \\
\cdots\ \mathbf{D}_{2d}(/\mathbf{D}_2) \downarrow \mathbf{C}_{2v} \\
\cdots\ \mathbf{D}_{2d}(/\mathbf{D}_{2d}) \downarrow \mathbf{C}_{2v}
\end{array}
\qquad (9.14)
$$

The first matrix in the left-hand side is the SMT of $\mathbf{D}_{2d}$ by $\mathbf{C}_{2v}$. The second matrix
is the inverse of the mark table for the $\mathbf{C}_{2v}$ group. The ith row of the matrix of
the right-hand side affords $\beta_k^{(ij)}$'s, where k runs over 1 to 5. Thereby, we have the
subductions

$$
\mathbf{D}_{2d}(/\mathbf{C}_1) \downarrow \mathbf{C}_{2v} = 2\mathbf{C}_{2v}(/\mathbf{C}_1)
$$

for the first row;

$$
\mathbf{D}_{2d}(/\mathbf{C}_2) \downarrow \mathbf{C}_{2v} = 2\mathbf{C}_{2v}(/\mathbf{C}_2)
$$

for the 2nd row; and so on. These subductions are collected in the $\downarrow \mathbf{C}_{2v}$ column
of Table 9.2.

The two methods for constructing the subduction tables are suitable to computer manipulation and can be easily programmed by any programming languages.
The programming may be a good exercise for readers. Subduction tables for other
point groups can be constructed in this way. Several tables are collected in Appendix C.

9.3   Chemical Meaning of Subduction

In the preceding section, we have introduced the concept "subduction" algebraically. We here explain the concept from a geometrical or chemical point of

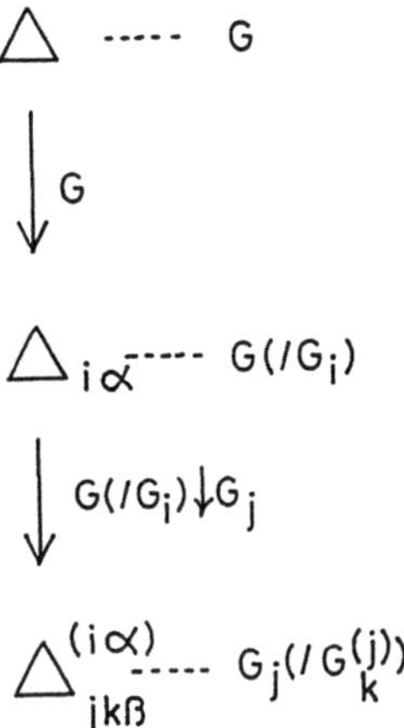

Figure 9.1: Division and subdivision of atoms in a molecule

view. Let Δ be a set of atoms contained in a given chemical compound (Fig 9.1). The set is divided into several sets of equivalent atoms, which are recognized as orbits: $\Delta_{i\alpha}$ ($\alpha = 1, 2, \ldots, \alpha_i$; $i = 1, 2, \ldots, s$). This process is characterized by the equation, $\sum_{i=1}^{s} \alpha_i \mathbf{G}(/\mathbf{G}_i)$ (Chapter 5). Each coset representation $\mathbf{G}(/\mathbf{G}_i)$ governs each one of the $\Delta_{i\alpha}$-orbits, where the subscript $i\alpha$ designates the dependence upon an α-th CR $\mathbf{G}(/\mathbf{G}_i)$.

An appropriate substitution of several atoms creates a new compound of $\mathbf{G}_j$ symmetry. Obviously, the $\mathbf{G}_j$ group is a subgroup of $\mathbf{G}$. Since we have partitioned the atoms of the original compound into the orbits, we can examine the orbits one by one. According to Theorem 9.1, each orbit ($\Delta_{i\alpha}$) is subdivided by the effect of $\mathbf{G}(/\mathbf{G}_i)\downarrow \mathbf{G}_j$ to create suborbits,

$$\Delta_{jk\beta}^{(i\alpha)} \quad (\beta = 1, 2, \ldots, \beta_k^{(ij)}), \tag{9.15}$$

each of which is governed by $\mathbf{G}_j(/\mathbf{G}_k^{(j)})$. Since the degree of $\mathbf{G}(/\mathbf{G}_i)$ is $| \mathbf{G} | / | \mathbf{G}_i |$ $(= | \Delta_{i\alpha} |)$ and that of $\mathbf{G}_j(/\mathbf{G}_k^{(j)})$ is $| \mathbf{G}_j | / | \mathbf{G}_k^{(j)} |$ $(=| \Delta_{jk\beta}^{(i\alpha)} |)$, eq. 9.7 affords

$$\frac{| \mathbf{G} |}{| \mathbf{G}_i |} = \sum_{k=1}^{v_j} \beta_k^{(ij)} \frac{| \mathbf{G}_j |}{| \mathbf{G}_k^{(j)} |}. \tag{9.16}$$

Example **9.3** Let us illustrate the above idea by using adamantane-2,6-dione (**1**) of $\mathbf{D}_{2d}$ symmetry as an example (Fig. 9.2). For simplicity's sake, we take 12 hydrogen atoms ($\circ$) into consideration. As shown in Chapter 8, these are divided into two

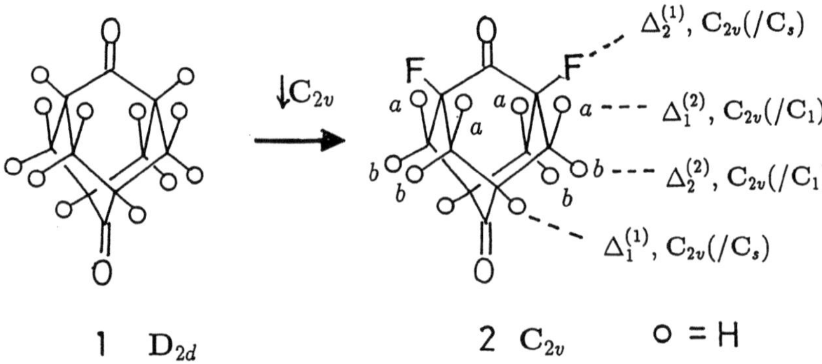

Figure 9.2: Division and subdivision in adamantane-2,6-dione

orbits, Δ_1 (four methine protons) and Δ_2 (eight methylene protons). They are subject to $D_{2d}(/C_s)$ and $D_{2d}(/C_1)$, respectively.

Consider that two hydrogen atoms of the Δ_1 orbit are substituted by fluorine atoms, as shown in Fig. 9.2. This substitution reduces the original D_{2d} symmetry into C_{2v}, producing 1,3-difluoroadamantane-2,6-dione (2). In the light of Method I or II (Chapter 6), we find four suborbits,

$$\Delta_1^{(1)} = H_2 \text{ (methine)} \quad \text{subject to } C_{2v}(/C_s),$$
$$\Delta_2^{(1)} = F_2 \quad \text{subject to } C_{2v}(/C_s),$$
$$\Delta_1^{(2)} = H_4^a \quad \text{subject to } C_{2v}(/C_1), \text{ and}$$
$$\Delta_2^{(2)} = H_4^b \quad \text{subject to } C_{2v}(/C_1).$$

The conversion of 1 into 2 can be explained in the light of the subduction concept. Table 9.2 affords

$$D_{2d}(/C_s) \downarrow C_{2v} = C_{2v}(/C_s) + C_{2v}(/C_s')$$

for Δ_1. This equation means that the $C_{2v}(/C_s)$ CR governs the $\Delta_1^{(1)}$ orbit and that the $C_{2v}(/C_s')$ governs the $\Delta_2^{(1)}$ orbit. This chemical explanation is an illustration of the process described in Example 9.1. On the other hand, the $D_{2d}(/C_1)$ orbit (Δ_2) is controlled by the subduction represented by

$$D_{2d}(/C_1) \downarrow C_{2v} = 2C_{2v}(/C_1).$$

The right-hand side of this equation indicates that the two suborbits generated ($\Delta_1^{(2)}$ and $\Delta_2^{(2)}$) are both subject to $C_{2v}(/C_1)$.

9.4 Unit Subduced Cycle Indices

Since the degree of $G_j(/G_k^{(j)})$ is

$$d_{jk} = |\, G_j \,| \,/\, |\, G_k^{(j)} \,|, \tag{9.17}$$

the corresponding $G_j/G_k^{(j)}$ set has a length of d_{jk}. Then, we assign a variable $s_{d_{jk}}$ to the $G_j/G_k^{(j)}$ set. Note that the $G_j/G_k^{(j)}$ set corresponds to each suborbit represented by eq. 9.15. We refer to this fact by saying that the $G_j(/G_k^{(j)})$ CR governs the suborbit. Since the suborbit has a length of d_{jk}, the variable $(s_{d_{jk}}^{(i\alpha)})$ can be also regarded as to be assigned to the suborbit, *i.e.* $\Delta_{jk\beta}^{(i\alpha)}$ ($\beta = 1, 2, \ldots, \beta_k^{(ij)}$). The superscript $(i\alpha)$ indicates the dependence upon the orbit $(\Delta_{i\alpha})$.

Theorem 9.1[1], in turn, indicates that the set of $G_j/G_k^{(j)}$'s constructs the SR $G(/G_i){\downarrow}G_j$ and that the set of suborbits $(\Delta_{jk\beta}^{(i\alpha)})$ constructs the orbit $(\Delta_{i\alpha})$. Accordingly, we define a unit subduced cycle index (USCI) as follows.

Definition 9.2 *A* unit subduced cycle index (USCI) *is defined by the equation,*

$$Z(G(/G_i) \downarrow G_j; s_d) = \prod_{k=1}^{v_j} (s_{d_{jk}})^{\beta_k^{(ij)}} \tag{9.18}$$

for each $G(/G_i) \downarrow G_j$, *where the subscript* d_{jk} *is expressed by*

$$d_{jk} = |\, G_j \,| \,/\, |\, G_k^{(j)} \,| \,. \tag{9.19}$$

Since the variable $s_{d_{jk}}$ is also regarded as to be assigned to the suborbit $(\Delta_{jk\beta}^{(i\alpha)})$, the USCI (eq. 9.18) can be considered to be assigned to $\Delta_{i\alpha}$.

Since $\beta_k^{(ij)}$ is dependent upon the group-subgroup relationships concerning G and G_i as well as concerning G_j and $G_k^{(j)}$, it is calculated to be constant. Hence, a tabulation of such USCIs is a valuable task. Once subduction tables have been constructed, tables of the USCIs are easily available. We obtain the following table by starting from Table 9.2. We will give discussions on the bottom row later (see also Corollary 5.2 in Chapter 5). Other tables of USCIs are collected in Appendix D.

The number of suborbits produced by $G(/G_i) \downarrow G_j$ is calculated to be

$$\beta_{ij} = \sum_{k=1}^{v_j} \beta_k^{(ij)}. \tag{9.20}$$

Each β_{ij} value can be calculated by summing up the powers of the corresponding USCI collected in a table of USCIs. Introduction of eq. 9.13 into eq. 9.20 affords an alternative method of calculating β_{ij}.

Table 9.3: Unit subduced cycle indices for D_{2d}

	$\downarrow C_1$	$\downarrow C_2$	$\downarrow C_2'$	$\downarrow C_s$	$\downarrow S_4$	$\downarrow C_{2v}$	$\downarrow D_2$	$\downarrow D_{2d}$
$D_{2d}(/C_1)$	s_1^8	s_2^4	s_2^4	s_2^4	s_4^2	s_4^2	s_4^2	s_8
$D_{2d}(/C_2)$	s_1^4	s_1^4	s_2^2	s_2^2	s_2^2	s_2^2	s_2^2	s_4
$D_{2d}(/C_2')$	s_1^4	s_2^2	$s_1^2 s_2$	s_2^2	s_4	s_4	s_2^2	s_4
$D_{2d}(/C_s)$	s_1^4	s_2^2	s_2^2	$s_1^2 s_2$	s_4	s_2^2	s_4	s_4
$D_{2d}(/S_4)$	s_1^2	s_1^2	s_2	s_2	s_1^2	s_2	s_2	s_2
$D_{2d}(/C_{2v})$	s_1^2	s_1^2	s_2	s_1^2	s_2	s_1^2	s_2	s_2
$D_{2d}(/D_2)$	s_1^2	s_1^2	s_1^2	s_2	s_2	s_2	s_1^2	s_2
$D_{2d}(/D_{2d})$	s_1	s_1	s_1	s_1	s_1	s_1	s_1	s_1
$\sum_{i=1}^s \overline{m}_{ji}$	1/8	1/8	1/4	1/4	1/4	0	0	0

Lemma 9.2 (The number of suborbits).

$$\beta_{ij} = \sum_{l=1}^{v_j} \widehat{M}_{il}^{(j)} \left(\sum_{k=1}^{v_j} \overline{m}_{lk}^{(j)} \right). \tag{9.21}$$

The term appearing in the last parentheses is the sum of each $G_k^{(j)}$ row in the inverse of the mark table of the G_j group. This sum is positive if and only if the $G_k^{(j)}$ subgroup is cyclic (Chapter 16).[2, 3] The number of suborbits, β_{ij}, can be proved to be equal to the number of (G_i, G_j)-double cosets.[5]

Example 9.4 In continuation of Example 9.2, we calculate the β_{ij}'s for $D_{2d}(/G_i) \downarrow C_{2v}$ $(G_i \leq D_{2d})$ by means of Lemma 9.2.

$$
\begin{array}{ccccc}
C_1 & C_2 & C_s & C_2' & C_{2v}
\end{array}
$$
$$
\begin{pmatrix}
8 & 0 & 0 & 0 & 0 \\
4 & 4 & 0 & 0 & 0 \\
4 & 0 & 0 & 0 & 0 \\
4 & 0 & 2 & 2 & 0 \\
2 & 2 & 0 & 0 & 0 \\
2 & 2 & 2 & 2 & 2 \\
2 & 2 & 0 & 0 & 0 \\
1 & 1 & 1 & 1 & 1
\end{pmatrix}
\begin{pmatrix}
1/4 \\
1/4 \\
1/4 \\
1/4 \\
0
\end{pmatrix}
=
\begin{pmatrix}
2 \\
2 \\
1 \\
2 \\
1 \\
2 \\
1 \\
1
\end{pmatrix}
\begin{array}{l}
\cdots\ D_{2d}(/C_1) \downarrow C_{2v} \\
\cdots\ D_{2d}(/C_2) \downarrow C_{2v} \\
\cdots\ D_{2d}(/C_2') \downarrow C_{2v} \\
\cdots\ D_{2d}(/C_s) \downarrow C_{2v} \\
\cdots\ D_{2d}(/S_4) \downarrow C_{2v} \\
\cdots\ D_{2d}(/C_{2v}) \downarrow C_{2v} \\
\cdots\ D_{2d}(/D_2) \downarrow C_{2v} \\
\cdots\ D_{2d}(/D_{2d}) \downarrow C_{2v}
\end{array}
$$

The first matrix in the left-hand side is the SMT of Example 9.2. The second matrix contains $\sum_{k=1}^{v_j} \overline{m}_{lk}^{(j)}$ derived from the inverse of the mark table for the C_{2v} group. Each value of the matrix of the right-hand side affords β_{ij}, which is the number of suborbits. Each of these vaules is equal to the sum of the corresponding row of eq. 9.14 (Example 9.2).

9.5 Unit Subduced Cycle Indices with Chirality Fittingness

Each $G_j/G_k^{(j)}$ set governed by the coset representation $(G_j(/G_k^{(j)}))$ has a chirality fittingness (Chapter 8).

$$\begin{cases} G_j(/G_k^{(j)}) : \text{achiral part} \quad \cdots \quad \text{homospheric} \\ G_j(/G_k^{(j)}) : \text{neutral part} \quad \cdots \quad \text{hemispheric} \\ G_j(/G_k^{(j)}) : \text{prochiral part} \quad \cdots \quad \text{enantiospheric} \end{cases}$$

According to this classification, we assign

$$\begin{cases} \text{a variable } a_{d_{jk}} \text{ to the } G_j/G_k^{(j)} \text{ set if this is an achiral part;} \\ \text{a variable } b_{d_{jk}} \text{ to the } G_j/G_k^{(j)} \text{ set if this is a neutral part; and} \\ \text{a variable } c_{d_{jk}} \text{ to the } G_j/G_k^{(j)} \text{ set if this is a prochiral part,} \end{cases}$$

where the symbol (d_{jk}) is defined by eq. 9.17. Then we obtain the following definition of a unit subduced cycle index with chirality fittingness (USCI-CF). This definition is also based on Theorem 9.1.[1]

Definition 9.3 *A unit subduced cycle index with chirality fittingness (USCI-CF) is defined by the equation,*

$$ZC(G(/G_i) \downarrow G_j; \$_d) = \prod_{k=1}^{\nu_j} (\$_{d_{jk}})^{\beta_k^{(ij)}} \tag{9.22}$$

for each $G(/G_i) \downarrow G_j$, *where*

$$\begin{aligned} \$ &= a \quad \text{for an achiral part,} \\ \$ &= b \quad \text{for a neutral part,} \\ &\text{and} \\ \$ &= c \quad \text{for a prochiral part.} \end{aligned}$$

The subscript d_{jk} is expressed by

$$d_{jk} = |\, G_j \,| \, / \, |\, G_k^{(j)} \,| . \tag{9.23}$$

Since the variable $\$_{d_{jk}}$ is also regarded as being assigned to the suborbit $(\Delta_{jk\beta}^{(i\alpha)})$, the USCI-CF (eq. 9.22) is also considered to be assigned to $\Delta_{i\alpha}$.

Such USCI-CFs can be precalculated in the same line as USCIs. Once subduction tables have been constructed, tables of the USCI-CFs are also available. We obtain Table 9.4 by starting from Table 9.2. Other tables of USCI-CFs are collected in Appendix E. It should be noted that two isomorphic groups have the essentially same USCI table; however their USCI-CF tables are not always the same. For example, the USCI-CF table of C_{2v} group is different from that of D_2 group, although the two groups are isomorphic.

Table 9.4: Unit subduced cycle indices with chirality fittingness for D_{2d}

	$\downarrow C_1$	$\downarrow C_2$	$\downarrow C_2'$	$\downarrow C_s$	$\downarrow S_4$	$\downarrow C_{2v}$	$\downarrow D_2$	$\downarrow D_{2d}$
$D_{2d}(/C_1)$	b_1^8	b_2^4	b_2^4	c_2^4	c_4^2	c_4^2	b_4^2	c_8
$D_{2d}(/C_2)$	b_1^4	b_1^4	b_2^2	c_2^2	c_2^2	c_2^2	b_2^2	c_4
$D_{2d}(/C_2')$	b_1^4	b_2^2	$b_1^2 b_2$	c_2^2	c_4	c_4	b_2^2	c_4
$D_{2d}(/C_s)$	b_1^4	b_2^2	b_2^2	$a_1^2 c_2$	c_4	a_2^2	b_4	a_4
$D_{2d}(/S_4)$	b_1^2	b_1^2	b_2	c_2	a_1^2	c_2	b_2	a_2
$D_{2d}(/C_{2v})$	b_1^2	b_1^2	b_2	a_1^2	c_2	a_1^2	b_2	a_2
$D_{2d}(/D_2)$	b_1^2	b_1^2	b_1^2	c_2	c_2	c_2	b_1^2	c_2
$D_{2d}(/D_{2d})$	b_1	b_1	b_1	a_1	a_1	a_1	b_1	a_1
$\sum_{i=1}^{s} \overline{m}_{ji}$	$1/8$	$1/8$	$1/4$	$1/4$	$1/4$	0	0	0

9.6   Desymmetrization Lattice

Group-subgroup lattices are useful to understand group-theoretical concepts.[4] A *desymmetrization lattice* (DL) for an orbit is a group-subgroup lattice that contains the data of the subduction of the orbit. Figure 9.3 illustrates a full desymmetrization lattice with the subduction data of the $D_{2d}(/C_s)$ orbit. Figure 9.4 depicts a desymmetrization lattice with USCIs and USCI-CFs. These lattices clarify what subsymmetries are realized by desymmetrizing the orbit.

3

First, we discuss a desymmetrization with achiral ligands only. *In general, a subgroup (subsymmetry) does not exist, if we do not find any division of an orbit by examining the subgroup and any of its supergroups in the lattice.* This case is solved by using a full DL or a DL with USCIs. For example, consider the substitution of an allene skeleton (**3**) which has four positions subject to the CR ($D_{2d}(/C_s)$). Figure 9.3 shows that there emerge no divisions in the conversion of $D_{2d}(/C_s)$ into $S_4(/C_1)$; therefore an S_4-molecule does not exist in this series. This conclusion is also verified by examining USCIs, where the symbol ($\circ$) denotes the existence of the subsymmetry and $\times$ the non-existence. In a similar way, we conclude that a C_2-molecule does not exist; on the other hand, a C_2'-molecule exists. In conclusion, there exist only four subsymmetries, *i.e.*, C_{2v} (*e.g.*4), C_s

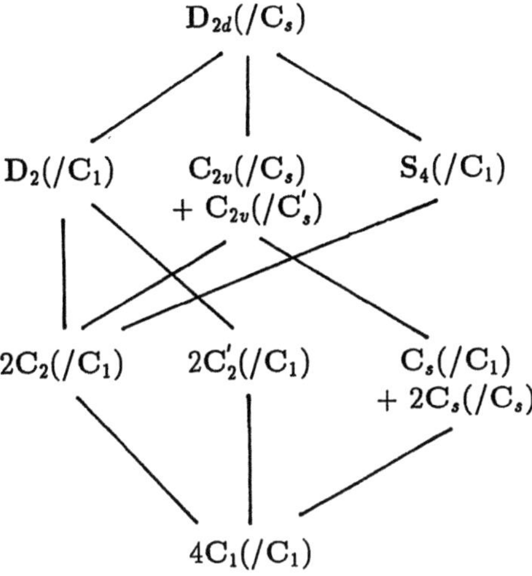

Figure 9.3: Desymmetrization lattice for $\mathbf{D}_{2d}(/\mathbf{C}_s)$

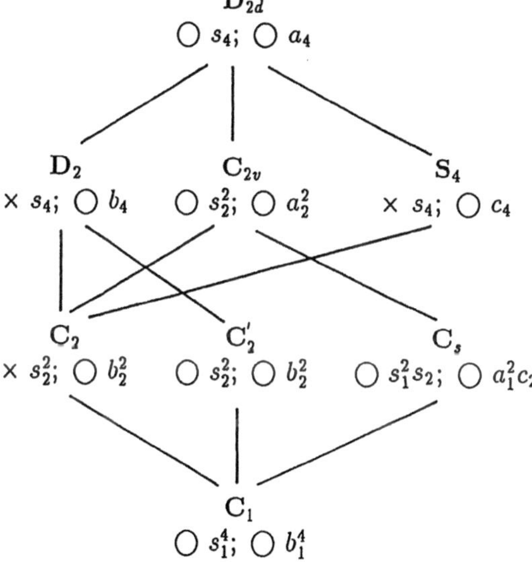

Figure 9.4: Desymmetrization lattice with USCIs and USCI-CFs for $\mathbf{D}_{2d}(/\mathbf{C}_s)$

($e.g.$5), C_2' ($e.g.$6) and C_1 ($e.g.$7), in this series.

$$\begin{array}{cccc} \text{4} \;\; C_{2v} & \text{5} \;\; C_s & \text{6} \;\; C_2' & \text{7} \;\; C_1 \end{array}$$

If we permit chiral ligands as well as achiral ligands, we use the full DL and the DL with USCI-CFs. In general, *a subgroup (subsymmetry) does not exist, if its chirality fittingness (or USCI-CF) is identical with that of any supergroup.* Figure 9.4 indicates that all of the subsymmetries exist under this condition. We depict an S_4- (8), a D_2- (9) and a C_2-molecule (10).

$$\begin{array}{ccc} \text{8} \;\; S_4 & \text{9} \;\; D_2 & \text{10} \;\; C_2 \end{array}$$

$$p, \overline{p} = \text{chiral} \qquad A = \text{achiral}$$

Bibliography

[1] S. Fujita, *Theor. Chim. Acta* **76**, 247 (1989).

[2] A. Kerber and K.-J. Thürlings, *Combinatorial Theory*, D. Jugnickel and K. Vedder Eds. Lecture Notes in Math. 969, Springer Berlin, pp. 191–211 (1982).

[3] S. Fujita, *J. Math. Chem.*, **5**, 99 (1990).

[4] S. Fujita, *J. Chem. Educ.*, **63**, 744 (1986).

[5] S. Fujita, submitted for publication.

Chapter 10

Prochirality [1]

We have discussed the concept of *chirality fittingness* (Chapter 8). In terms of this concept, we re-examine the concept of "prochirality", which has been proposed by Hanson.[1, 2] Hanson's definition is based upon permutation-group theory and plays down point-group theory, though the latter is essential to discuss stereochemistry. The prochirality concept has been restricted to molecular systems in which two ligands or faces are differentiated to create a chiral compound. By integrating the permutation- and point-group theories, the present method gives a general approach that treats two or more ligands successfully.

We have also developed another concept of *subduction of coset representations* in Chapter 9. If we combine the two concepts developed by us, a more detailed result is available for discussing prochirality. The prochirality is shown to be one of the symmetrical properties inherent in an enantiospheric orbit.

10.1 Desymmetrization of Enantiospheric Orbits

We here examine an enantiospheric orbit that is subject to the coset representation $G(/G_i)$. Note that G is an achiral group and G_i is a chiral group.

Let G^{max} be the subgroup that contains all of the proper rotations of G. We first discuss such desymmetrizations that are formulated as $G(/G_i) \downarrow G^{max}$. The corresponding coset decomposition contains a transversal $\{t_1, t_2, \ldots, t_r\}$, in which $r/2$ elements are proper rotations and the remaining $r/2$ elements are improper rotations (see Chapter 9). When we designate the latter half by the symbol u_α ($\alpha = 1, 2, \ldots, r/2$), we can rewrite the coset decomposition as follows.

$$G = (G_i t_1 + G_i t_2 + \ldots + G_i t_{r/2}) + (G_i u_1 + G_i u_2 + \ldots + G_i u_{r/2}). \quad (10.1)$$

It should be noted that $u_\alpha = t_\alpha u_1$ ($\alpha = 1, 2, \ldots, r/2$) and that u_1 ($\in G$) is σ_h, σ_v, σ_d, i, or $S_n(n = 4p)$, according to G. As shown in Chapters 8 and 9, each coset

---

[1]Reprinted in part with permission from S. Fujita, *J. Am. Chem. Soc.*, **112**, 3390–3397 (1990). ©(1990) American Chemical Society.

in eq. 10.1 corresponds to a ligand Ω_k ($\in \Omega$) in such a manner that $G_i t_k \longleftrightarrow \Omega_k$ ($1 \leq k \leq r/2$) and $G_i u_k \longleftrightarrow \Omega_k$ ($1 \leq k \leq r/2$). Thereby, Ω is divided into two subsets, $i.e.$,

$$\Omega^{(+)} = \{\Omega_1, \Omega_2, \ldots, \Omega_{r/2}\}, \tag{10.2}$$
$$\Omega^{(-)} = \{\Omega_{r/2+1}, \Omega_{r/2+2}, \ldots, \Omega_r\}. \tag{10.3}$$

In the light of the coset decomposition (eq. 10.1), we obtain a coset representation that contains permutations,

$$\pi_g = \begin{pmatrix} G_i t_1 & \ldots & G_i t_{r/2} & G_i t_1 u_1 & \ldots & G_i t_{r/2} u_1 \\ G_i t_1 g & \ldots & G_i t_{r/2} g & G_i t_1 u_1 g & \ldots & G_i t_{r/2} u_1 g \end{pmatrix}, \tag{10.4}$$

for $\forall g \in G$.

Starting from eq. 10.1, we have the coset decomposition G^{max} by G_i,

$$G^{max} = G_i t_1 + G_i t_2 + \ldots + G_i t_{r/2}, \tag{10.5}$$

because G_i is also a subgroup of G^{max}. Obviously, the subgroup G^{max} is a normal subgroup of G, where $|G^{max}| = |G|/2$.

From the purmutations represented by eq. 10.4, we select permutations which correspond to $\forall g \in G^{max}$. Thereby, we obtain the chiral subduction $G(/G_i) \downarrow G^{max}$. Since $g (\in G^{max})$ is a proper rotation, the left half of each permutation is concerned with proper rotations only (or equivalently with $\Omega^{(+)}$) and the right half with improper rotations only (or equivalently with $\Omega^{(-)}$). Thus, there emerge two orbits. By careful inspection, we find that the two halves are identical with each other from a viewpoint of permutation. Since eq. 10.5 indicates that the left half of the permutation (eq. 10.4) belongs to $G^{max}(/G_i) \downarrow G^{max}$ ($=G^{max}(/G_i)$), the subduced representation ($G(/G_i) \downarrow G^{max}$) contains at least one $G^{max}(/G_i)$ CR. Since the original $G(/G_i)$ CR is transitive, the left and right halves of eq. 10.4 are mixed up on the action of G. This means that local symmetry of the left half is the same as or conjugate to that of the right half. Because the former is G_i, we arrive at the following lemma.

Lemma 10.1 *Let G be an achiral point group. Consider an enantiospheric orbit governed by $G(/G_i)$, where G_i is a chiral subgroup of G. Let G^{max} be a maximal chiral subgroup of G. Then, there are two possibilities.*

$$\text{Case (a):} \quad G(/G_i) \downarrow G^{max} = 2G^{max}(/G_i) \tag{10.6}$$
$$\text{Case (b):} \quad G(/G_i) \downarrow G^{max} = G^{max}(/G_i) + G^{max}(/G_i'), \tag{10.7}$$

where G_i and G_i' are conjugate in G but not conjugate in G^{max}.

We have pointed out the chirality fittingness of an enantiospheric orbit in Theorem 8.1 (Chapter 8). There are two ways of packing the enantiospheric orbits, *i.e.*, a paired achiral packing and a compensated chiral packing. First, we discuss a paired achiral packing. Equation 10.6 means that a half of A_r in the orbit $(G(/G_i))$ are substituted by $B_{r/2}$ to give a chiral G^{max}-molecule. This operation is illustrated by the following scheme.

$$
\begin{array}{c}
\mathbf{G} - molecule \\
\boxed{A \mid A \mid \ldots \mid A} \\
\boxed{A \mid A \mid \ldots \mid A}
\end{array}
\;\xrightarrow{\;\downarrow\; \mathbf{G}^{max}\;}\;
\begin{array}{c}
\mathbf{G}^{max} - molecule \\
\boxed{B \mid B \mid \ldots \mid B} \\
\boxed{A \mid A \mid \ldots \mid A}
\end{array}
\tag{10.8}
$$

or

$$
\begin{array}{c}
\mathbf{G} - molecule \\
\boxed{A \mid A \mid \ldots \mid A} \\
\boxed{A \mid A \mid \ldots \mid A}
\end{array}
\;\xrightarrow{\;\downarrow\; \mathbf{G}^{max}\;}\;
\begin{array}{c}
\mathbf{G}^{max} - molecule \\
\boxed{A \mid A \mid \ldots \mid A} \\
\boxed{B \mid B \mid \ldots \mid B}
\end{array}
\tag{10.9}
$$

wherein A and B represent achiral substituents. The two chiral molecules produced are enantiomeric to each other.

Second, we deal with a compensated chiral packing. We use the symbols Q and Q' to designate a pair of antipodal ligands. Equation 10.6 means that $r/2$ of Q's in the orbit $(G(/G_i))$ are substituted by $r/2$ of B's to give a chiral G^{max}-molecule. This operation is illustrated by the following scheme.

$$
\begin{array}{c}
\mathbf{G} - molecule \\
\boxed{Q \mid Q \mid \ldots \mid Q} \\
\boxed{Q' \mid Q' \mid \ldots \mid Q'}
\end{array}
\;\xrightarrow{\;\downarrow\; \mathbf{G}^{max}\;}\;
\begin{array}{c}
\mathbf{G}^{max} - molecule \\
\boxed{B \mid B \mid \ldots \mid B} \\
\boxed{Q' \mid Q' \mid \ldots \mid Q'}
\end{array}
\tag{10.10}
$$

or

$$
\begin{array}{c}
\mathbf{G} - molecule \\
\boxed{Q \mid Q \mid \ldots \mid Q} \\
\boxed{Q' \mid Q' \mid \ldots \mid Q'}
\end{array}
\;\xrightarrow{\;\downarrow\; \mathbf{G}^{max}\;}\;
\begin{array}{c}
\mathbf{G}^{max} - molecule \\
\boxed{Q \mid Q \mid \ldots \mid Q} \\
\boxed{B \mid B \mid \ldots \mid B}
\end{array}
\tag{10.11}
$$

The two chiral molecules produced are again enantiomeric to each other.

The following example illustrates the two possibilities described in Lemma 10.1.

Example **10.1** Let us consider adamantane-2,6-dione (**1**). Among the 12 hydrogen atoms (○) in **1**, we take the eight methylene hydrogen atoms into consideration. The substitution of four hydrogen atoms produces **2a**, in which ○'s and ●'s construct distinct $D_2(/C_1)$ orbits (Fig. 10.1). This is an example of case (a), which is algebraically represented by

$$
D_{2d}(/C_1) \downarrow D_2 = 2D_2(/C_1).
\tag{10.12}
$$

If ○ and ● are interchanged, the resulting molecule (**2b**) is enantiomeric to **2a**.

An example of case (b) is the conversion of **3** into **4a** (or **4b**). The four methylene carbons of **3** construct an orbit governed by $D_{2d}(/C_2')$. This orbit is

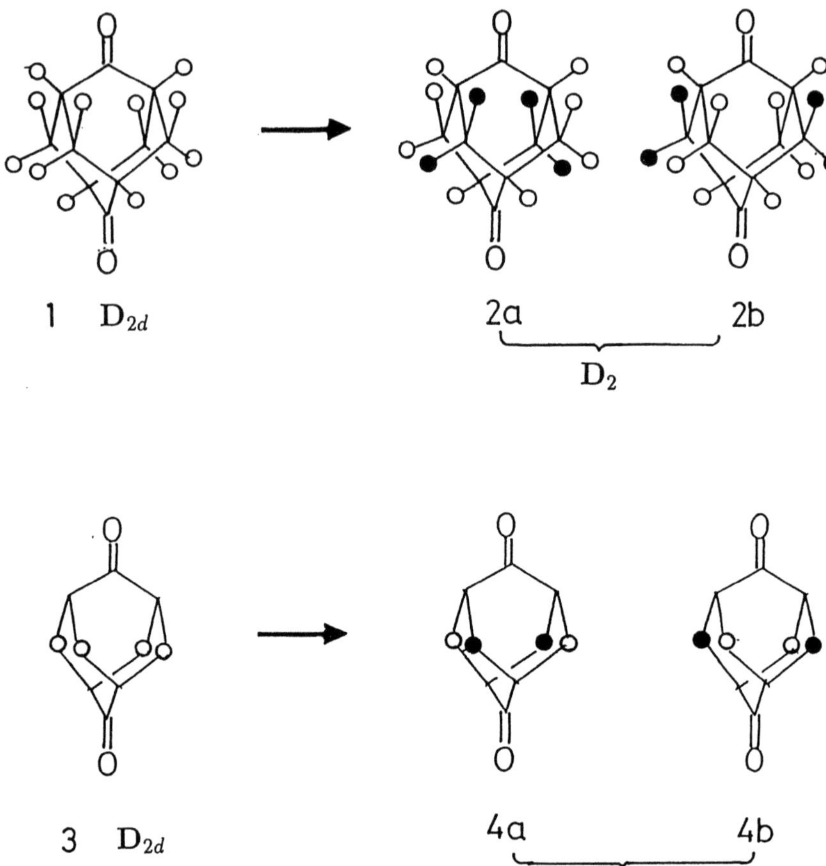

Figure 10.1: D$_2$ molecules derived from a D$_{2d}$ molecule

divided into a $\mathbf{D}_2(/\mathbf{C}_2')$ orbit and $\mathbf{D}_2(/\mathbf{C}_2'')$ orbit in each of the desymmetrized compounds (**4a** and **4b**). This operation is expressed by

$$\mathbf{D}_{2d}(/\mathbf{C}_2') \downarrow \mathbf{D}_2 = \mathbf{D}_2(/\mathbf{C}_2') + \mathbf{D}_2(/\mathbf{C}_2'). \tag{10.13}$$

In both cases (a) and (b), the original $\mathbf{G}(/\mathbf{G}_i)$ orbit is a single enantiospheric orbit that is divided into two halves on the action of $\mathbf{G}^{max}$ (Chapter 8). The halves correspond to the two orbits given in the right-hand side of eq. 10.6 or eq. 10.7.

10.2 Prochirality

In the preceding section, we have restricted our discussion within the irreversible effect of desymmetrization. However, Lemma 10.1 covers a wide variety of desymmetrizations of enantiospheric orbits, whether irreversible or reversible.

The two halves of an enantiospheric $\mathbf{G}(/\mathbf{G}_i)$ orbit do not separate energetically within an achiral environment, since the two belong to the same equivalence class on the action of $\mathbf{G}$ point group. Consider that the enantiospheric orbit is placed under a chiral environment. Such chiral environments may be an attack by a chiral reagent, a solvation by a chiral solvent and so on. The chiral environment differentiates the two parts energetically. This effect is schematically represented by the following expression (consider first *a paired achiral packing*).

$$\begin{array}{|c|c|c|c|}\hline A^Q & A^Q & \dots & A^Q \\\hline A & A & \dots & A \\\hline\end{array} \;,\;\; \text{or} \;\; \begin{array}{|c|c|c|c|}\hline A & A & \dots & A \\\hline A^Q & A^Q & \dots & A^Q \\\hline\end{array} \;,$$
$$\underbrace{}_{r/2} \qquad\qquad \underbrace{}_{r/2}$$

where A^Q represents the ligand A perturbed by a chiral reagent (Q). These perturbed molecules (*i.e.*, reaction intermediates) no longer belong to the original $\mathbf{G}$ symmetry. They are obviously diastereomeric with each other; therefore, they are energetically different. If the left intermediate is energetically preferred, the process represented by eq. 10.8 occurs preferably. If the right one is energetically preferred, scheme 10.9 produces the other enantiomer.

With respect to *a compensated chiral packing*, similar discussions can be applied. Let the symbol Q^R denote a chiral ligand Q perturbed by the action of a chiral reagent (R). Then, we have the following intermediate,

$$\begin{array}{|c|c|c|c|}\hline Q^R & Q^R & \dots & Q^R \\\hline Q' & Q' & \dots & Q' \\\hline\end{array} \;, \tag{10.14}$$
$$\underbrace{}_{r/2}$$

if this is energetically preferred. It follows that the process represented by eq. 10.10 occurs.

The discussions in this section are restricted within Case (a) of Lemma 10.1. However, they are applicable to Case (b) without any modification. Hence, Lemma 10.1 can be generalized to a theorem.

Theorem 10.1 (Fujita[3]) *An enantiospheric orbit is capable of separating into two hemispheric orbits of the same length under a chiral environment, whether the change is reversible or irreversible.*

A *prochiral* compound is defined as an achiral compound that has at least one enantiospheric orbit. Since the prochiral compound may possess homospheric orbits along with enantiospheric orbits, we have two type of prochiral compounds; Type I has enantiospheric orbits only, and Type II has enantionspheric and homospheric orbits. An achiral compound that contains homospheric orbits only is called a *para-achiral* compound, since there exist no direct methods of converting the compound into a chiral one. In conclusion, organic (and inorganic) compounds are classified into the following categories.

$$
\left\{
\begin{array}{l}
\text{achiral} \left\{
\begin{array}{lll}
\text{prochiral-I} & \cdots & \text{enantiospheric} \\
\text{prochiral-II} & \cdots & \text{enantiospheric + homospheric} \\
\text{para-achiral} & \cdots & \text{homospheric}
\end{array}
\right. \\
\text{chiral} \qquad\qquad \cdots \quad \text{hemispheric}
\end{array}
\right.
$$

Among these categories, the para-achiral molecules will be discussed in Chapter 11. The prochirality of the present definition differs from Hanson's original definition.[1] This point will be discussed in Chapter 21.

10.3 Further Desymmetrization of Enantiospheric Orbits

In this section, we deal with Case (a) of Lemma 10.1 only. Consider the SR $\mathbf{G}^{max}(/\mathbf{G}_i) \downarrow \mathbf{G}_j$ where $\mathbf{G}_j$ is a subgroup of $\mathbf{G}^{max}$. If we apply Theorem 9.1 (Chapter 9) to the SR, we obtain following equation.

$$\mathbf{G}^{max}(/\mathbf{G}_i) \downarrow \mathbf{G}_j = \sum_{k=1}^{v_j} \gamma_k^{(ij)} \mathbf{G}_j(/\mathbf{G}_k^{(j)}), \tag{10.15}$$

where the multiplicities ($\gamma_k^{(ij)}$'s) are non-negative integers. Since $\mathbf{G}_i$ and $\mathbf{G}_j$ are subgroups of $\mathbf{G}$ and of $\mathbf{G}^{max}$, eqs. 10.6 and 10.15 afford the following theorem.

Theorem 10.2 *Suppose that an enantiospheric orbit follows Case (a) of Lemma 10.1. Then, any chiral subduction, $\mathbf{G}(/\mathbf{G}_i) \downarrow \mathbf{G}_j$, is represented by*

$$\mathbf{G}(/\mathbf{G}_i) \downarrow \mathbf{G}_j = \sum_{k=1}^{v_j} (2\gamma_k^{(ij)}) \mathbf{G}_j(/\mathbf{G}_k^{(j)}), \tag{10.16}$$

wherein $\mathbf{G}$ contains proper and improper rotations; and its subgroups, $\mathbf{G}_i$ and $\mathbf{G}_j$, consist of proper rotations only.

As a result, any $G_k^{(j)}$'s are proper rotations and any $G_j(/G_k^{(j)})$'s represent hemispheric orbits.

This theorem is exemplified by $D_{2d}(/C_1)\!\downarrow\!D_2$, $D_{2d}(/C_1)\!\downarrow\!C_2$, $D_{2d}(/C_1)\!\downarrow\!C_2'$, and $D_{2d}(/C_1)\!\downarrow\!C_1$ as discussed in Chapter 9. The tables collected in Appendix C contain further examples of this theorem.

Suppose that G contanis u_1 satisfying $u_1u_1 = I$ (an identity) as an improper rotation and that we can construct a point group $\widetilde{G}_j$ ($\leq G$) that contains G_j as the maximum chiral subgroup; i.e., $\widetilde{G}_j = \{g, u_1g \mid \forall g \in G_j\}$, where u_1 ($\in G$) is σ_h, σ_v, σ_d, or i.[2] Let us examine a subduction represented by $G(/G_i)\!\downarrow\!\widetilde{G}_j$, where $G(/G_i)$ is enantiospheric. For all proper rotations (g), the concrete form of this subduction is represented by eq. 10.4. For all improper rotations (u_1g), we obtain

$$\pi_{u_1g} = \begin{pmatrix} G_it_1 & \dots & G_it_{r/2} & G_it_1u_1 & \dots & G_it_{r/2}u_1 \\ G_it_1u_1g & \dots & G_it_{r/2}u_1g & G_it_1g & \dots & G_it_{r/2}g \end{pmatrix}, \qquad (10.17)$$

since $(G_it_\alpha u_1)(u_1g) = G_it_\alpha(u_1u_1)g = G_it_\alpha g$. Note that eq. 10.17 holds for $\forall g \in G_j$ and that $G(/G_i)\!\downarrow\!\widetilde{G}_j$ contains the permutations (eqs. 10.4 and 10.17). In the permutation (eq. 10.17), the kth column ($k \leq r/2$) is always associated with $(r/2+k)$th column. This fact holds for the permutation (eq. 10.4), which constructs the $G^{max} \downarrow G_j$ representation. The relationship between these permutations (eqs. 10.4 and 10.17) indicates the correpondence between the orbits of $G(/G_i)\!\downarrow\!\widetilde{G}_j$ and those of $G(/G_i) \downarrow G_j$. Hence, eq. 10.15 can be extended into a theorem.

Theorem 10.3 *Suppose that an enantiospheric orbit is in accord with Case (a) of Lemma 10.1. If we can construct $\widetilde{G}_j$ as above, an achiral subduction of a prochiral part, $G(/G_i) \downarrow \widetilde{G}_j$, is represented by*

$$G(/G_i) \downarrow \widetilde{G}_j = \sum_{k=1}^{v_j} (\gamma_k^{(ij)})\widetilde{G}_j(/G_k^{(j)}), \qquad (10.18)$$

where $G_k^{(j)}$ is chiral subgroups. The coset representations, $\widetilde{G}_j(/G_k^{(j)})$, appearing in the right-hand side of eq. 10.18 is prochiral parts and the corresponding orbits are all enantiospheric.

This theorem is exemplified by comparing the $D_{2d}(/C_1)\!\downarrow\!C_{2v}$ column with $D_{2d}(/C_1)\!\downarrow\!C_2$ column of the subduction table for D_{2d}, where G_i is equal to C_1, C_2, C_2', or D_2. Note that the summation in eq. 10.18 is concerned with the SSG of G_j, not with the SSG of $\widetilde{G}_j$. In eq. 10.18, $\widetilde{G}_j(/G_k^{(j)})$ may be equivalent to $\widetilde{G}_j(/G_{k'}^{(j)})$ if $G_k^{(j)}$ is conjugate to $G_{k'}^{(j)}$ as subgroups of $\widetilde{G}_j$ and if they are not conjugate as subgroups of G_j. For example, compare the subduction $T_d(/C_2)\!\downarrow\!D_{2d} = D_{2d}(/C_2) + 2D_{2d}(/C_2')$ with $T_d(/C_2)\!\downarrow\!D_2 = D_2(/C_2) + D_2(/C_2') + D_2(/C_2'')$.

---

[2] In order for $\widetilde{G}_j$ be a group, it has to satisfy $u_1G_j = G_ju_1$ and $u_1u_1 \in G_j$.

Equations 10.16 and 10.18 mean that *the chiral subduction of an enantiospheric orbit creates even-number of hemispheric orbits*, which appear in pairs in the original enantiospheric orbit. If we consider sequential subductions, *i.e.*, $\mathbf{G} \Longrightarrow \widetilde{\mathbf{G}}_j \Longrightarrow \mathbf{G}_j$,[3] the corresponding change of an enantiospheric orbit is represented by

$$\mathbf{G}(/\mathbf{G}_i)(a\ part) \Longrightarrow \widetilde{\mathbf{G}}_j(/\mathbf{G}_k^{(j)}) \Longrightarrow 2\mathbf{G}_j(/\mathbf{G}_k^{(j)}).$$

The above result can be generalized to the case for which eq. 10.16 holds. The following scheme explains the generalized operation.

$$
\begin{array}{c}
\mathbf{G} - molecule \\
\begin{array}{|c|c|c|c|c|c|c|}
\hline
A & \cdots & A & A & \cdots & A & \cdots & A \\
\hline
A & \cdots & A & A & \cdots & A & \cdots & A \\
\hline
\end{array}
\end{array}
\ \overset{\downarrow \mathbf{G}_j}{\Longrightarrow}\
\begin{array}{c}
\geq \mathbf{G}_j - molecule \\
\begin{array}{|c|c|c|c|c|c|c|}
\hline
A & \cdots & A & A & \cdots & A & \cdots & A \\
\hline
A & \cdots & B & B & \cdots & B & \cdots & A \\
\hline
\end{array}
\end{array}. \qquad (10.19)
$$

The left-hand side represents an enantiospheric orbit with a paired achiral packing, which is subject to $\mathbf{G}(/\mathbf{G}_i)$. The part lying between two $\|$'s contains $2 \times |\mathbf{G}_j| / |\mathbf{G}_k^{(j)}|$ of As. Suppose that a half of this part is occupied by B. This operation creates two hemispheric orbits shown in the right-hand side in agreement with Theorem 10.3. Each of the hemispheric orbits is subject to $\mathbf{G}_j(/\mathbf{G}_k^{(j)})$.

In a case in which we can tentatively consider $\widetilde{\mathbf{G}}_j$ in the subduction process (*cf.* Theorem 10.3), the part lying between two $\|$'s of eq. 10.19 is subject to $\widetilde{\mathbf{G}}_j(/\mathbf{G}_k^{(j)})$. The relationship between $\widetilde{\mathbf{G}}_j(/\mathbf{G}_k^{(j)})$ and $\mathbf{G}_j(/\mathbf{G}_k^{(j)})$ is essentially the same as that between $\mathbf{G}(/\mathbf{G}_i)$ and $\mathbf{G}^{max}(/\mathbf{G}_i)$. Hence, the discussion at eq. 10.6 also holds for this case.

An alternative replacement creates the antipode (the mirror image) of the above molecule, as expressed by the following scheme.

$$
\begin{array}{c}
\mathbf{G} - molecule \\
\begin{array}{|c|c|c|c|c|c|c|}
\hline
A & \cdots & A & A & \cdots & A & \cdots & A \\
\hline
A & \cdots & A & A & \cdots & A & \cdots & A \\
\hline
\end{array}
\end{array}
\ \overset{\downarrow \mathbf{G}_j}{\Longrightarrow}\
\begin{array}{c}
\geq \mathbf{G}_j - molecule \\
\begin{array}{|c|c|c|c|c|c|c|}
\hline
A & \cdots & B & B & \cdots & B & \cdots & A \\
\hline
A & \cdots & A & A & \cdots & A & \cdots & A \\
\hline
\end{array}
\end{array}. \qquad (10.20)
$$

The compound represented by eq. 10.19 is enantiomeric to the one represented by eq. 10.20. In other words, they are the mirror images of each other.

The same discussion is also applicable to the case of a compensated chiral packing of an enantiospheric orbit. The following two ways of packing produce a pair of antipodes.

$$
\begin{array}{c}
\mathbf{G} - molecule \\
\begin{array}{|c|c|c|c|c|c|c|}
\hline
Q & \cdots & Q & Q & \cdots & Q & \cdots & Q \\
\hline
Q' & \cdots & Q' & Q' & \cdots & Q' & \cdots & Q' \\
\hline
\end{array}
\end{array}
\ \overset{\downarrow \mathbf{G}_j}{\Longrightarrow}\
\begin{array}{c}
\geq \mathbf{G}_j - molecule \\
\begin{array}{|c|c|c|c|c|c|c|}
\hline
Q & \cdots & Q & Q & \cdots & Q & \cdots & Q \\
\hline
Q' & \cdots & B & B & \cdots & B & \cdots & Q' \\
\hline
\end{array}
\end{array} \qquad (10.21)
$$

---

[3] We tentatively presume that the process of $\mathbf{G} \Longrightarrow \widetilde{\mathbf{G}}_j$ produces a $\widetilde{\mathbf{G}}_j$ molecule. Generally speaking, however, this process produces a molecule having a symmetry higher than $\widetilde{\mathbf{G}}_j$.

and

$$
\begin{array}{c}
\text{G} - molecule \\
\boxed{\begin{array}{c|c|c|c|c|c} Q & \cdots & Q & Q & \cdots & Q & \cdots & Q \\ \hline Q' & \cdots & Q' & Q' & \cdots & Q' & \cdots & Q' \end{array}}
\end{array}
\;\;\xRightarrow{\;\downarrow \text{G}_j\;}\;\;
\begin{array}{c}
\geq \text{G}_j - molecule \\
\boxed{\begin{array}{c|c|c|c|c|c} Q & \cdots & B & B & \cdots & B & \cdots & Q \\ \hline Q' & \cdots & Q' & Q' & \cdots & Q' & \cdots & Q' \end{array}}
\end{array}.
$$

$$(10.22)$$

The above discussions are summarized in the form of a theorem.

Theorem 10.4 *Suppose an enantiospheric orbit is subduced chirally in the light of eq. 10.16. If two modes of subduced packings of an appropriate orbit $(\text{G}_j(/\text{G}_k^{(j)}))$ are represented by eqs. 10.19 and 10.20 (or by eqs. 10.21 and 10.22), they are enantiometric to each other.*

This theorem indicates that a prochiral compound, which has at least one enantiospheric orbit, can be converted into a chiral compound. It is to be noted that the resulting enantiomeric pair has not yet been proved to have G_j symmetry itself. The pair has been known as far to belong to a chiral subgroup H ($\text{G} \geq \text{H} \geq \text{G}_j$) which contains G_j as a subgroup. The determination of H is open to further investigation. If we assume that the resulting $\text{G}_j(/\text{G}_k^{(j)})$ is a faithful representation of G_j, the molecule created belongs to G_j symmetry.

For illustrating Theorem 10.4, let us examine an enantiospheric orbit ($\text{D}_{2d}(/\text{C}_1)$) appearing in the compound (**1**) of D_{2d} symmetry. The subduction of this orbit into D_2 is represented by eq. 10.12. This subduction is realized by compounds **2a** and **2b**, which belong to D_2 symmetry (Fig. 10.1). They are enantiomeric to each other.

The subduction of the orbit into C_2 is represented by

$$\text{D}_{2d}(/\text{C}_1) \downarrow \text{C}_2 = 4\text{C}_2(/\text{C}_1). \tag{10.23}$$

Figure 10.2 illustrates this derivation. The operation producing **5a** is possible if we differentiate two ligands marked with a from others. The selection of two ligands marked with b creates a compound homomeric to **5a**. The two selections are equivalent energetically, even if we employ chiral reagents. Another selection (c) produces **5b**, which is enantiomeric to **5a**. In the compound (**5a**), the original $\text{D}_{2d}(/\text{C}_1)$ is divided into four orbits marked by a, b, c, and d, each of which is subject to $\text{C}_2(/\text{C}_1)$.

Figure 10.3 shows a desymmetrization into C_2'. This case corresponds to the subduction represented by

$$\text{D}_{2d}(/\text{C}_1) \downarrow \text{C}_2' = 4\text{C}_2'(/\text{C}_1). \tag{10.24}$$

This subduction creates another pair of C_2 molecules (**6a** and **6b**). The process producing **6a** is possible if we distinguish the methylene hydrogens (marked with a) from others. The corresponding homomer can be created by selecting

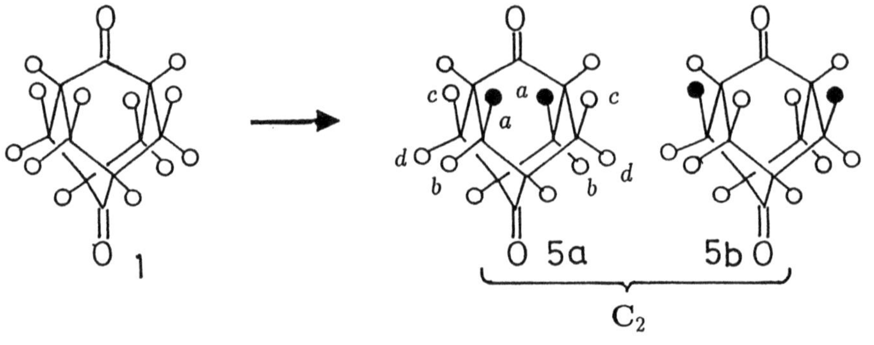

Figure 10.2: C_2 molecules derived from a D_{2d} molecule

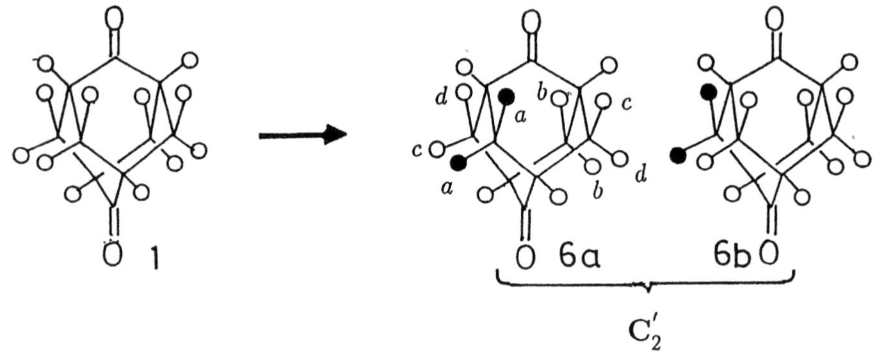

Figure 10.3: C_2' molecules derived from a D_{2d} molecule

the methylene hydrogens marked with *b*. These two processes are energetically equivalent with and without chiral perturbation. Note that the four hydrogens (*a* an *b*) belong to the same half of the original $\mathbf{D}_{2d}(/\mathbf{C}_1)$ orbit of **1**. The remaining methylene hydrogens construct the other half of the original $\mathbf{D}_{2d}(/\mathbf{C}_1)$ orbit, which is concerned with the creation of **6b** and its homomer.

10.4 Chiral syntheses

In the preceding sections, we have discussed prochirality as an attribute of an enantiospheric orbit. Although various asymmteric syntheses have been reported,[4, 5] they are based on the principle that a chiral reagent differentiates only two functional groups (or faces). The present prochirality concept is applicable to cases that contains two or more functional groups. We use the term *chiral syntheses* in place of "asymmetric syntheses", because we deal here with potential cases that produce molecules of higher chirality. In a chiral synthesis, a chiral reagent having several chiral units (Q) attacks a substrate molecule at its functional groups of enantiospheric orbits. This process has been formulated by considering a substituent perturbed chirally (A^Q). Consider an enantiospheric orbit subject to a prochiral part, $\mathbf{G}(/\mathbf{G}_i)$. If a chiral reagent attacks the substrate on one half of the orbit, there emerges an intermediate of $\mathbf{G}^{max}$ symmetry as follows in accord with Lemma 10.1 (see also Section 10.2).

$$
\begin{array}{c}
\mathbf{G} - substrate \\
\begin{array}{|c|c|c|c|}
\hline
A & A & \cdots & A \\
\hline
A & A & \cdots & A \\
\hline
\end{array}
\end{array}
\Longrightarrow
\begin{array}{c}
\mathbf{G}^{max} - intermediate \\
\begin{array}{|c|c|c|c|}
\hline
A^Q & A^Q & \cdots & A^Q \\
\hline
A & A & \cdots & A \\
\hline
\end{array}
\end{array}
\Longrightarrow
\begin{array}{c}
\mathbf{G}^{max} - product \\
\begin{array}{|c|c|c|c|}
\hline
B & B & \cdots & B \\
\hline
A & A & \cdots & A \\
\hline
\end{array}
\end{array}
. \quad (10.25)
$$

The intermediate may have a lower symmetry than $\mathbf{G}^{max}$ if A^Q is incompatible to the local symmetry ($\mathbf{G}^{max}$); however, the chiral nature maintains in this case.

On the other hand, an attack on the opposite half of the enantiospheric orbit produces an intermediate of $\mathbf{G}^{max}$ symmetry. This process is illustrated by

$$
\begin{array}{c}
\mathbf{G} - substrate \\
\begin{array}{|c|c|c|c|}
\hline
A & A & \cdots & A \\
\hline
A & A & \cdots & A \\
\hline
\end{array}
\end{array}
\Longrightarrow
\begin{array}{c}
\mathbf{G}^{max} - intermediate \\
\begin{array}{|c|c|c|c|}
\hline
A & A & \cdots & A \\
\hline
A^Q & A^Q & \cdots & A^Q \\
\hline
\end{array}
\end{array}
\Longrightarrow
\begin{array}{c}
\mathbf{G}^{max} - product \\
\begin{array}{|c|c|c|c|}
\hline
A & A & \cdots & A \\
\hline
B & B & \cdots & B \\
\hline
\end{array}
\end{array}
\quad (10.26)
$$

Because the two $\mathbf{G}^{max}$-intermediates are diastereomeric, one of them is energetically predominant over the other. Experiments or quantum-theoretical calculations are necessary to determine which intermediate is energetically preferred. If the first intermediate (eq. 10.25) is tentatively preferred, the corresponding enantiomer is produced preferably.

When we use a chiral reagent having chiral parts of opposite chirality (Q'), we find that the reagent attacks the opposite half compared with the intermediate (eq.

10.25). The following intermediate (eq. 10.27) is energetically preferred, inasmuch as eq. 10.25 is preferred. Note that the intermediate of eq. 10.27 is enantiomeric to that of eq. 10.25.

$$
\begin{array}{ccc}
\mathbf{G}-substrate & \mathbf{G}^{max}-intermediate & \mathbf{G}^{max}-product \\
\boxed{A\mid A\mid \ldots \mid A} & \boxed{A\mid A\mid \ldots \mid A} & \boxed{A\mid A\mid \ldots \mid A} \\
\boxed{A\mid A\mid \ldots \mid A} \Longrightarrow & \boxed{A^{Q'}\mid A^{Q'}\mid \ldots \mid A^{Q'}} \Longrightarrow & \boxed{B\mid B\mid \ldots \mid B}
\end{array} \qquad (10.27)
$$

Thereby, eq. 10.27 is energetically preferred than the following counterpart that contains an intermediate enantiomeric to that of eq. 10.26.

$$
\begin{array}{ccc}
\mathbf{G}-substrate & \mathbf{G}^{max}-intermediate & \mathbf{G}^{max}-product \\
\boxed{A\mid A\mid \ldots \mid A} & \boxed{A^{Q'}\mid A^{Q'}\mid \ldots \mid A^{Q'}} & \boxed{B\mid B\mid \ldots \mid B} \\
\boxed{A\mid A\mid \ldots \mid A} \Longrightarrow & \boxed{A\mid A\mid \ldots \mid A} \Longrightarrow & \boxed{A\mid A\mid \ldots \mid A}
\end{array} \qquad (10.28)
$$

The process (eq. 10.27) produces an enantiomer of the compound generated by eq. 10.25. Chiral syntheses of this type is called a $\mathbf{G}(/\mathbf{G}_i) \downarrow \mathbf{G}^{max}$-differentiating reactions, since their processes are characterized by Lemma 10.1.

Figure 10.4 illustrates a potential experiment in which an enantiospheric orbit is subject to $\mathbf{C}_s(/\mathbf{C}_1)$. The starting material (**7**) of the $\mathbf{C}_s$ symmetry has two CH_2OH ligands. We regard each ligand as A in the lump.[4] Then the two A's are subject to $\mathbf{C}_s(/\mathbf{C}_1)$, belonging to an enantiospheric orbit. In order to obtain a chiral product, we must differentiate only two ligands belonging to the orbit. Suppose that a chiral reagent (Q) attacks either of ligands. Then, such an interaction as $CH_2OH\cdots Q$ (A^Q) appears; the attack produces a chiral environment. Either one of such possible intermediates (**8a** or **8b**) is energetically predominant and yields a final product of $\mathbf{C}_1$.

For an example of this type, Fuji[6] reported an asymmetric protonation of a dicarboxylate, in which two carboxylate groups belonging to an enantiospheric orbit are differentiated. These reactions have been classified into enantiotopic-group-differentiating reactions.[7] In the light of the present notation, they are characterized as $\mathbf{C}_s(/\mathbf{C}_1)\downarrow\mathbf{C}_1$-ligand-differentiating reactions.[5]

To the best of our knowledge, all of the asymmetric syntheses reported so far are associated with $\mathbf{C}_s(/\mathbf{C}_1)$ from the present point of view. However, Theorem 10.2 and the accompanying discussions imply possibilities of more general chiral syntheses to which no attention has ever been paid. Consider an enantiospheric orbit (subject to $\mathbf{G}(/\mathbf{G}_i)$) which is subduced chirally in the light of eq. 10.16. One of the resulting $\mathbf{G}_j(/\mathbf{G}_k^{(j)})$'s is a representation of $\mathbf{G}_j$. Suppose that a chiral

---

[4]This treatment is formulated more strictly in terms of the concept "proligand". See Chapter 21.

[5]An enantiospheric orbit of the faces of a carbonyl group or of the related functional group, which is subject to the $\mathbf{C}_s(/\mathbf{C}_1)$ representation, undergoes a chiral synthesis. Many examples have been reported and reviewed.[4, 8] These reactions have been classified into enantiotopic-face-differentiating reactions.[7] They are characterized more specifically as $\mathbf{C}_s(/\mathbf{C}_1)\downarrow\mathbf{C}_1$-face-differentiating reactions in terms of the present terminology.

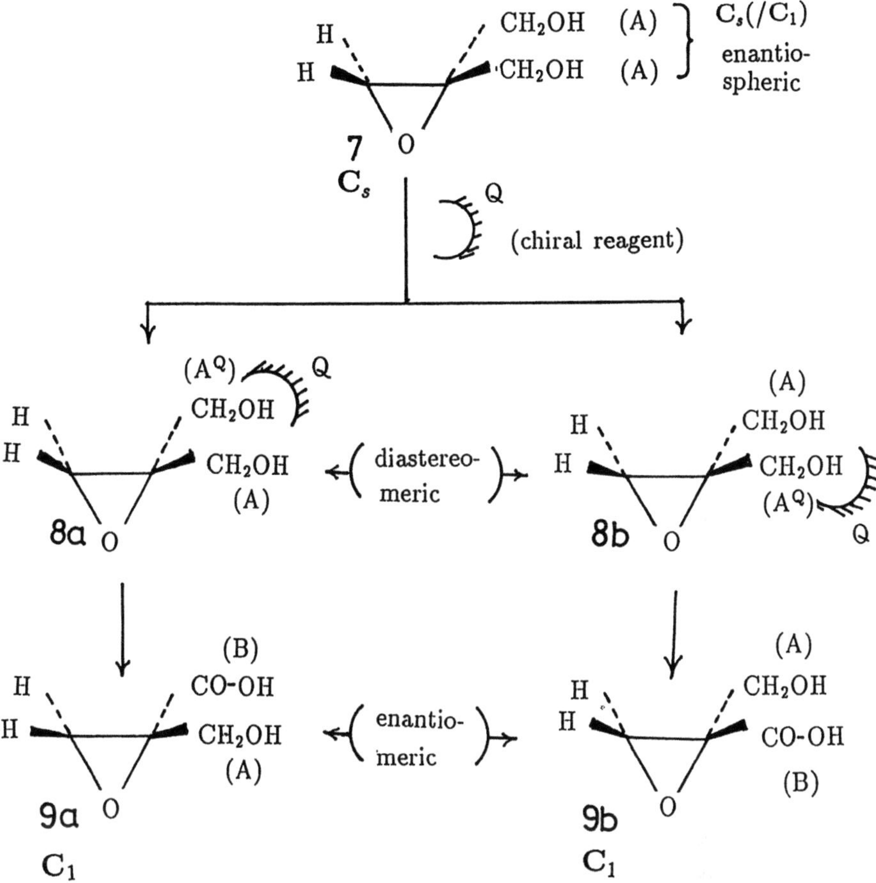

Figure 10.4: A potential chiral synthesis generating C_1 molecule from a C_s molecule

reagent, which has $\mid G_j \mid / \mid G_k^{(j)} \mid$ chiral units (Q), attacks such a $G_j(/G_k^{(j)})$ part. The following two processes afford diastereomeric intermediates.[6] Although careful experiments or theoretical calculations are necessary to determine which, either one of the intermediates has lower activation energy and is preferred to give either of enantiomeric products.

$$
\mathbf{G} - substrate \quad\Longrightarrow\quad \geq \mathbf{G}_j - intermediate
$$

$$
\begin{array}{|c c|c|c c|c|c c|}
\hline
A & \dots & A & A & \dots & A & \dots & A \\
\hline
A & \dots & A & A & \dots & A & \dots & A \\
\hline
\end{array}
\Longrightarrow
\begin{array}{|c c|c|c c|c|c c|}
\hline
A & \dots & A & A & \dots & A & \dots & A \\
\hline
A & \dots & A^Q & A^Q & \dots & A^Q & \dots & A \\
\hline
\end{array}
$$

$$
\geq \mathbf{G}_j - molecule
$$

$$
\Longrightarrow
\begin{array}{|c c|c|c c|c|c c|}
\hline
A & \dots & A & A & \dots & A & \dots & A \\
\hline
A & \dots & B & B & \dots & B & \dots & A \\
\hline
\end{array}
\qquad (10.29)
$$

and

$$
\mathbf{G} - substrate \quad\Longrightarrow\quad \geq \mathbf{G}_j - intermediate
$$

$$
\begin{array}{|c c|c|c c|c|c c|}
\hline
A & \dots & A & A & \dots & A & \dots & A \\
\hline
A & \dots & A & A & \dots & A & \dots & A \\
\hline
\end{array}
\Longrightarrow
\begin{array}{|c c|c|c c|c|c c|}
\hline
A & \dots & A^Q & A^Q & \dots & A^Q & \dots & A \\
\hline
A & \dots & A & A & \dots & A & \dots & A \\
\hline
\end{array}
$$

$$
\geq \mathbf{G}_j - molecule
$$

$$
\Longrightarrow
\begin{array}{|c c|c|c c|c|c c|}
\hline
A & \dots & B & B & \dots & B & \dots & A \\
\hline
A & \dots & A & A & \dots & A & \dots & A \\
\hline
\end{array}
\qquad (10.30)
$$

Let us examine the C_{2v}-molecule (**10**), in which four CH_2OH ligands construct an orbit subject to $C_{2v}(/C_1)$. This orbit is enantiospheric and active to a chiral synthesis. Figure 10.5 indicates a model of chiral synthesis which produces a C_2-molecule from a C_{2v}-molecule. This process corresponds to the subduction represented by $C_{2v}(/C_1) \downarrow C_2 = 2C_2(/C_1)$. We can consider a potential chiral reagent that has two chiral units (Q) in a diagonal positions. Since the resulting intermediates (**11a** and **11b**) are diastereomeric, the energetically predominant intermediate yields either of enantiomeric products (**12a** or **12b**). We refer to this reaction as a $C_{2v}(/C_1)\downarrow C_2$-ligand-differentiating reaction.

Figure 10.6 illustrates a model of chiral synthesis that produces a C_1-molecule from a C_{2v}-molecule. This process corresponds to the subduction represented by $C_{2v}(/C_1) \downarrow C_1 = 4C_1(/C_1)$. We can consider a chiral reagent that has one chiral unit (Q). The intermediates (**13a** and **13b**) afford the respective products that are enantiomeric (**14a** and **14b**). Since **13a** and **13b** are diastereomeric, the energetically predominant one yields the corresponding product (**14a** or **14b**). We refer to this reaction as a $C_{2v}(/C_1) \downarrow C_1$-ligand-differentiating reaction.

A compensated chiral packing of an enantiospheric orbit undergoes another type of chiral synthesis. This can be explained in the same line as above and will be a good exercise to readers.

---

[6]The symmetry of the intermediates are G_j or higher symmetry if A^Q satisfies the local symmetry.

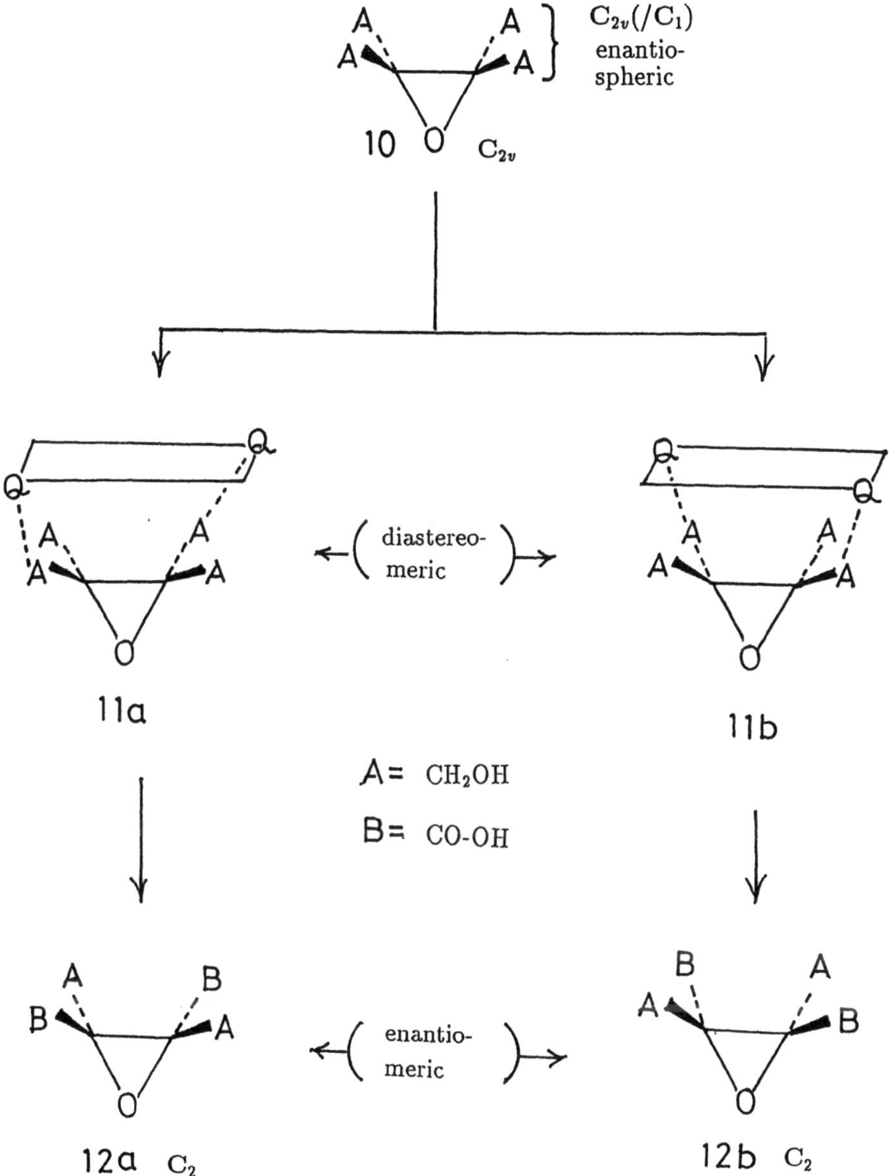

Figure 10.5: A potential chiral synthesis generating C_2 molecule from a C_{2v} molecule

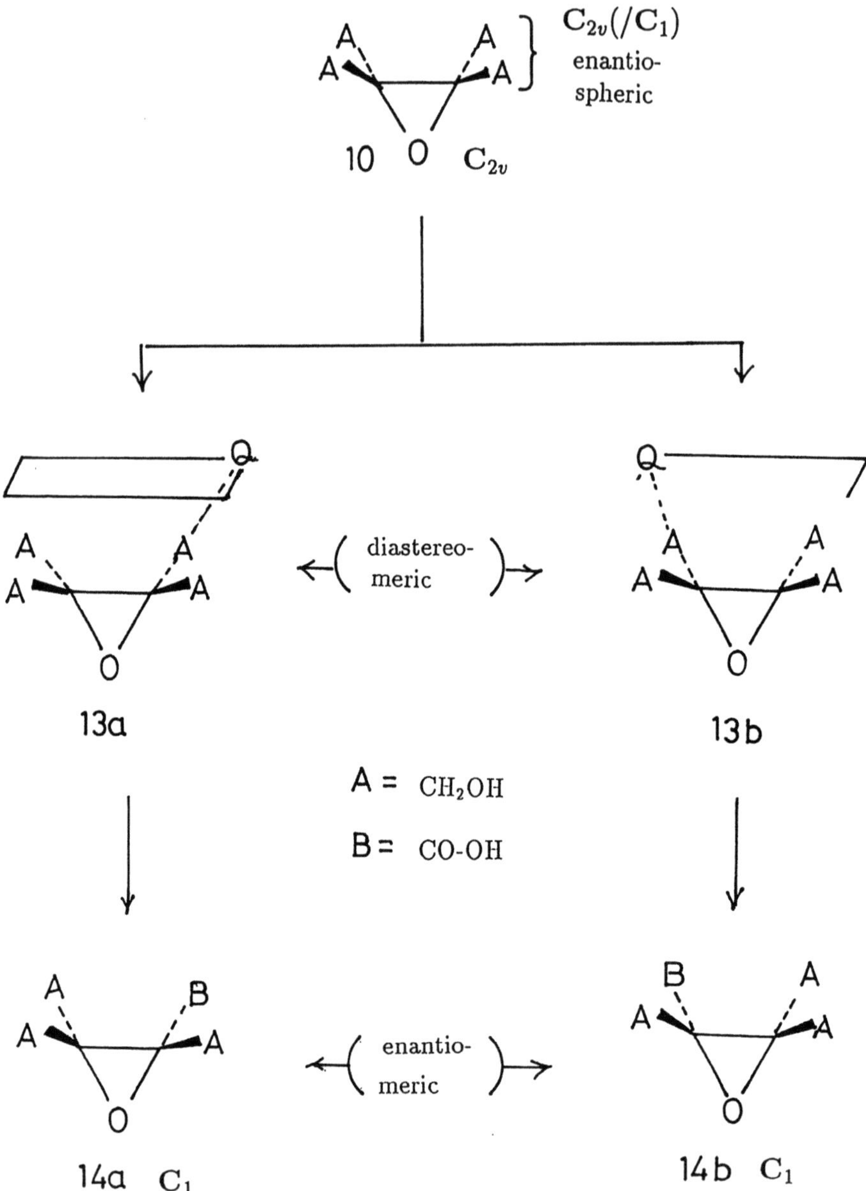

Figure 10.6: A potential chiral synthesis generating C_2 molecule from a C_{2v} molecule

Bibliography

[1] K. R. Hanson, *J. Am. Chem. Soc.*, **88**, 2731 (1966).

[2] H. Hirschmann and K. R. Hanson, *J. Org. Chem.*, **36**, 3293 (1971).

[3] S. Fujita, *J. Am. Chem. Soc.*, **112**, 3390 (1990).

[4] a) J. D. Morrison and H. S. Mosher, *Asymmetric Organic Reactions*, Prentic-Hall, Englewood Ciffs (1971). b) K. B. Kagan and J. C. Fiaud, *Top. Stereochem.*, **10**, 175 (1978). c) R. Bentley, *Molecular Asymmetry in Biology*, Vol. 1, Academic, New York-London (1969); Vol. 2, Academic, New York-London (1970). d) K. Fuji, *Yuki Gosei Kagaku Kyokaishi*, **44**, 623 (1986).

[5] E. L. Eliel, *Stereochemistry of Carbon Compounds*, McGraw-Hill, New York-London (1962).

[6] K. Fuji, M. Node, S. Terada, M. Murata, H. Nagasawa, *J. Am. Chem. Soc.*, **107**, 6404 (1985).

[7] Y. Izumi, *Angew. Chem. Int. Ed. Engl.*, **10**, 871 (1971).

[8] K. Yamamoto, *Yuki Gosei Kagaku Kyokaishi*, **47**, 122 (1989).

Chapter 11

Desymmetrization of Para-Achiral Compounds[1]

A homospheric orbit does not split away with and without a chiral environment. This fact implies that there is no direct method for converting a para-achiral molecule into a chiral one. However, a multistep conversion is available to solve this problem. Mislow and Siegel discussed the desymmetrization of achiral objects that contain one or more symmetry elements of the second kind (σ, i, or S_4) and proposed the concept of (pro)^p-chirality.[1] Later, Halevi added some extension to this concept.[2] This section deals with these issues from the present viewpoint.

11.1 Chiral Subduction of Homospheric Orbits

The subduction of coset representations introduced in Chapter 9 is a foundation of manipulating desymmetrization.[3] First, we examine direct desymmetrization of homospheric orbits into a chiral point group, although this process is chemically impossible. Consider $G(/G_i) \downarrow G_j$, in which $G(/G_i)$ correponds to a homospheric orbit and the subgroup G_j contains proper rotations only. This process is called a *chiral subduction* of a homospheric orbit. In this case, both G and G_i contain proper and improper rotations. Let G^{max} be the maximum subgroup of G that contains proper rotations only. And let G_i^{max} be the maximum subgroup of G_i that contains proper rotations only. Thereby, we have coset decompositions,

$$G = G_i t_1 + G_i t_2 + \ldots + G_i t_r \tag{11.1}$$

and

$$G^{max} = G_i^{max} t_1 + G_i^{max} t_2 + \ldots + G_i^{max} t_r. \tag{11.2}$$

Note that we can select a common transversal $\{t_1, t_2, \ldots, t_r\}$. This fact implies the following theorem.

---

[1]This chaper is based on the article published in S. Fujita, *J. Am. Chem. Soc.*, **112**, 3390–3397 (1990).

Theorem 11.1 *Any chiral subduction of a homospheric orbit, i.e.* $G(/G_i) \downarrow G_j$, *is represented by*

$$G(/G_i) \downarrow G_j = G^{max}(/G_i^{max}) \downarrow G_j, \tag{11.3}$$

where G_j is a chiral subgroup of G (and also of G^{max}).

This equation means that the $G(/G_i)$ orbit (homospheric) and the $G^{max}(/G_i^{max})$ orbit (hemispheric) behave in the same manner on the process of the subduction ($\downarrow G_j$). Chemically speaking, this theorem indicates that there is no direct method for converting such homospheric orbits into hemispheric ones. Consider that a set (orbit) of atoms (or ligands) in a given molecule is governed simultaneously by $G(/G_i)$ and $G^{max}(/G_i^{max})$. This simultaneous government is possible if the orbit is homospheric (Theorem 11.1). This fact implies that the homospheric orbit behaves in the same manner if it is placed under a chiral or achiral environment.

In terms of Theorem 9.1 (Chapter 9), the coset decomposition represented by eq. 11.2 produces

$$G^{max}(/G_i^{max}) \downarrow G_j = \sum_{k=1}^{v_j} \gamma_k^{(ij)} G_j(/G_k^{(j)}). \tag{11.4}$$

Equations 11.3 and 11.4 provide the following equation.

$$G(/G_i) \downarrow G_j = \sum_{k=1}^{v_j} \gamma_k^{(ij)} G_j(/G_k^{(j)}), \tag{11.5}$$

wherein G and G_i contain proper and improper rotations; and G_j contains proper rotations only. Comparison between eqs. 11.4 and 11.5 indicates that the respective right-hand sides are identical with each other. This identity indicates Theorem 11.1 concretely.

If we compare eq. 11.5 with Theorem 10.2 (Chapter 10), they are different in their coefficients. The coefficient (2) appearing in Theorem 10.2 indicates the possibility of chiral syntheses in the case of an enantiospheric orbit. On the other hand, eq. 11.5 shows that there exists no possibility of chiral syntheses.

In a special case in which G_j is equal to G_{max}, eq. 11.5 is converted into

$$G(/G_i) \downarrow G^{max} = G^{max}(/G_i^{max}). \tag{11.6}$$

If u_1 is selected to be an improper rotation (rotoreflection) that satisfies $G_i = G_i^{max} + G_i^{max} u_1$, eq. 11.1 is converted into the coset decomposition represented by

$$G = (G_i^{max} t_1 + G_i^{max} t_2 + \ldots + G_i^{max} t_r) + (G_i^{max} u_1 + G_i^{max} u_2 + \ldots + G_i^{max} u_r), \tag{11.7}$$

wherein t_α is proper rotations selected as representatives, and $u_\alpha = u_1 t_\alpha$ ($\alpha = 1, 2, \ldots, r/2$). For any $g \in G_j$ (a chiral subgroup), we have

$$\begin{aligned}
G_i t_\alpha g &= G_i^{max} t_\alpha g + G_i^{max} u_1 t_\alpha g \\
&= G_i^{max} t_\alpha g + G_i^{max} u_\alpha g.
\end{aligned}$$

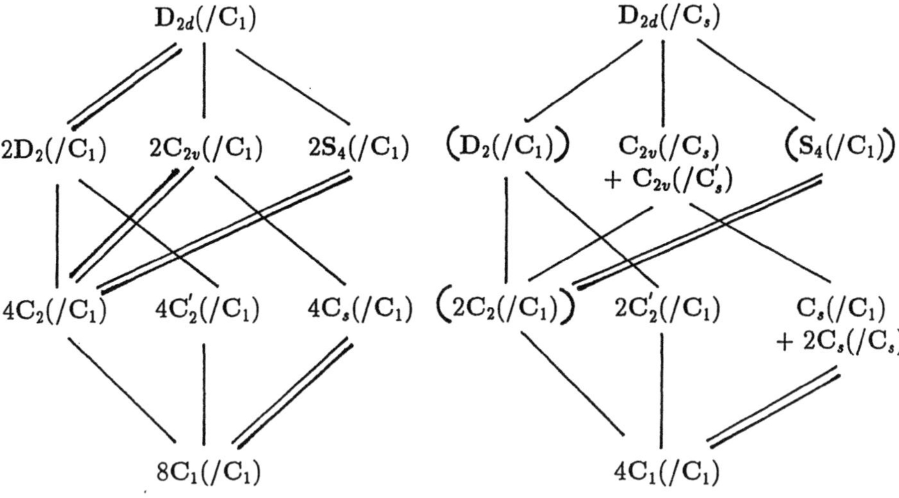

Figure 11.1: Desymmetrization lattice for a $D_{2d}(/C_1)$ orbit (left) and a $D_{2d}(/C_s)$ orbit (right)

This correspondence means that eqs. 11.1 and 11.7 create the relationship,

$$G(/G_i^{max}) \downarrow G_j = 2G(/G_i) \downarrow \dot{G}_j. \tag{11.8}$$

11.2 Desymmetrization of Homospheric Orbits

We have already introduced a *desymmetrization lattice* for an orbit in Chapter 9. Such desymmetrization lattices are quite useful to discuss existence and non-existence of subgroups. Here, we take no account of a case in which any new chiral units are added.[2] Figures 11.1 shows desymmetrization lattices for a $D_{2d}(/C_1)$ (left) and a $D_{2d}(/C_s)$ orbit (right), respectively.

Comparison between these lattices provides useful information on symmetrical properties of the D_{2d} point group. Suppose that we do not find any division of an orbit, when we examine a given group and any of its supergroups in these lattices. Then, the group does not exist mathchemically.[3] We have already shown that S_4 subgroup is absent in the $D_{2d}(/C_s)$ series (Fig. 11.1, right). According

---

[2] If any chiral units are added, we should take a slightly different approach, in which chirality fittingness shoud be taken into consideration. See Chapter 9.

[3] A "mathchemically" possible process is defined as a process which is allowed by mathematical-chemical consideration. This process can be realized experimentally by some method, whether being difficult or not.

to the same reason, D_2 and C_2 do not exist in this series. Molecules of the other subsymmetries are possible to exist from a mathchemical point of view.

Other coset representations for D_{2d} group also constitute the corresponding desymmetrization lattices. Thereby, we conclude the existence or non-existence of molecules of a specific subsymmetry. Table 11.1 summarizes the results, in which T denotes existence of a molecule and F denotes the non-existence. When we examine a compound having two or more orbits, the corresponding rows of Table 11.1 are combined in a LOGICAL-OR fashion (*i.e.*, TT = T, TF = FT = T and FF = F). The result of T indicates the existence of such a molecule, F corresponds to the non-existence.

Table 11.1: Existence of molecules produced by desymmetrization of D_{2d}

	C_1	C_2	C_2'	C_s	S_4	C_{2v}	D_2	D_{2d}
$D_{2d}(/C_1)$	T	T	T	T*	T*	T*	T	T*
$D_{2d}(/C_2)^\dagger$	F	T	F	F	T*	T*	T	T*
$D_{2d}(/C_2')$	T	F	T	T*	F	F	T	T*
$D_{2d}(/C_s)$	T	F	T	T*	F	T	F	T
$D_{2d}(/S_4)^\dagger$	F	F	F	F	T	F	F	T
$D_{2d}(/C_{2v})$	F	F	F	F	F	T	F	T
$D_{2d}(/D_2)^\dagger$	F	F	F	F	F	F	T	T
$D_{2d}(/D_{2d})$	F	F	F	F	F	F	F	T

† Forbidden coset representation.

* Prochiral.

It should be noted that the symbol F indicates the non-existence of a single orbit. If a molecule has two or more orbits, we should examine all of these orbits by a desymmetrization lattice which has the data of these orbits taken from a subduction table. More conveniently, we apply Table 11.1 to such problems. This solution is illustrated by the following example.

Example **11.1** Let us work out a potential conversion of adamantane-2,6-dione (**1**) into a D_2-molecule (Fig. 11.2). We have already discussed the same conversion in Example 10.1 (Chapter 10), in which we took account of the 8 methylene hydrogen atoms only. We here deal with all of the 12 hydrogen atoms of **1**. The molecule (**1**) has a $D_{2d}(/C_s)$ orbit (methine protons; Δ_1) and a $D_{2d}(/C_1)$ orbit (methylene protons; Δ_2), as shown in Example 5.1 (Chapter 5). We find the value for the $D_{2d}(/C_s)$ as F (false) and that for $D_{2d}(/C_1)$ as T (true) in Table 11.1 and combine them in a LOGICAL-OR fashion, *i.e.*, FT = T. Hence, we conclude that a D_2-molecule such as **2** is possible to exist. See also Fig. 11.1.

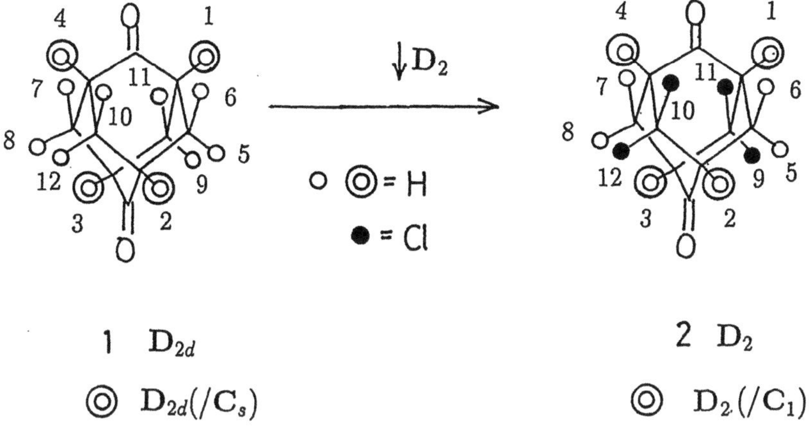

Figure 11.2: A hypothetical process corresponding to $\downarrow$D₂

The concrete form of a permutation representation consists of eight permutations as follows.

$$
\begin{array}{rcll}
 & & \Delta_1 & \Delta_2 \\
I & \sim & (1)(2)(3)(4) & (5)(6)(7)(8)(9)(10)(11)(12) \\
C_{2(1)} & \sim & (1\ 2)(3\ 4) & (5\ 6)(7\ 8)(9\ 10)(11\ 12) \\
C_{2(2)} & \sim & (1\ 3)(2\ 4) & (5\ 7)(6\ 8)(9\ 11)(10\ 12) \\
C_{2(3)} & \sim & (1\ 4)(2\ 3) & (5\ 8)(6\ 7)(9\ 12)(10\ 11) \\
\sigma_{d(1)} & \sim & (1)(2\ 3)(4) & (5\ 9)(6\ 11)(7\ 10)(8\ 12) \\
S_4 & \sim & (1\ 2\ 4\ 3) & (5\ 10\ 8\ 11)(6\ 12\ 7\ 9) \\
S_4^3 & \sim & (1\ 3\ 4\ 2) & (5\ 11\ 8\ 10)(6\ 9\ 7\ 12) \\
\sigma_{d(2)} & \sim & (1\ 4)(2)(3) & (5\ 12)(6\ 10)(7\ 11)(8\ 9) \\
\end{array}
$$

The left part concerning the vertical double line is associated with the $\mathbf{D}_{2d}(/\mathbf{C}_s)$ CR and the right is concerned with the $\mathbf{D}_{2d}(/\mathbf{C}_1)$ CR. We restrict this representation within the elements of its chiral subgroup, *e.g.*, $\mathbf{D}_2$. This restriction affords the subduction of the two CRs by $\mathbf{D}_2$. We then obtain 4 permutations expressed by

$$
\begin{array}{ccc}
 & \Delta_1 & \Delta_1^{(2)} \; \Big| \; \Delta_2^{(2)} \\
I \sim & (1)(2)(3)(4) & (5)(6)(7)(8) \; \Big| \; (9)(10)(11)(12) \\
C_{2(1)} \sim & (1\ 2)(3\ 4) & (5\ 6)(7\ 8) \; \Big| \; (9\ 10)(11\ 12) \\
C_{2(2)} \sim & (1\ 3)(2\ 4) & (5\ 7)(6\ 8) \; \Big| \; (9\ 11)(10\ 12) \\
C_{2(3)} \sim & (1\ 4)(2\ 3) & (5\ 8)(6\ 7) \; \Big| \; (9\ 12)(10\ 11)
\end{array}
$$

This set of permutations constitutes a subduced representation by D_2. The subduction affords no effect on the $D_{2d}(/C_s)$ part, which corresponds to the equation, $D_{2d}(/C_s) \downarrow D_2 = D_2(/C_1)$ (*cf.* eq. 11.6). Although the $D_2(/C_1)$ orbit cannot exist as a single orbit in the $D_{2d}(/C_s)$ series (Table 11.1), it can exist if a molecular environment is suitable, as shown in **2**.

In the light of the subduction, the $D_{2d}(/C_1)$ part is divided into two orbits shown by a single vertical line. This conversion is expressed by the equation, $D_{2d}(/C_1) \downarrow D_2 = 2D_2(/C_1)$. See Example 10.1 (Chapter 10).

The geometric meaning of the subduction is illustrated in Fig. 11.2. The numbering of the 12 positions corresponds to the integers contained in Δ_1 and Δ_2. The four hydrogen atoms at the bridgehead positions (marked with $\odot$) construct the orbit (Δ_1). This orbit is not divided in the subduction process; thus, the $D_{2d}(/C_s)$ orbit in **1** is converted into the $D_2(/C_1)$ orbit in **2**. The orbit (Δ_2) is further divided into $\Delta_1^{(2)}$ and $\Delta_2^{(2)}$. When we consider 4 hydrogens ($\circ$) for the $\Delta_1^{(2)}$ orbit and 4 chlorine atoms ($\bullet$) for $\Delta_2^{(2)}$, we find that the resulting compound (**2**) belongs to D_2 symmetry.

11.3 Chemoselective and Stereoselective Processes

Although there exist no direct methods to convert para-achiral compounds into chiral compounds, some indirect methods are available to do this task. This section discusses these indirect methods.

In Chapter 9, we have generally discussed the processes of desymmetrization, especially chiral syntheses. Desymmetrizing conversions are classified on the basis of the requirement of chiral environments. If a conversion requires any additional chiral environment (*e.g.*, an attack by a chiral reagent), this process is referred to as a stereoselective process. Otherwise, the process is called a chemoselective process. If we take account of the chirality/achirality of a product, we obtain the following classification. The term "chiral syntheses" are synonymous with stereoselective chiral processes.

$$
\left\{
\begin{array}{ll}
\text{chemoselective achiral process :} & \text{achiral} \rightarrow \text{achiral} \\
\text{chemoselective chiral process :} & \text{chiral} \rightarrow \text{chiral} \\
\text{stereoselective chiral process :} & \text{achiral} \rightarrow \text{chiral}
\end{array}
\right.
$$

The chemoselective achiral process is a conversion of an achiral compound of G symmetry into an achiral one of its subsymmetry (G_j). This process requires no chiral environments.[4]

The chemoselective chiral process is a conversion of a chiral compound into a chiral derivative of the same or lower symmetry. This process requires no *additional* chiral perturbation other than the intrinsic chirality of the starting compound. However, chemists' convension have referred to this process as one of asymmetric syntheses.[5]

The stereoselective chiral process (chiral synthesis) is a conversion of a prochiral compound into a chiral derivative. This reaction requires chiral environment. A double straight line denotes this type of process in Fig. 11.1.[6] Table 11.1 contains the symbol (T^*) that denotes the presence of an enantiospheric orbit capable of undertaking a stereoselective chiral process.

In order to convert a para-achiral compound into a chiral molecule, we take at least two steps:

1. We first apply a chemoselective achiral process to this compound, affording a prochiral intermediate. This process involves a conversion of a homospheric orbit into an enantiospheric orbit.

2. Then we use a chiral reagent onto the intermediate in a subsequent stereoselective chiral process. This step contains a conversion of the enantiospheric orbit into hemispheric orbits.

Example **11.2** Let us begin with a D_{2d} molecule (**1**) having a $D_{2d}(/C_s)$ orbit as well as a $D_{2d}(/C_1)$ orbit. In contrast to Examples 10.1 (Chapter 10) and 11.1 (Chapter 11), the present treatment is concerned with the $D_{2d}(/C_s)$ orbit only. Figure 11.3 illustrates chiral molecules *via* C_s (**3**). This orbit contains four methine protons. A chemoselective process can convert this compounds into **3**, which is in accordance with the equation, $D_{2d}(/C_s) \downarrow C_s - C_s(/C_1) + 2C_s(/C_s)$. This is a simple monofluorination. Since four methines of **1** are equivalent, there emerges a single monofluorinated product. Although the original $D_{2d}(/C_s)$ orbit is homospheric, one of the resultiong orbits is subject to $C_s(/C_1)$. This orbit is capable of accepting a chiral reagent. Hence, stereoselective monochlorination of **3** gives **5**, if the attack from the direction (c) by a chiral chlorinating agent is

---

[4]The present discussion takes account of mathchemical possibility of such conversions. Note that experimental methods for realizing these conversions are left out of consideration. In general, a chemoselective achiral process yields a set of configurational isomers. They are regarded to be isolable by some method rather than chiral methods, even if difficult.

[5]Conversions of this type are called diastereo-differentiating asymmetric syntheses.

[6]For simplicity's sake, we deal only with atom(ligand)-differentiating reactions in this paper. However, there conceptually exist bond(edge)-differentiating and face-differentiating reactions. They can be treated in the same way as described here. This hypothetical differentiation suggests a new type of chiral sythesis.

presumed to be energetically preferred rather than the attack from the direction (d).

An alternative stereoselective chiral reaction is possible. If we consider the two carbonyl groups in **1**, the corresponding four faces construct a $\mathbf{D}_{2d}(/\mathbf{C}_s)$ orbit. This orbit is converted into two $\mathbf{C}_s(/\mathbf{C}_1)$ orbits during the process of desymmetrizing **1** to **3**. The corresponding subduction is also represented by $\mathbf{D}_{2d}(/\mathbf{C}_s)\downarrow \mathbf{C}_s = \mathbf{C}_s(/\mathbf{C}_1) + 2\mathbf{C}_s(/\mathbf{C}_s)$. Since the two faces of the top carbonyl group in **3** is subject to the $\mathbf{C}_s(/\mathbf{C}_1)$ (enantiospheric), either one of the faces(a or b) is attacked preferablly by a chiral reagent to give **4** if this is energetically preferable. The two faces of the bottom carbonyl in **3** are distinctly governed by the two $\mathbf{C}_s(/\mathbf{C}_s)$'s. Hence, an attack on the one face is energetically different from another attack on the other face, even under achiral environments. These chemoselective attacks produces two molecules that are different but belong to the same point group ($\mathbf{C}_s$) as that of the starting molecule (**3**). A more complicated case has been discussed in our paper.[3]

In connection with the preceding analysis, we should make a remark on "(pro)^p-chirality". Mislow *et al.* [1] have difined as (pro)^p-chirality any finite, achiral object that can be desymmetrized into a chiral object by at most p stepwise replacement of a point by a differently labeled one. Although this statement has correctly indicated the dependence of the (pro)^p-chirality upon the individual object, their discussion has implied the dependence upon point groups rather than upon individual objects. Thus, they has closed their discussion by a comment, "According to our scheme, desymmetrization of an object with $\mathbf{T}_d$ symmetry yields an object that can belong to only one of four subsymmetries ($\mathbf{C}_{3v}$, $\mathbf{C}_{2v}$, $\mathbf{C}_s$, or $\mathbf{C}_1$)."[1] This conclusion, however, is correct only for the case of an object having a $\mathbf{T}_d(/\mathbf{C}_{3v})$ orbit in terms of the present approach.

Halevi[2] has pointed out that the p value depends upon a route considered. Since his discussion takes no account of the intransitivity of atoms, his extension is also insufficient to be applied to general problems. In addition, the definitions of (pro)^p-chirality, both the original and the revised, have implicitly postulated that an asymmetric synthesis can only be accomplished by differentiating two enantiotopic groups or faces. However, the present discussions shows that stereoselective chiral reactions can occur at various enantiospheric orbits with two or more ligands. Hence, the concepts of *enantiosphericity* and *prochirality* in the present sense are preferred than the (pro)^p-chirality. Moreover, the classification into type I and II prochiralities is useful to discuss stereoselective reactions.

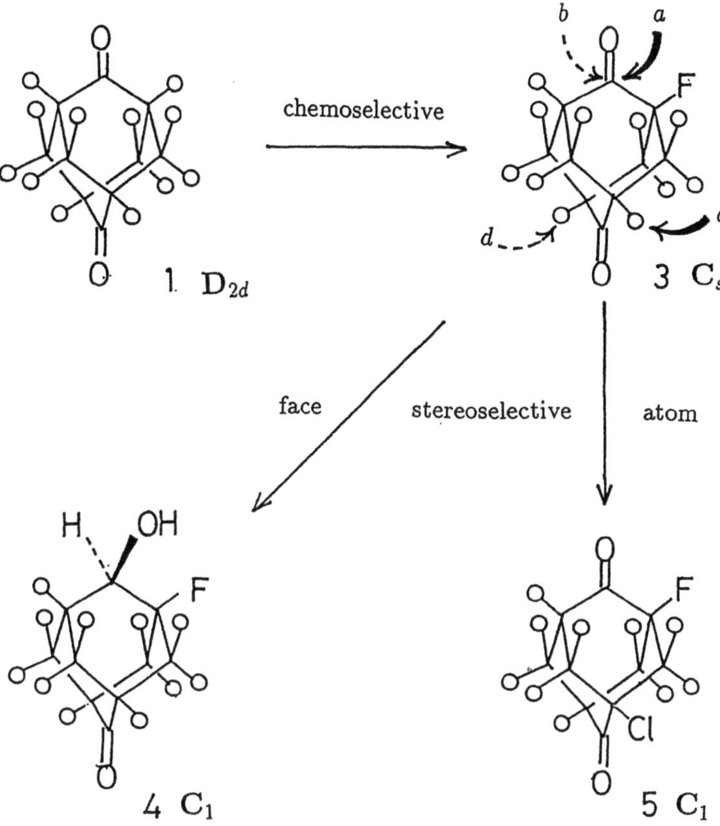

Figure 11.3: Potential derivation of chiral molecules from a D_{2d} molecule *via* $D_{2d}(/C_s)\downarrow C_s$

Bibliography

[1] K. Mislow, J. Siegel, *J. Am. Chem. Soc.*, **106**, 3319 (1984).

[2] E. A. Halevi, *J. Chem. Res. (S)*, 206 (1985).

[3] S. Fujita, *J. Am. Chem. Soc.*, **112**, 3390 (1990).

Chapter 12

Topicity and Stereogenicity [1]

Since the terminology in stereochemistry has been to a great extent empirically developed, it has not escaped some ambiguity. Thus, there have appeared many reviews for clarifying or getting rid of such ambiguity.[1]–[7] The term *stereogenic* was originally proposed by McCasland[8] in order to avoid confusions provided by so-called "asymmetric" and "pseudoasymmetric" carbon atoms. The term stereogenic has been utilized as a key concept in the revised CIP-system for describing chirality and related matters.[9] Later, Mislow and Siegel[10] have extended the definition of a stereogenic atom so as to manipulate a stereogenic element (unit); this is defined as an element (uint) bearing several groups of such nature that an interchange of two groups produces a stereoisomer. They also discussed the conceptual distinction between the stereogenicity and chirotopicity. The term stereogenic has been adopted by several authors[11] in place of the terms "asymmetric" and "chiral". In the present chapter, we redefine topicity terms in the light of chirality fittingness of orbits and then discuss stereogenicity.

12.1 Topicity Based On Chirality Fittingness of an Orbit

Terms concerning *topicity* were proposed by Mislow and Raban[12] and have been widely adopted in chemical and biochemical fields. These terms were classified by means of a flow chart based on pairwise relations.[13] In order to discuss the stereogenicity, we shall formulate the topicity in a stricter fashion.

Chirality fittingness of an orbit. In Chapter 8,[14, 15] we have discussed the chirality fittingness.

$$
\text{chirality fittingness (sphericity)} \quad \begin{cases} \text{homospheric} \\ \text{enantiospheric} \\ \text{hemispheric} \end{cases}
$$

[1]Reprinted in part with permission from S. Fujita, *Tetrahedron*, **46**, 5943–5954 (1990). ©(1990) Pergamon Press PLC.

We regard a molecule as a three-dimensional object that consists of a skeleton and ligands,[2] where the substitution positions of the skeleton are replaced by the ligands. A set of equivalent positions in the skeleton is regarded as an orbit governed by a CR ($G(/G_i)$), where ligands on the orbit are subject to a global symmetry (G) and a local symmetry (G_i) through the CR. It is dependent upon a problem at issue what part of the molecule is selected as a skeleton and what parts are regarded as ligands.

Let us work out adamantanone (**1**) that belongs to C_{2v} symmetry. By the action of the C_{2v} point group, the 14 positions of **1** are divided into five orbits (Δ_1 to Δ_5). For instance, the Δ_1 orbit contains 2 bridgehead positions ($\{1,2\}$). Since this orbit is subject to the CR ($C_{2v}(/C_s)$), it is determined to be homospheric.[3] The Δ_2 orbit consists of $\{3,4,5,6\}$ and is governed by CR ($C_{2v}(/C_1)$). This orbit is therefore enantiospheric. Table 12.1 summarizes such orbits appearing in **1** as well as their governing CRs and chirality fittingness.

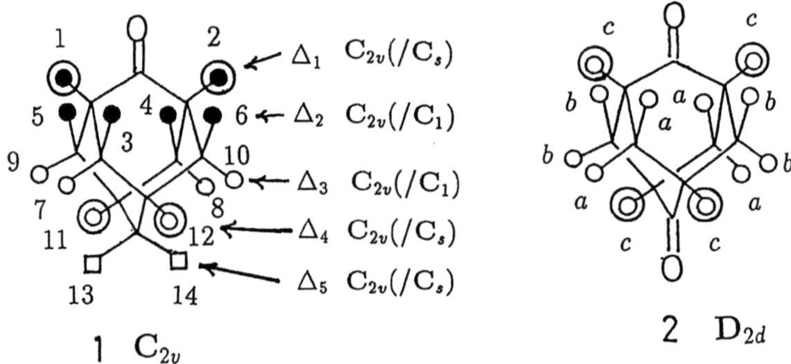

Table 12.1: Orbits and coset representations in 1

Orbit	Positions	Members (Ligands)	coset representation	Chirality fittingness (Sphericity)
Δ_1	$\{1,2\}$	H_2	$C_{2v}(/C_s)$	homospheric
Δ_2	$\{3,4; 5,6\}$	H_4	$C_{2v}(/C_1)$	enantiospheric
Δ_3	$\{7,8; 9,10\}$	H_4	$C_{2v}(/C_1)$	enantiospheric
Δ_4	$\{11,12\}$	H_2	$C_{2v}(/C_s')$	homospheric
Δ_5	$\{13,14\}$	H_2	$C_{2v}(/C_s)$	homospheric

If we consider proper rotations only among the symmetry operations of an

---

[2]We use the term "ligand" to comprise both atoms and groups.

[3]This assignment needs a table of marks for the C_{2v} group. For the procedure of such assignment, see Chapter 6.

achiral point group $\mathbf{G}$, we have a chiral point group $\mathbf{G}^{max}$, where $\mid \mathbf{G} \mid = 2 \mid \mathbf{G}^{max} \mid$. According to this restriction,[4] a homospheric orbit $(\mathbf{G}(/\mathbf{G}_i))$ is converted into a hemispheric orbit $\mathbf{G}^{max}(/\mathbf{G}_i^{max})$, where the length of the former orbit is equal to that of the latter (Chapter 11). This fact indicates that the homospheric orbit is superposed to itself by proper rotations as well as by improper rotations. On the other hand, the restriction divides an enantiospheric orbit into two hemispheric orbits $\mathbf{G}^{max}(/\mathbf{G}_i)$, each length of which is one half of that of the original orbit. This fact means that the enantiospheric orbit can be divided into two halves,[5] which cannot be superposed to each other by proper rotations only, but can be interchanged with each other by improper rotations. For instance, the Δ_2 orbit of **1** can be divided into two halves, *i.e.* {3,4} and {5,6}, if we only take proper rotations into consideration. Such subdivisions are represented by a semicolon in each couple of braces, as shown in Table 12.1.

Definition of topicity. These discussions indicate that each position is characterized as a member of a homo-, enantio-, or hemi-spheric orbit. For some purposes, we further examine the membership of each position in either (left-handed or right-handed) half of an enantiospheric orbit. As a result, the present approach conceptually needs not introduce such terms as concerning "topicity", which refer to a relation between two positions. In other words, we can discuss such relations by characterizing the two positions as members of the same orbit or of different orbits. We can, for example, recognize the relationship among positions {3, 4, 5, and 6} without the topicity terms; thus, by indicating that they have the same membership in an enantiospheric orbit (Δ_2). However, the terms concerning topicity have long been used in organic chemistry and are so familar to organic chemists that we cannot now discontinue their usage. Hence, it is impartial as well as convenient to use the terms after redefinition from the present point of view. We here define the following terms in order to designate relationship between two stereochemical objects.

$$\left\{ \begin{array}{l} \text{homotopic} \quad \left\{ \begin{array}{l} \text{holotopic} \\ \text{hemitopic} \end{array} \right. \\ \\ \text{enantiotopic} \\ \text{diastereotopic} \\ \text{heterotopic} \end{array} \right.$$

The term *homotopic* is defined as being the same membership of two positions (atoms, faces, or ligands)

(1) in a homospheric orbit,

---

[4] In the real 3-dimensional world, any improper rotation is a rather hypothetical operation. However, if we neglect improper rotations, we cannot characterize the relationship between the two halves of an enantiospheric orbit.

[5] These two halves can be classified as left-handed and right-handed by means of an appropriate criterion; this criterion is an issue to be examined elsewhere.

(2) in either half of an enantiospheric orbit, or

(3) in a hemispheric orbit.[6]

For instance, the 1- and 2-positions of **1** are related to each other in a homotopic fashion by the first criterion, since they belong to the homospheric orbit (Δ_1) governed by $C_{2v}(/C_s)$. The 3- and 4-positions of **1** are homotopic by the second criterion, since they construct one half of the Δ_2 orbit that is enantiospheric because of the corresponding CR ($C_{2v}(/C_1)$). It should be noted that the case (1) is different from the other two cases in the sense that two positions of the former relation (1) are superimposable by a proper rotation as well as by an improper rotation; on the other hand two positions of the latter relation ((2) or (3)) are superimposable by a proper rotation but not by any improper rotations. In order to explicate this difference, we propose the term *holotopic* to denote the case (1). We introduce the term *hemitopic* to denote the cases (2) and (3), if necessary. Thus, the present term homotopic involves holotopic and hemitopic.

When one position is a member of one half of an enantiospheric orbit and the other position is a member of the other half, the relationship between the two positions is defined as *enantiotopic*. For instance, the relationship between the 3- and 5-positions of **1** and that between the 3- and 6-positions are different, but both are enantiotopic. We can also use this term to designate such two halves of an enantiospheric orbit. Thereby, a set of {3,4} is enantiotopic to another set of {5,6} in the Δ_2 orbit of the molecule (**1**).

The term *diastereotopic* is used to characterize a relation between two positions, where the one belongs to an appropriate orbit subject to a CR and the other position is a member of another orbit subject to the same CR. For example, the relationship between the 3- and 7-positions of **1** is diastereotopic, since the 3-position belongs to a $C_{2v}(/C_1)$ orbit (Δ_2) and the 7-position is a member of another $C_{2v}(/C_1)$ orbit (Δ_3). By extension, the Δ_2 orbit is diastereotopic to the Δ_3 orbit, because both of the orbits are governed by the same CR ($C_{2v}(/C_1)$).

Two orbits of a molecule in a diastereotopic relation should be meaningly related to each other, if the symmetry of the molecule is hypothetically converted into a higher one. In connection with **1**, for instance, consider adamantane-2,6-dione (**2**) of D_{2d} symmetry, in which 8 methylene positions (o) construct an orbit subject to CR $D_{2d}(/C_1)$. Since this orbit is enantiospheric, the 8 positions are divided into two halves (*a* and *b*), which are enantiotopic to each other. A hypothetical monoreduction of **2** into **1** results in separation into two enantiospheric orbits (Δ_2 and Δ_3 in **1**). This example gives a sound foundation to the differentiation of a diastreotopic relation from a heterotopic relation in some cases.

In the light of the present definition, one may conclude that the relationship

---

[6]Previously, we overlooked this case in the definition of homotopicity. See Ref. [15].

between Δ_4 and Δ_5 in **1** is diastereotopic, because they are subject to the same type of homospheric orbits $(C_{2v}(/C_s),C_{2v}(/C_s'))$. This misunderstanding can be avoided, since there is no hypothetical operations that make them equivalent. The same conclusion can be obtained if the topic term is used only to compare two objects of the same constitution or of similar constitutions. Thereby, Δ_4 (methine hydrogens) and Δ_5 (methylene hydrogens) belong to distinct categories and should not be compared; hence, the two orbits are heterotopic (see below).

However, there still exist borderline cases between diastereotopic and heterotopic objects. In other words, the dintinction between diastereotopic and heterotopic depends upon a skeleton selected. It should be noted that the term "diastereotopic" originates in the history of organic stereochemistry in which two-dimensional constitutional isomers are taken into primary consideration and three-dimensional isomers are then taken account of. If we consider a molecule as a three-dimensional object from the beginning, the term "diastereotopic" would be unnecessary.

The term *heterotopic* is used to refer to cases in which two positions to be examined are different in their assigned orbits and in their CRs. For example, the relationship between the 3- and 1-positions of **1** is heterotopic, since the former position belongs to the $C_{2v}(/C_1)$ orbit (Δ_2) and the latter is a member of the $C_{2v}(/C_s)$ orbit (Δ_1).

The above discussions have shown the ambiguity between "diastereotopic" and "heterotopic". This ambiguity can be averted by using the sphericity terms primarily and the topicity terms subsidiarily.

12.2 Stereogenicity

We define a *stereogenic transposition on a skeleton* as an operation that interchanges two ligands so as to produce stereoisomers.[7] If such a transposition produces no stereoisomers, this operation is defined as *non-stereogenic*. A skeleton is referred to as being stereogenic if it involves at least one stereogenic transposition.[8] In other words, a stereogenic skeleton can undergo non-stereogenic transpositions along with stereogenic ones. The present discussion focuses on (non-)stereogenic transformations, but not on stereogenic skeletons. For simplicity of discussion, we take no account of transpositions within a ligand. In addition, we tentatively suppose that a skeleton is rigid (or invariant) with respect to bond rotations (see Chapter 20). In this section, we introduce several new concepts to clarify the stereogenicity.

---

[7]The term "stereoisomer" is here used to denote *isoskeletal isomers*, which are defined as two molecules having an identical skeleton and an identical set of ligands but different modes of substitution. The latter term is similar to, but conceptually distinct from "permutational isomers", which was introduced by Ugi *et al.*[16] This point will be discussed elsewhere.

[8]This definition stems from Mislow's one on a stereogenic unit. See Ref. [10].

Substitution equivalent. Consider a monohydroxy derivative (**3**) as an example (Figure 12.1). We interchange the hydroxyl group at the 3-position and the hydrogen atom at the 7-postition. This transposition produces an isomer (**4**); hence, this is a stereogenic transposition.

In order to make clear the nature of this transformation, we introduce another formulation. Thus, the molecule (**3**) is regarded as a derivative produced by a substitution of the hydroxyl group at the 3-position of compound **1**, while **4** is derived by the OH-subsitution of **1** at the 7-position. This means that the transposition concerning **3** and **4** is equivalent to a set of the distinct substitution processes of the common intermediate (**1**), which we call a *a common skeleton*. We call the latter set of processes a *substitution equivalent* of the former transposition.

The symmetry of such common skeletons should be the same as or higher than those of original molecules. Among possible skeletons, we select a common skeleton of the highest symmetry. In this example, we replace the 3-OH of **3** (or the 7-OH of **4**) by a hydrogen atom. Then we regard the resulting adamantanone (**1**) having 14 positions as a common skeleton of C_{2v} symmetry. Taking account of such a substitution equivalent provides us with a convenient method of determining whether a transposition is stereogenic or not. The substitution equivalent is closely related to systematic enumerations by unit subduced cycle indices, where a derivative is produced by placing a set of ligands on the positions of a skeleton (Chapters 15-19). A pair of distinct replacements in the enumeration corresponds to a pair of molecules participating a transposition. Hence, we can determine stereogenicity by examining whether the pair of replacements in a substitution equivalent generate two distinct molecules or not.

Stereogenic transpositions. In accord with the topicities described in the preceding section, there are four types of transpositions; homotopic, enantiotopic, diastereotopic, and heterotopic transpositions. As shown below, these topicities are concerned with common skeletons, not with molecules to be examined.

The example illustrated in Figure 12.1 is a *diastereotopic transposition*. The 3- and 7-positions of the common skeleton (**1**) of C_{2v} symmetry belong to the respective $C_{2v}(/C_1)$ orbits (Δ_2 and Δ_3); that is, they are diastereotopic to each other. Hence, this transposition is a diastereotopic transposition. In general, a substitution equivalent for a diastereotopic transposition produces two distinct stereoisomers; hence this transposition is stereogenic.

A *heterotopic transposition* is illustrated by an interconversion between **3** and **5** (Fig. 12.2). The former molecule (**3**) belongs to C_1 symmetry and the latter (**5**) has C_s symmetry. We can also select **1** as a common skeleton in this case, where we pay our attention to the 3- and 1-positions. The 3- and 1-positions of the

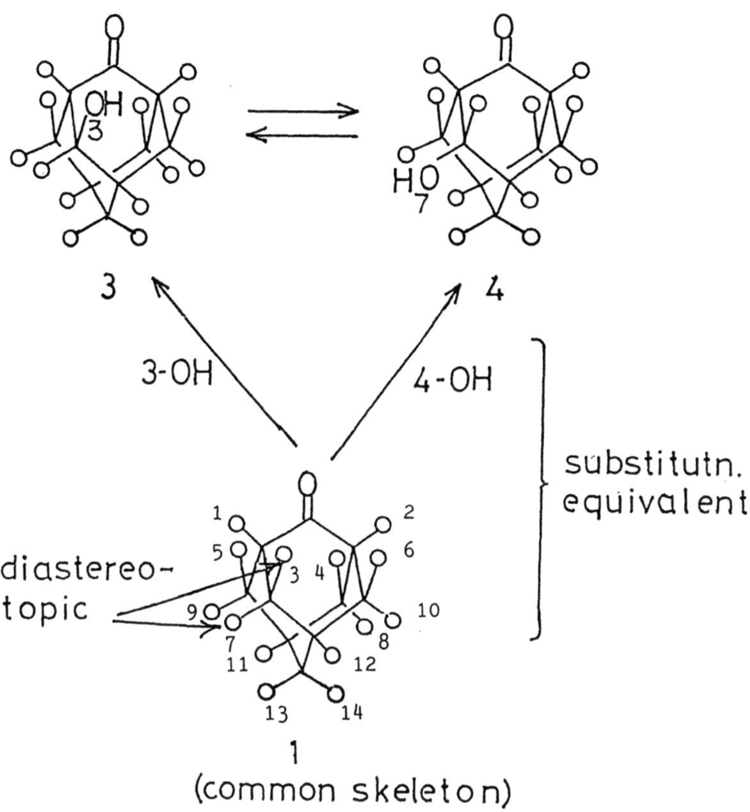

Figure 12.1: Substitution equivalent for a stereogenic transposition

The term diastereotopic is concerned with the common skeleton (1), not with the molecules (3 and 4). The 3-position of 1 is a member of a $C_{2v}(/C_1)$ orbit, i.e. $\Delta_2 = \{3,4;5,6\}$; the 7-position of 1 belongs to another $C_{2v}(/C_1)$ orbit, i.e. $\Delta_3 = \{7,8;9,10\}$.

common skeleton (1) are concluded to be heterotopic, since the 3-position belongs
to the $C_{2v}(/C_1)$ orbit (Δ_2), while the 1-position is involved in the $C_{2v}(/C_s)$ orbit
(Δ_1). Obviously, a heterotopic transposition is in general stereogenic to produce
an isoskeletal isomer.

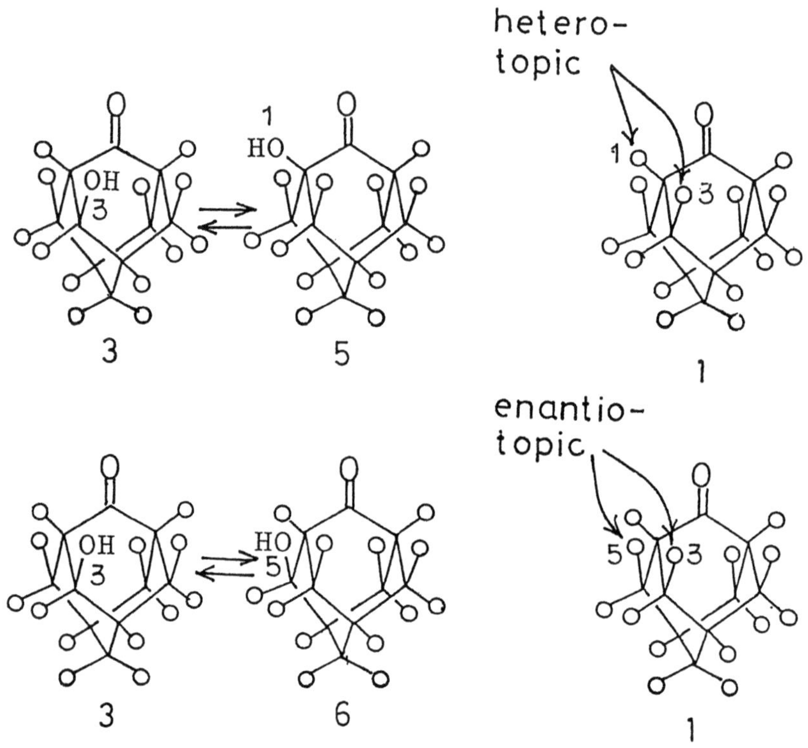

Figure 12.2: Heterotopic and enantiotopic transpositions as stereogenic ones

The rightmost figures depict a common skeleton appeared in substitution equivalents. The
terms, heterotopic and enantiotopic, come from the common skeleton.

The interconversion between **3** and **6** (the enantiomer of **3**) is an *enantiotopic
transposition*, since the positions 3 and 5 of the common skeleton (1) are enantiotopic to each other. In other words, these positions belong to the same $C_{2v}(/C_1)$
(Δ_2), but separately occupy the respective halves of this orbit. Thereby, this process is stereogenic to produce a pair of enantiomers.

Non-stereogenic transpositions. Figure 12.3 depicts homotopic transpositions, all of which are non-stereogenic. Homotopic transpositions are subdivided into three categories in accordance with three types of homotopicities.

The molecule (**3**) is converted into **3a** by interchanging 3-OH and 4-H. The latter molecule (**3a**) is superimposable to the former (**3**) by a proper rotation (a C_2 operation). We use the term *homomeric* to denote this type of relation between two molecules. In the light of this definition, we describe that this transposition is non-stereogenic, since it generates a homomer (**3a**) of the starting molecule (**3**). If we consider **1** as a common skeleton in the corresponding substitution equivalent, the 3- and 4-positions belong to one half of an enantiospheric orbit (Δ_2 in Table 12.1) and are homotopic in the light of the first criterion of the above definition. Hence, this transposition is a homotopic transposition.

The second type of homotopic transposition is exemplified by the interconversion between **7** and its homomer (**7a**). In the corresponding common skeleton (**1**), the 1- and 2-positions are the members of a homospheric orbit (Δ_1 governed by $C_{2v}(/C_s)$), which are homotopic to each other by means of the second criterion of the definition. Hence, this process is homotopic transposition and non-stereogenic.

The interconversion between **8** and its homomer (**8a**) is one of the third-type homotopic transpositions, becuase the corresponding common skeleton is a C_2 molecule (**9**), in which the 1- and 2-positions construct a hemispheric orbit governed by the CR $C_2(/C_1)$.

Any two ligands of the same membership, no matter which of the three types concerns, are equivalent on proper rotations. Hence, the replacement of the one member and that of the other member can be easily proved to produce molecules homomeric to each other. In general, any homotopic transposition produces homomers; hence, the homotopic transposition is non-stereogenic. Because the present discussion is concerned with all possible cases, we end up with a simple theorem: *a homotopic transposition is non-stereogenic; the other types of transpositions are all stereogenic.*

homotopic transposition $\qquad\cdots$ non-stereogenic

$$\left.\begin{array}{l}\text{enantiotopic transposition}\\ \text{diasterotopic transposition}\\ \text{heterotopic transposition}\end{array}\right\}\ \cdots\ \text{stereogenic}$$

So-called "pseudoasymmetric carbon atoms". A test using the topicity of a common skeleton in a substitution equivalent is a convenient method of checking stereogenicity. Two chiral stereoisomers of 2,3,4-trihydroxyglutaric acid, **10** and **11**, have already been discussed.[10] From the present point of view, the two

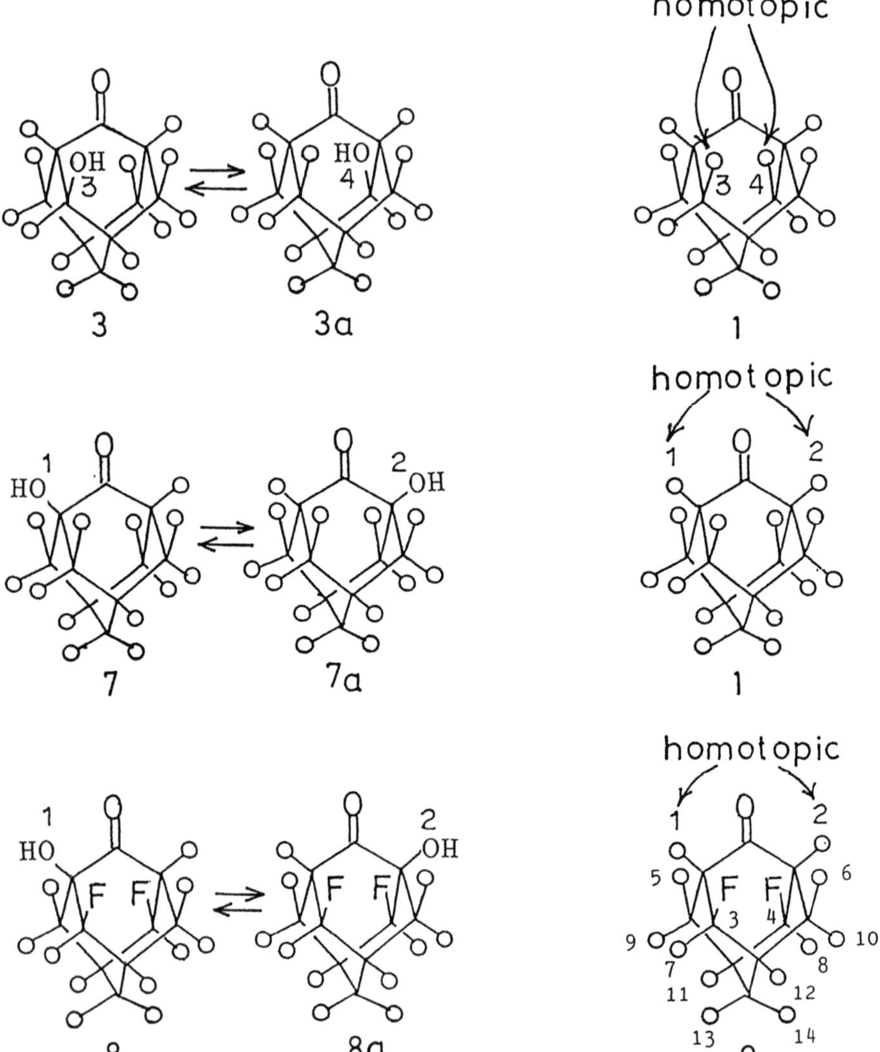

Figure 12.3: Homotopic transpositions as non-stereogenic ones

Note that $\Delta_2 = \{3,4;5,6\}$ is an enantiospheric orbit governed by $C_{2v}(/C_1)$ and $\Delta_1 = \{1,2\}$ is a homospheric orbit governed by $C_{2v}(/C_s)$ in the common skeleton (1). In the common skeleton (9), $\Delta_1 = \{1,2\}$ is a hemispheric orbit governed by $C_2(/C_1)$.

isomers are represented by **12** and **13**, in which Q and $\overline{Q}$ denote a pair of antipodal ligands (CH(OH)COOH). For a transposition on the central carbon of **12**, we consider a common skeleton (**14**) of C_2 symmetry (Figure 12.4), in which the resulting two positions construct a hemispheric orbit governed by $C_2(/C_1)$. This fact indicates that this transposition is a homotopic transposition, which is concluded to be non-stereogenic. Similarly, a transposition on **13** is also homotopic and hence non-stereogenic.

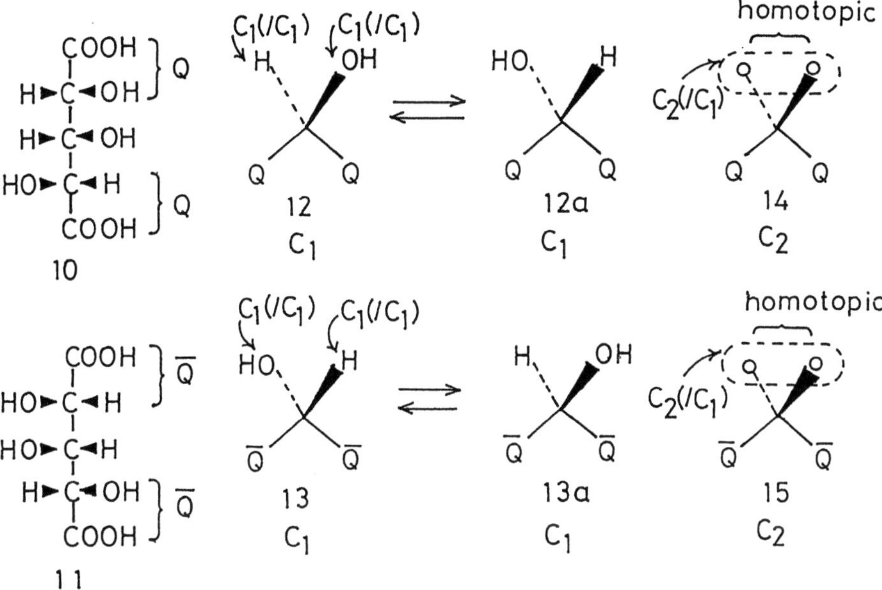

Figure 12.4: Chiral 2,3,4-trihydroxyglutaric acids

The rightmost figures (**14** and **15**) depict common skeletons, where each of the $C_2(/C_1)$ orbits is hemispheric.

Let us next discuss the two diastereomers (**16** and **17**) of 2,3,4-trihydroxyglutaric acid, in which the central carbon atom has been dubbed "pseudoasymmetric" (Figure 12.5). These molecules are formulated by means of the present procedure to be **18** and **19**, each of which can be considered to belong to C_s symmetry. For the purpose of discussing a transposition on the central carbon of **18** or **19**, we consider a common skeleton (**20**) of C_s symmetry. Note that the hypothetical removal of H and OH keeps the C_s symmetry invariant. The resulting two positions belong separately to one-membered $C_s(/C_s)$ orbits; that is, they are diastereotopic. Hence, this transposition is concluded to be a diastereotopic one, which is stereogenic, bringing out an interconversion between the diastreomers (**18** and **19**).

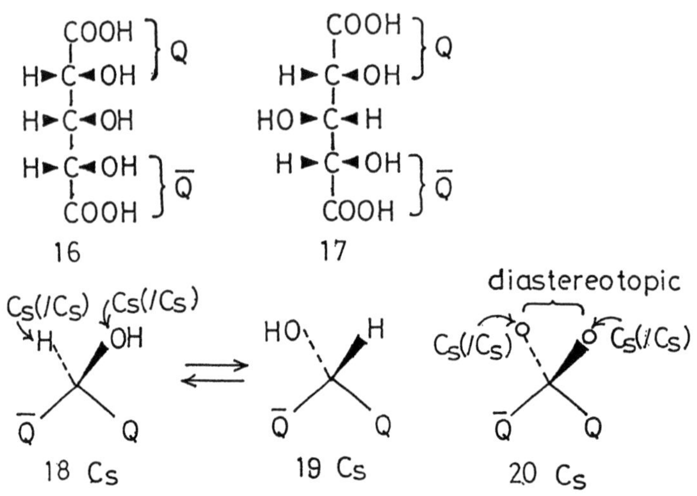

Figure 12.5: Meso 2,3,4-trihydroxyglutaric acids

The rightmost figure (**20**) depicts a common skeleton, where each of the $C_s(/C_s)$ orbits is homospheric.

So-called "chirotopicity *vs.* stereogenicity". We have already discussed that the concept "chirotopicity" proposed by Mislow *et al.* [10] can be replaced by the chirality fittingness (sphericity) of an orbit.[15] We proposed the term *homospheric* in place of "achirotopic" and the terms *enantiospheric* and *hemispheric* in place of "chirotopic". It should be emphasized that these two systems of terminology are conceptually distinct. That is to say, the present terms concerning the sphericity (chirality fittingness) are the attributes of an orbit, whereas the terms regarding the chirotopicity are the attributes of an individual ligand.[9] In other words, the

---

[9] We use the term "topic" in order to denote a *relation*, but not to designate an *attribute*. The terms homotopic

sphericity is concerned with a molecule as a whole in the form of such orbits; on the other hand, the chirotopicity focuses on a part of the molecule. Although it was emphasized that chirotopicity and stereogenicity are independent attributes,[10, 4] the standpoint based on the chirotopicity is still insufficient to differentiate the attributes.

From the present point of view, the so-called chirotopicity is described by the sphericity of an orbit appearing in *the skeleton of a molecule at issue.* For example, the 3-OH ligand of **3** is a member of a $C_1(/C_1)$ orbit, since this molecule belongs to C_1 symmetry (asymmetric). The symbol $C_1(/C_1)$ indicates that this orbit is hemispheric. In the chirotopicity convention, C_1 in the parentheses indicates that this ligand is chirotopic.

On the other hand, stereogenicity is concerned with an orbit appearing in *a common skeleton* that is considered in a substitution equivalent. For example, the 3- and 11-positions construct one half of the $C_{2v}(/C_1)$ orbit (Δ_2) (*i.e.* they are homotopic) in the common skeleton **1** for the transposition between 3-OH and 11-H of **3**. Then, we have concluded that this homotopic transposition is non-stereogenic. It should again be emphasized that this homotopic relation stems from the sphericity of the common skeleton (**1**) and is not concerned with the molecule to be examined (**3**).

---

etc. agree with this criterion, since they are concerned with relations. On the other hand, the terms achirotopic and chirotopic do not satisfy this criterion, because they are used to designate attributes. The conventional usage of the topic are concluded to mix up these two distinct categories.

Bibliography

[1] J. K. O'loane, *Chem. Rev.*, **80**, 41 (1980).

[2] E. L. Eliel, *Top. Curr. Chem.*, **105**, 1 (1982).

[3] H. Hirschmann, K. R. Hanson, *Top. Stereochem.*, **14**, 183 (1983).

[4] D. J. Brand, J. Fischer, *J. Chem. Educ.*, **64**, 1035 (1987).

[5] J. Jonas, *Collect. Czech. Chem. Commun.*, **53**, 2675 (1988).

[6] S. A. Kaloustian, M. K. Kaloustian, *J. Chem. Educ.*, **52**, 56 (1975).

[7] E. L. Eliel, *J. Chem. Educ.*, **57**, 52 (1980).

[8] G. E. McCasland, *A New General System for the Naming of Stereoisomers*, Chemical Abstracts Service, Columbus, Ohio (1953), p. 2.

[9] V. Prelog, G. Helmchen, *Angew. Chem. Int. Ed. Engl.*, **21**, 567 (1982).

[10] K. Mislow, J. Siegel, *J. Am. Chem. Soc.*, **106**, 3319 (1984).

[11] (a) R. Tripathy, R. W. Franck, K. D. Onan, *J. Am. Chem. Soc.*, **110**, 3257 (1988). (b) F. A. L. Anet, D. J. O'Leary, *J. Am. Chem. Soc.*, **111**, 8935 (1989).

[12] K. Mislow, M. Raban, *Top. Stereochem.*, **1**, 1 (1967).

[13] K. Mislow, *Bull. Soc. Chim. Belg.*, **86**, 595 (1977).

[14] S. Fujita, *Bull. Chem. Soc. Jpn.*, **63**, 315 (1990).

[15] S. Fujita, *J. Am. Chem. Soc.*, **112**, 3390 (1990).

[16] (a) I. Ugi, D. Marquarding, H. Klusacek, G. Gokel, P. Gillespie, *Angew. Chem. Int. ed. Engl.*, **9**, 703 (1970). (b) I. Ugi, J. Bugundji, R. Kopp, D. Marquarding, *Perspectives in Theoretical Stereochemisty*, Lecture Notes in Chemistry Series 36, Springer-Verlag, Heidelberg (1984).

Chapter 13

Counting Orbits

The purpose of this chaper is to introduce the Pólya-Redfield theorem, which has been widely used for enumeration of chemical compounds. This theorem is of fundamental importance, not only because of its wide applicability but also because of its conceptual depth.[1, 2] Many mathematical textbooks have dealt with the topics in various manners.[3]–[7] Chemical enumerations have been reviewed in several books[8]–[10] and articles.[11]–[13] Recent advances of chemical enumerations have been reported.[14] We have reported enumerations of organic reactions after the formulation of imaginary transition structures.[15]

13.1 The Cauchy-Frobenius Lemma

The issue of this section is a derivation of so-called Burnside's lemma (the Cauchy-Frobenius lemma). We start from Lemma 5.3 that has been proved in Chapter 5.

Theorem 13.1 (Cauchy-Frobenius) *Let* $\mathbf{P}$ *be a permutation group on* Δ, *which is partitioned into* r *orbits;* $\Delta_1, \Delta_2, \ldots, \Delta_r$. *Then the number* (r) *is given by the equation,*

$$r = \frac{1}{|\mathbf{P}|} \sum_{p \in \mathbf{P}} n_{(p)}, \tag{13.1}$$

where $n_{(p)}$ *denotes the number of fixed elements in* Δ *on the action of a permutation* $p \in \mathbf{P}$.

Proof. We count all of the elements that are fixed by all of the permutations of $\mathbf{P}$. First, we examine each permutation and count fixed points to afford $n_{(p)}$. This is repeated over all of the permutations. As a result, we have $\sum\limits_{p \in \mathbf{P}} n_{(p)}$.

Alternatively, we count the number $(\tilde{n}_{(\delta)})$ of permutations that keep each element $(\delta \in \Delta)$ invariant. This is repeated over all of the elements of Δ. As a result, we obtain $\sum_{\delta \in \Delta} \tilde{n}_{(\delta)}$. It follows that

$$\sum_{p \in \mathbf{P}} n_{(p)} = \sum_{\delta \in \Delta} \tilde{n}_{(\delta)}. \tag{13.2}$$

Suppose that δ is an element of an orbit Δ_i ($i = 1, 2, \ldots$, or r). Since $\tilde{n}_{(\delta)}$ is the number of permutations that keep δ invariant, this is equal to the order of the stabilizer $(\mathbf{P}_\delta)$ of δ. The latter is given by Lemma 5.3 (Chapter 5) as follows.

$$\tilde{n}_{(\delta)} = |\,\mathbf{P}_\delta\,| = |\,\mathbf{P}\,|\,/\,|\,\Delta_i\,|\,. \tag{13.3}$$

Since this equation holds for all δ of Δ_i, we have

$$\sum_{\delta \in \Delta_i} \tilde{n}_{(\delta)} = |\,\mathbf{P}\,|\,/\,|\,\Delta_i\,| \times |\,\Delta_i\,| = |\,\mathbf{P}\,| \quad \text{(for each } \Delta_i\text{)}. \tag{13.4}$$

When we sum up this over all i ($i = 1, 2, \ldots, r$), we have the total number,

$$\sum_{\delta \in \Delta} \tilde{n}_{(\delta)} = \sum_{i=1}^{r} \sum_{\delta \in \Delta_i} \tilde{n}_{(\delta)} = r\,|\,\mathbf{P}\,|\,. \tag{13.5}$$

Equations 13.2 and 13.5 give eq. 13.1 to be proved.

Theorem 13.1 is equivalent to Corollary 5.2 (Chapter 5), although they are different in their explicit forms. Correlation of them will be discussed later.

Example **13.1** Let us re-examine Example 5.2 (Chapter 5), which have dealt with the 12 hydrogen atoms of adamantane-2,6-dione ($\mathbf{D}_{2d}$ symmetry). The concrete form of a permutation group on the set of 12 hydrogen atoms consists of eight permutations as follows.

$\mathbf{D}_d$		permutation		$n_{(p)}$
I	$\sim$ (1)(2)(3)(4)	(5)(6)(7)(8)(9)(10)(11)(12)	$\cdots$	12
$C_{2(1)}$	$\sim$ (1 2)(3 4)	(5 6)(7 8)(9 10)(11 12)	$\cdots$	0
$C_{2(2)}$	$\sim$ (1 3)(2 4)	(5 7)(6 8)(9 11)(10 12)	$\cdots$	0
$C_{2(3)}$	$\sim$ (1 4)(2 3)	(5 8)(6 7)(9 12)(10 11)	$\cdots$	0
$\sigma_{d(1)}$	$\sim$ (1)(2 3)(4)	(5 9)(6 11)(7 10)(8 12)	$\cdots$	2
S_4^3	$\sim$ (1 3 4 2)	(5 11 8 10)(6 9 7 12)	$\cdots$	0
S_4	$\sim$ (1 2 4 3)	(5 10 8 11)(6 12 7 9)	$\cdots$	0
$\sigma_{d(2)}$	$\sim$ (1 4)(2)(3)	(5 12)(6 10)(7 11)(8 9)	$\cdots$	2
			$\sum_{p \in \mathbf{P}} n_{(p)}$ $\cdots$	16

Since $|\,\mathbf{P}\,|$ is equal to 8, we have $r = 16/8 = 2$. This is equal to the value obtained in Example 5.2 (Chapter 5).

We have already given a comment on homomorphism in Chapter 5. In order to make further discussions, we should examine the homomorphism in a stricter fashion.

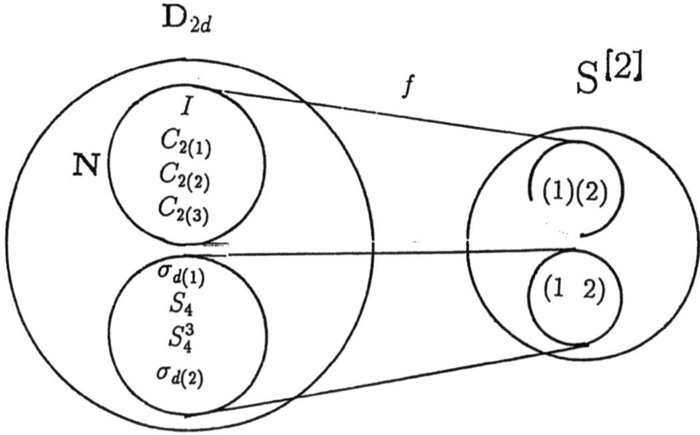

Figure 13.1: Kernel of homomorphism

Definition 13.1 (Kernel of Homomorphism) *Let f be a homomorphism of a finite group $\mathbf{G}$ onto a permutation group $\mathbf{P}$. The set $\mathbf{N}$ of the elements of $\mathbf{G}$ that are mapped onto I (identity) of $\mathbf{P}$ is called the kernel of the homomorphism.*

$$\mathbf{N} = f^{-1}(I) = \{g \mid f(g) = I \in \mathbf{P}, g \in \mathbf{G}\}. \tag{13.6}$$

Figure 13.1 illustrates this concept by using the mapping of $\mathbf{D}_{2d}$ onto $\mathbf{S}^{[2]}$.

Theorem 13.2 *The kernel $\mathbf{N}$ defined in Def. 13.1 is a normal group of $\mathbf{G}$.*

Proof. (1) For $\forall g, g' \in \mathbf{N}$, we have $f(g) = f(g') = I \in \mathbf{P}$.

Because of $f(g)f(g^{-1}) = f(gg^{-1}) = f(I) = I$ and $f(g) = I$, we obtain $f(g^{-1}) = I(\in \mathbf{P})$. It follows that $g^{-1} \in \mathbf{N}$.

From the equation, $f(gg') = f(g)f(g') = I \in \mathbf{P}$, we obtain $gg' \in \mathbf{N}$.

These equations conclude that $\mathbf{N}$ is a group.

(2) We have

$$f(g^{-1}\mathbf{N}g) = f(g^{-1})f(\mathbf{N})f(g) = f(g^{-1})f(g) = f(g^{-1}g) = f(I) = I. \tag{13.7}$$

This means that $g^{-1}\mathbf{N}g$ is also a kernel, *i.e.* $g^{-1}\mathbf{N}g = \mathbf{N}$. That is to say, $\mathbf{N}$ is a normal subgroup.

Theorem 13.3 *Let f be a homomorphism of $\mathbf{G}$ onto $\mathbf{P}$, where $\mathbf{N}$ is the kernel of the homomorphism. We can regard f as a mapping from a factor group $\mathbf{G}/\!/\mathbf{N}$ onto $\mathbf{P}$. Then, $\mathbf{G}/\!/\mathbf{N} \cong \mathbf{P}$.*

Proof. (1) Let $\mathbf{P}$ denote a permutation group: $\mathbf{P}=\{p_1,p_2,\ldots,p_m\}$. For $p_i \in \mathbf{P}$, we have an appropriate t_i ($\in \mathbf{G}$) that satisfies $f^{-1}(p_i) = t_i$. Then, we obtain

$$f(\mathbf{N}t_i) = f(\mathbf{N}f^{-1}(p_i)) = f(\mathbf{N})f(f^{-1}(p_i)) = f(f^{-1}(p_i)) = p_i. \tag{13.8}$$

It follows that we obtain a mapping $p_i \to \mathbf{N}t_i$. In a similar way, we have a mapping $p_j \to \mathbf{N}t_j$ by replacing p_i by p_j.

(2) We will prove that $\mathbf{N}t_i \cap \mathbf{N}t_j = \emptyset$ if $p_i \neq p_j$. Assume that $\mathbf{N}t_i$ and $\mathbf{N}t_j$ have a common element a, when the equation $p_i \neq p_j$ holds. Then, we have $a = gt_i$ and $a = g't_j$ for $\exists g, g' \in \mathbf{N}$. As a result, we obtain

$$f(a) = f(gt_i) = f(g)f(t_i) = f(t_i) = f(f^{-1}(p_i)) = p_i$$
$$f(a) = f(g't_j) = f(g)f(t_j) = f(t_j) = f(f^{-1}(p_j)) = p_j$$

This means that $p_i = p_j$. This result is against the original assumption. It follows that $\mathbf{N}t_i \cap \mathbf{N}t_j = \emptyset$.

(3) According to the preceding result, we have a partition represented by

$$\mathbf{G} = \mathbf{N}t_1 + \mathbf{N}t_2 + \cdots + \mathbf{N}t_m, \tag{13.9}$$

where the equation $f^{-1}(p_i) = t_i \in \mathbf{G}$ holds for $p_i \in \mathbf{P}$ ($i = 1, 2, \ldots, m$). This result indicates that p_i corresponds to $\mathbf{N}t_i$ in one-to-one fashion. It follows that $\mathbf{G}/\!/\mathbf{N} \cong \mathbf{P}$.

Theorem 13.4 (Generalization of the Cauchy-Frobenius lemma) *A finite group* $\mathbf{G}$ *is homomorphically associated with a permutation group* $\mathbf{P}$ *on* $\Delta = \{1, 2, \ldots, n\}$ *in the manner that* $g_\alpha \in \mathbf{G} \to p(g_\alpha) \in \mathbf{P}$. *On the action of* $\mathbf{G}$, *the* Δ *set is partitioned into* r *orbits:* $\Delta_1, \Delta_2, \ldots, \Delta_r$. *Then, the number* ($r$) *of the orbits is calculated by*

$$r = \frac{1}{|\mathbf{G}|} \sum_{g_\alpha \in \mathbf{G}} n_{p(g_\alpha)}, \tag{13.10}$$

where $n_{p(g_\alpha)}$ *denotes the number of fixed points of* Δ *on the action of* $p(g_\alpha)$.

Proof. Suppose that $\mathbf{P}=\{p_1,p_2,\ldots, p_m\}$. Let $\mathbf{N}$ be the kernel of the homomorphic mapping $f : \mathbf{G} \to \mathbf{P}$. Theorem 13.3 provides

$$\mathbf{G} = \mathbf{N}f^{-1}(p_1) + \mathbf{N}f^{-1}(p_2) + \cdots + \mathbf{N}f^{-1}(p_m). \tag{13.11}$$

Hence, we have

$$|\mathbf{G}| = m\,|\mathbf{N}| = |\mathbf{P}|\,|\mathbf{N}|. \tag{13.12}$$

For $\forall g_\alpha \in \mathbf{N}f^{-1}(p_i)$, there exists an appropriate $g(\in \mathbf{N})$ that satisfies $g_\alpha = gf^{-1}(p_i)$. As a result, we obtain

$$p(g_\alpha) = f(g_\alpha) = f(gf^{-1}(p_i)) = f(g)f(f^{-1}(p_i)) = p_i.$$

This means that all of the $|\mathbf{N}|$ elements contained in $\mathbf{N}f^{-1}(p_i)$ are mapped to p_i. Hence, we have

$$\sum_{g_\alpha \in \mathbf{G}} n_{(p(g_\alpha))} = \sum_{p_i \in \mathbf{P}} |\mathbf{N}| \, n_{(p_i)} = |\mathbf{N}| \sum_{p_i \in \mathbf{P}} n_{(p_i)} \tag{13.13}$$

Theorem 13.1 affords

$$r = \frac{1}{|\mathbf{P}|} \sum_{p_i \in \mathbf{P}} n_{(p_i)}. \tag{13.14}$$

When eqs. 13.12 an 13.13 are introduced into eq. 13.14, we obtain

$$r = \frac{1}{|\mathbf{P}||\mathbf{N}|} \sum_{g_\alpha \in \mathbf{G}} n_{(p(g_\alpha))} = \frac{1}{|\mathbf{G}|} \sum_{g_\alpha \in \mathbf{G}} n_{(p(g_\alpha))}. \tag{13.15}$$

This is the equation to be proved.

Theorem 13.4 (or Theorem 13.1) indicates the same number of fixed points as calculated in Corollary 5.2 (Chapter 5) despite of their different formulations. The former theorem stems from the formulation based on conjugacy classes, while the latter corollary is obtained by starting from conjugate subgroups. We will show the relationship between the two formulations in Chapter 16.

13.2 Configurations

In this section, we formulate a chemical compound (or a 3D object) as a configuration that is suitable to mathematical manipulation. Let us consider derivation of allene (1), in which four hydrogens are regarded as substitution positions (Fig. 13.2).

If an appropriate set of ligands ($X^{\theta_1}Y^{\theta_2}$; $\theta_1 + \theta_2 = 4$) are substituted for the positions, we have 16 configurations (f_1 to f_{16}) shown in Fig. 13.2 ($\circ$ = X and $\bullet$ = Y). The term $X^{\theta_1}Y^{\theta_2}$ characterizes each configuration; it can be regarded as a molecular formula in a chemical sense. Among the 16 configurations, consider configurations having X^3Y^1 (f_2 to f_5). They are superimposable on the action of a permutation group $\mathbf{P}$, which is associated with $\mathbf{D}_{2d}(/\mathbf{C}_s)$ governing the four positions of the allene skeleton (1). Chemically speaking, they are identical with each other.[1] As for X^2Y^2, we have six configurations (f_6 to f_{11}). The four configurations (f_6 to f_9) are superimposable; the remaining two (f_{10} and f_{11}) construct another set.

Consider the set of the configurations: $\mathbf{F} = \{f_1, f_2, \ldots, f_{16}\}$. Let us apply a permutation $p_2 = (1\ \ 2)(3\ \ 4)$ to the allene skeleton. This operation is the renumbering of the four positions. Figure 13.3 illustrates the p_2 permutation of the skeleton and the effect on the configuration (f_3), providing the configuration (f_4). This operation is repeated over all configurations. Then, we have a permutation,[2]

---

[1] These configurations belong to $\mathbf{C}_s$ point group. The equivalency of these configurations is here concerned with the permutation group of the allene skeleton.

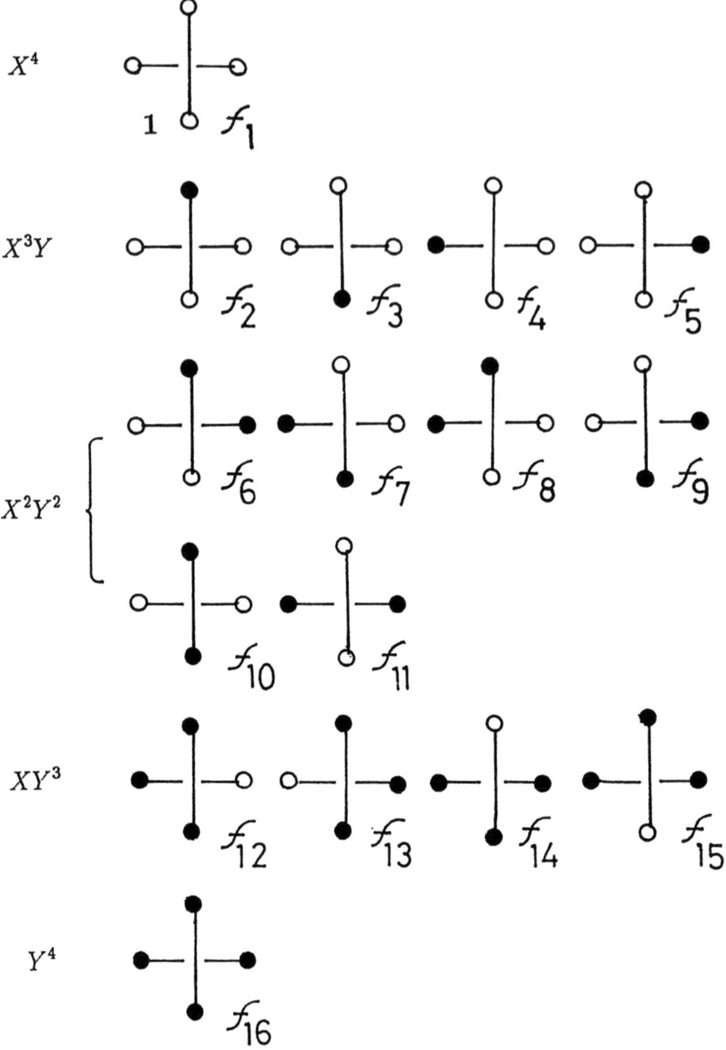

Figure 13.2: An allene skeleton (1) and its derivatives. All configurations are depicted.

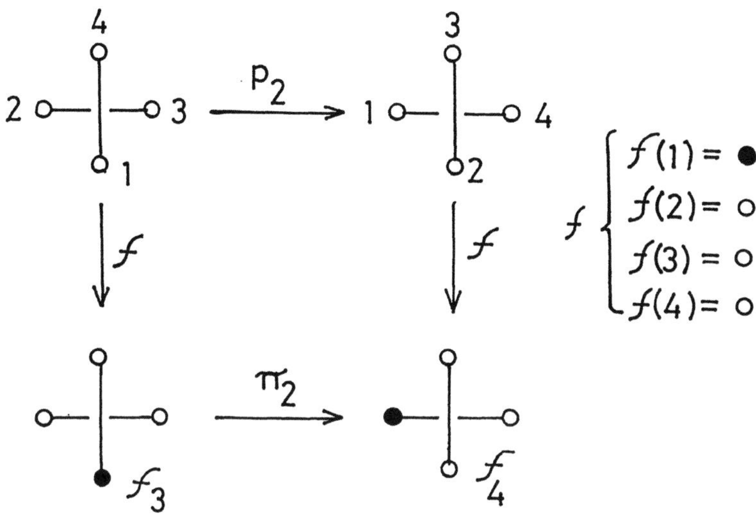

Figure 13.3: A permutation of configurations

$$\pi_2 = \begin{pmatrix} f_1 & f_2 & f_3 & f_4 & f_5 & f_6 & f_7 & f_8 & f_9 & f_{10} & f_{11} & f_{12} & f_{13} & f_{14} & f_{15} & f_{16} \\ f_1 & f_5 & f_4 & f_3 & f_2 & f_6 & f_7 & f_9 & f_8 & f_{11} & f_{10} & f_{14} & f_{15} & f_{12} & f_{13} & f_{16} \end{pmatrix}, \quad (13.16)$$

which corresponds to p_2. When we perform each symmetry operation to **F** by applying it to Δ, we have eight permutations. They are listed in Table 13.1, where the second row of each permutation is shown for simplicity's sake. Thereby, we obtain a permutation group,

$$\Pi = \{\pi_1, \pi_2, \ldots, \pi_8\}, \quad (13.17)$$

which is homomorphic to **P**.[3]

In order to make our discussions more precise, we here define *configurations*. Let Δ be a domain that consists of $|\Delta|$ elements called positions:

$$\Delta = \{\delta_1, \delta_2, \ldots, \delta_{|\Delta|}\},$$

Let **P** (order $|\mathbf{P}|$) be a permutation group that acts on the domain (Δ). Let **X** be a codomain (or range) that contains $|\mathbf{X}|$ elements called *ligands* (figures):

$$\mathbf{X} = \{X_1, X_2, \ldots, X_{|\mathbf{X}|}\}.$$

---

[2]It should be proved that π_2 *etc.* are permutations. The proof will be shown later.

[3]Note that **P** is associated with $\mathbf{D}_{2d}(/\mathbf{C}_s)$. The **P** is a permutation group, while the $\mathbf{D}_{2d}(/\mathbf{C}_s)$ is a permutation representation. They are conceptually distinct. However, they have the same form, since the $\mathbf{D}_{2d}(/\mathbf{C}_s)$ is a faithful representation. In this case, we can equalize them, though such manipulation lacks mathematical strictness.

Table 13.1: Permutations of **F** and fixed configurations

$D_{2d}(/C_s)$	f_1	f_2	f_3	f_4	f_5	f_6	f_7	f_8	f_9	f_{10}	f_{11}	f_{12}	f_{13}	f_{14}	f_{15}	f_{16}
$p_1 \sim \pi_1 =$	f_1	f_2	f_3	f_4	f_5	f_6	f_7	f_8	f_9	f_{10}	f_{11}	f_{12}	f_{13}	f_{14}	f_{15}	f_{16}
	✓	✓	✓	✓	✓	✓	✓	✓	✓	✓	✓	✓	✓	✓	✓	✓
$p_2 \sim \pi_2 =$	f_1	f_5	f_4	f_3	f_2	f_6	f_7	f_9	f_8	f_{11}	f_{10}	f_{14}	f_{15}	f_{12}	f_{13}	f_{16}
	✓					✓	✓									✓
$p_3 \sim \pi_3 =$	f_1	f_4	f_5	f_2	f_3	f_7	f_6	f_8	f_9	f_{11}	f_{10}	f_{15}	f_{14}	f_{13}	f_{12}	f_{16}
	✓							✓	✓							✓
$p_4 \sim \pi_4 =$	f_1	f_3	f_2	f_5	f_4	f_7	f_6	f_9	f_8	f_{10}	f_{11}	f_{13}	f_{12}	f_{15}	f_{14}	f_{16}
	✓									✓	✓					✓
$p_5 \sim \pi_5 =$	f_1	f_2	f_3	f_5	f_4	f_8	f_9	f_6	f_7	f_{10}	f_{11}	f_{13}	f_{12}	f_{14}	f_{15}	f_{16}
	✓	✓	✓							✓	✓			✓	✓	✓
$p_6 \sim \pi_6 =$	f_1	f_5	f_4	f_2	f_3	f_9	f_8	f_6	f_7	f_{11}	f_{10}	f_{15}	f_{14}	f_{12}	f_{13}	f_{16}
	✓															✓
$p_7 \sim \pi_7 =$	f_1	f_4	f_5	f_3	f_2	f_8	f_9	f_7	f_6	f_{11}	f_{10}	f_{14}	f_{15}	f_{13}	f_{12}	f_{16}
	✓															✓
$p_8 \sim \pi_8 =$	f_1	f_3	f_2	f_4	f_5	f_9	f_8	f_7	f_6	f_{10}	f_{11}	f_{12}	f_{13}	f_{15}	f_{14}	f_{16}
	✓			✓	✓					✓	✓	✓	✓			✓
formula	X^4	X^3Y^1				X^2Y^2						X^1Y^3				Y^4

Let f be a function from the domain Δ to the codomain $(\mathbf{X})$, *i.e.*,

$$f : \Delta \to \mathbf{X} \tag{13.18}$$

wherein the functions are called *configurations*.

Theorem 13.5 *Consider a set of the configurations,*

$$\mathbf{F} = \{f_1, f_2, \ldots, f_\alpha, \ldots, f_\beta, \ldots, f_{|\mathbf{F}|}\}. \tag{13.19}$$

Then, any p belonging to $\mathbf{P}$ is associated with π_p represented by

$$\pi_p = \left(\begin{array}{cccc} f_1 & f_2 & \cdots & f_{|\mathbf{F}|} \\ f_1 p & f_2 p & \cdots & f_{|\mathbf{F}|} p \end{array} \right). \tag{13.20}$$

(1) This is a permutation on $\mathbf{F}$.
(2) Let Π be a set of all π_p ($p \in \mathbf{P}$). Then, Π is a homomorphism of $\mathbf{P}$.
(3) Moreover, $\mathbf{P} \cong \Pi$ (i.e. isomorphic).

Proof. (1) Assume that f_α and f_β, which are different to each other, are mapped into the same function f_γ by π_p. Then, we have $f_\gamma(\delta) = f_\alpha(p(\delta))$ and $f_\gamma(\delta) = f_\beta(p(\delta))$ for all δ ($\in \Delta$). It follows that $f_\alpha(\delta) = f_\beta(\delta)$ for all δ, because the set of $p(\delta)$'s is identical with the set of δ. This result is against the original assumption. Hence, π_p is a permutation on $\mathbf{F}$.

(2) It is sufficient to prove that $\pi_{qp} = \pi_q \pi_p$.

$$\pi_p = \begin{pmatrix} f_1(\delta) & f_2(\delta) & \cdots & f_{|\mathbf{F}|}(\delta) \\ f_1 p(\delta) & f_2 p(\delta) & \cdots & f_{|\mathbf{F}|} p(\delta) \end{pmatrix}$$

$$\pi_q = \begin{pmatrix} f_1 p(\delta) & f_2 p(\delta) & \cdots & f_{|\mathbf{F}|} p(\delta) \\ f_1 qp(\delta) & f_2 qp(\delta) & \cdots & f_{|\mathbf{F}|} qp(\delta) \end{pmatrix}$$

$$\pi_{qp} = \begin{pmatrix} f_1(\delta) & f_2(\delta) & \cdots & f_{|\mathbf{F}|}(\delta) \\ f_1 qp(\delta) & f_2 qp(\delta) & \cdots & f_{|\mathbf{F}|} qp(\delta) \end{pmatrix}$$

These three equations afford $\pi_{qp} = \pi_q \pi_p$.

(3) Let $\mathbf{N}$ be the kernel of the homomorphism created by (2). Then, we find any p ($\in \mathbf{N}$) that satisfies $f_\alpha(\delta) = f_\alpha(p(\delta))$ for all δ and all α. This is obviously I itself. Hence, $\mathbf{N} = \{I\}$. It follows that $\mathbf{P} \cong \Pi$.

13.3 The Pólya-Redfield Theorem

Orbits of configurations. Since the Π group is a permutation group on $\mathbf{F}$, we can consider orbits in $\mathbf{F}$. This type of orbits (orbits of configurations) will be defined later; however, the intuitive discussions above indicate that two configurations are equivalent if they are superimposable on the action of an appropriate permutation concerning the allene skeleton and that such an equivalence class constructs an orbit. Chemically speaking, such an orbit in $\mathbf{F}$ corresponds to one compound derived from the skeleton. By inspection of Table 13.1, we have six orbits, *i.e.*, $\mathbf{F}_1 = \{f_1\}$, $\mathbf{F}_2 = \{f_2, f_3, f_4, f_5\}$, $\mathbf{F}_3 = \{f_6, f_7, f_8, f_9\}$, $\mathbf{F}_4 = \{f_{10}, f_{11}\}$, $\mathbf{F}_5 = \{f_{12}, f_{13}, f_{14}, f_{15}\}$, and $\mathbf{F}_6 = \{f_{16}\}$. Each of these orbits represents one compound; thus there emerge six derivatives.

We use chemical terms and/or mathematical terms according to our issues. The following table shows correspondence between several pairs of such terms.

Mathematical term		Chemical term
figure	$\longleftrightarrow$	ligand
an orbit of configurations	$\longleftrightarrow$	a compound (derivative)
equivalent configurations	$\longleftrightarrow$	homomers
non-equivalent configurations with the same weight	$\longleftrightarrow$	isomers

The number of the orbits is calculated by means of Theorem 13.4. In Table 13.1, fixed configurations[4] are checked for counting them. The number of such fixed configurations is calculated for every row, as shown in the following tabulation.

---

[4] We sometimes refer to such configurations (or any other objects) as points in an abstract fashion. This convension is based on the idea that any set acted by a permutation group may not be differentiated if we take no account of the concrete meaning of the set.

$\mathbf{D}_{2d}$		$\mathbf{P}$		Π on $\mathbf{F}$	$n_{\pi(g_\alpha)}$
I	$\sim$	$p_1 = (1)(2)(3)(4)$	$\sim$	π_1	16
$C_{2(1)}$	$\sim$	$p_2 = (1\ \ 2)(3\ \ 4)$	$\sim$	π_2	4
$C_{2(2)}$	$\sim$	$p_3 = (1\ \ 3)(2\ \ 4)$	$\sim$	π_3	4
$C_{2(3)}$	$\sim$	$p_4 = (1\ \ 4)(2\ \ 3)$	$\sim$	π_4	4
$\sigma_{d(1)}$	$\sim$	$p_5 = (1)(2\ \ 3)(4)$	$\sim$	π_5	8
S_4^3	$\sim$	$p_6 = (1\ \ 3\ \ 4\ \ 2)$	$\sim$	π_6	2
S_4	$\sim$	$p_7 = (1\ \ 2\ \ 4\ \ 3)$	$\sim$	π_7	2
$\sigma_{d(2)}$	$\sim$	$p_8 = (1\ \ 4)(2)(3)$	$\sim$	π_8	8
				$\displaystyle\sum_{g_\alpha} n_{\pi(g_\alpha)}$	48

Since $|\,\mathbf{D}_{2d}\,|$ is equal to 8, the number of orbits is calculated to be $48/8 = 6$. This is equal to the number of derivatives that are derived from the allene skeleton (**1**).

Consider a subset of $\mathbf{F}$ that corresponds to the term $[\theta] = X^2Y^2$, *i.e.*, $\mathbf{F}^{[\theta]} = \{f_6, f_7, f_8, f_9, f_{10}, f_{11}\}$. The corresponding part of Table 13.1 provides a permutation group ($\Pi^{[\theta]}$) on $\mathbf{F}^{[\theta]}$, which is homomorphic to $\mathbf{P}$ and $\mathbf{D}_{2d}$.[5] By counting fixed configuration with a mark ($\sqrt{}$) in Table 13.1, the number of orbits in $\mathbf{F}^{[\theta]}$ is calculated to be

$$\frac{1}{8}(6 + 2 + 2 + 2 + 2 + 0 + 0 + 2) = 2$$

according to Theorem 13.4.

We here define *orbits of configurations* in a stricter fashion. Let Δ be a domain that consists of n elements called positions: $\Delta = \{\delta_1, \delta_2, \ldots, \delta_n\}$. Let $\mathbf{P}$ (order $|\,\mathbf{P}\,|$) be a permutation group that acts on the domain (Δ). Let $\mathbf{X}$ be a codomain (or range) that contains m elements called *ligands* (figures): $\mathbf{X} = \{X_1, X_2, \ldots, X_m\}$. Let f_α and f_β be two functions from the domain Δ to the codomain ($\mathbf{X}$), *i.e.*,

$$f_\alpha : \Delta \to \mathbf{X}$$
$$\text{and}$$
$$f_\beta : \Delta \to \mathbf{X},$$

wherein the functions are called *configurations*.

Definition 13.2 (Binary relation) *A relationship between f_α and f_β is defined as a binary relation ($f_\alpha \sim f_\beta$) if there exists p ($\in \mathbf{P}$) that satisfies the following equation,*

$$f_\alpha(\delta) = f_\beta(p(\delta)) \quad \text{for } \forall \delta \in \Delta. \tag{13.21}$$

Theorem 13.6 *The binary relation defined above is an equivalence relation. As a result, a set of the functions (configurations) are partitioned into equivalence classes (orbits) by this relation.*

---

[5]It should be proved that each element of $\mathbf{F}^{[\theta]}$ is a permutation. The proof will be shown later.

Proof. Let $\mathbf{F}$ be a set of all the functions (configurations) defined above.

(1) (Reflective law). Let f_α be a function of $\mathbf{F}$. For I (identity) of the permutation group $\mathbf{P}$, $f_\alpha(\delta) = f_\alpha(I(\delta))$ for $\forall \delta \in \Delta$. Therefore, $f_\alpha \sim f_\alpha$.

(2) (Symmetric law). Let f_α and f_β be functions of $\mathbf{F}$. Suppose that $\exists p(\in \mathbf{P})$ satisfies the binary relation $(f_\alpha \sim f_\beta)$. Then $f_\alpha(\delta) = f_\beta(p(\delta))$ for $\forall \delta \in \Delta$. If $\forall \delta$ is substituted by $p^{-1}(\delta)$, then we have $f_\alpha(p^{-1}(\delta)) = f_\beta(pp^{-1}(\delta)) = f_\beta(\delta)$, because $pp^{-1} = I$. Since p^{-1} also belongs to $\mathbf{P}$, this equation shows that $f_\beta \sim f_\alpha$.

(3) (Transitive law). When $f_\alpha \sim f_\beta$ and $f_\beta \sim f_\gamma$, the following equations are obtained in terms of Def. 13.2.

$$f_\alpha(\delta) = f_\beta(p(\delta)) \quad (\forall \delta \in \Delta) \quad \text{and}$$
$$f_\beta(\delta) = f_\gamma(p'(\delta)) \quad (\forall \delta \in \Delta),$$

wherein $\exists p$ and $\exists p'$ belongs to $\mathbf{P}$. Then we obtain: $f_\alpha(\delta) = f_\beta(p(\delta)) = f_\gamma(p'p(\delta))$. Since p belongs to the permutation group $(\mathbf{P})$, $p'p$ is also an element of $\mathbf{P}$. Hence, $f_\alpha \sim f_\gamma$.

By means of the equivalence relation described in Def. 13.2, the set of configurations $\mathbf{F}$ is partitioned into several orbits, which are called *orbits of configurations*. Chemically speaking, each of the orbits corresponds to one chemical compound. Hence, enumeration of compounds is replaced by enumeration of such orbits of configurations.

Generating functions. Compounds having the same molecular formula are called *isomers*. The next issue is to count such isomers. Let us return to Table 13.1. Each configuration f_α is associated with the corresponding term $X^{\theta_1}Y^{\theta_2}$, which is regarded as a molecular formula.[6] When we examine the first row, we have the number of fixed configurations that is itemized with respect to the molecular formulas, *i.e.*,

$$X^4 + 4X^3Y + 6X^2Y^2 + 4XY^3 + Y^4,$$

since I (or p_1) fixes all of the configurations (marked by the symbol $\sqrt{}$). By counting configurations marked with $\sqrt{}$ for each row of Table 13.1, we obtain the following generating functions.[7]

$$
\begin{aligned}
p_2: & \quad X^4 + 2X^2Y^2 + Y^4, \\
p_3: & \quad X^4 + 2X^2Y^2 + Y^4, \\
p_4: & \quad X^4 + 2X^2Y^2 + Y^4, \\
p_5: & \quad X^4 + 2X^3Y + 2X^2Y^2 + 2XY^3 + Y^4, \\
p_6: & \quad X^4 + Y^4,
\end{aligned}
$$

---

[6] The corresponding mathematical term is a "weight" of the function (configuration). We also use this term later.

[7] A generating function is a fundamental concept in combinatorics. Most textbooks cited as references explain this concept in detail.

Table 13.2: Generation functions for counting configurations from 1

D_{2d}		P		permutation on F	cycle structure	dummy variable	generating function
I	$\sim$	$p_1 = (1)(2)(3)(4)$	$\sim$	π_1	(1^4)	s_1^4	$(X+Y)^4$
$C_{2(1)}$	$\sim$	$p_2 = (1\ 2)(3\ 4)$	$\sim$	π_2	(2^2)	s_2^2	$(X^2+Y^2)^2$
$C_{2(2)}$	$\sim$	$p_3 = (1\ 3)(2\ 4)$	$\sim$	π_3	(2^2)	s_2^2	$(X^2+Y^2)^2$
$C_{2(3)}$	$\sim$	$p_4 = (1\ 4)(2\ 3)$	$\sim$	π_4	(2^2)	s_2^2	$(X^2+Y^2)^2$
$\sigma_{d(1)}$	$\sim$	$p_5 = (1)(2\ 3)(4)$	$\sim$	π_5	$(1^2 2)$	$s_1^2 s_2$	$(X+Y)^2(X^2+Y^2)$
S_4^3	$\sim$	$p_6 = (1\ 3\ 4\ 2)$	$\sim$	π_6	(4)	s_4	$1+X^4$
S_4	$\sim$	$p_7 = (1\ 2\ 4\ 3)$	$\sim$	π_7	(4)	s_4	$1+X^4$
$\sigma_{d(2)}$	$\sim$	$p_8 = (1\ 4)(2)(3)$	$\sim$	π_8	$(1^2 2)$	$s_1^2 s_2$	$(X+Y)^2(X^2+Y^2)$

$$p_7: \quad X^4 + Y^4, \text{ and}$$
$$p_8: \quad X^4 + 2X^3Y + 2X^2Y^2 + 2XY^3 + Y^4.$$

We can write down all of the configurations in this case, since there emerge only 16 configurations. However, general isomer enumerations must deal with more complicated cases that contain a vast number of configurations. The Pólya-Redfield theory provides us with a versatile method for counting configurations.

For example, let us examine $p_4 = (1\ 4)(2\ 3)$ acting on the allene skeleton (1). Consider the set of positions $\{1,4\}$, on which X or Y is placed. In order for the p_4 permutation to keep the resulting configuration invariant, the mode of substitution must be X^2 or Y^2. Hence, $s_2 \equiv X^2 + Y^2$ is a generating function; this is called a *ligand inventory*.[8] For the set $\{2,3\}$, the same ligand inventory is necessary. It follows that $s_2^2 = (X^2 + Y^2)^2$ is a generating function that gives the number of fixed configurations on the action of this permutation. Note that the s_2^2 variable corresponds to the cycle structure of the permutation, (2^2). Table 13.2 lists such examination over all of the permutations. When we expand the generating functions in Table 13.2, we can verify the polynomials shown above.

The present discussion is easily extended to general cases. Thus, we obtain the following lemma.

Lemma 13.1 *Let* $\mathbf{P}$ *be a permutation group on* $\Delta = \{\delta_1, \delta_2, \cdots, \delta_n\}$. *Suppose that the cycle structure of* p *(*$\in \mathbf{P}$*) is represented by* $(1^{\nu_1} 2^{\nu_2} \cdots n^{\nu_n})$, *where* $\sum_{i=1}^n i\nu_i = n$. *Each of the positions of* Δ *is occupied by a ligand that is selected from a codomain (or range),* $\mathbf{X} = \{X_1, X_2, \ldots, X_m\}$. *Consider that the* p *permutation keeps invariant a configuration having* $\theta_1\ X_1, \theta_2\ X_2, \cdots,$ *and* $\theta_m\ X_m$, *where*

$$[\theta]: \quad \theta_1 + \theta_2 + \cdots + \theta_m = n. \tag{13.22}$$

---

[8]The corresponding mathematical term is "figure inventory". The term "figure" corresponds to the chemical term "ligand".

Let the symbol A_θ denote the total number of such configurations that are fixed on the action of p. A generating function for calculating A_θ is represented by

$$\sum_{[\theta]} A_\theta X_1^{\theta_1} X_2^{\theta_2} \cdots X_m^{\theta_m} = s_1^{\nu_1} s_2^{\nu_2} \cdots s_n^{\nu_n}, \tag{13.23}$$

where the summation is concerned with all of the partitions ($[\theta]$) shown in eq. 13.22. The variables in the right-hand side are ligand inventories expressed by

$$s_d = X_1^d + X_2^d + \cdots + X_m^d. \tag{13.24}$$

The term $X_1^{\theta_1} X_2^{\theta_2} \cdots X_m^{\theta_m}$ appearing in eq. 13.23 is regarded as a molecular formula in a chemical sense. In order to simplify our discussion, we introduce the following definition.

Definition 13.3 (Weight of configuration) *Consider a domain Δ and a co-domain $\mathbf{X}$:*

$$\Delta = \{\delta_1, \delta_2, \ldots, \delta_n\},$$
$$\mathbf{X} = \{X_1, X_2, \ldots, X_m\},$$

where the domain Δ is subject to a permutation group $\mathbf{P}$. Let f_α be a function (configuration) from the domain Δ to the codomain ($\mathbf{X}$), i.e.,

$$f_\alpha : \Delta \to \mathbf{X}. \tag{13.25}$$

The function f_α represents a configuration having θ_1 X_1, θ_2 $X_2, \cdots$, and θ_m X_m, where

$$[\theta]: \quad \theta_1 + \theta_2 + \cdots + \theta_m = n.$$

Then, a weight of the configuration $W(f_\alpha)$ is defined by the expression,

$$W(f_\alpha) = X_1^{\theta_1} X_2^{\theta_2} \cdots X_m^{\theta_m}. \tag{13.26}$$

This definition will be extended to more general cases.

The following theorem can be easily proved by starting from Def. 13.3.

Theorem 13.7 *If two configurations $f_\alpha : \Delta \to \mathbf{X}$ and $f_\beta : \Delta \to \mathbf{X}$ defined in Def. 13.3 are equivalent, they have the same weight, i.e., $W(f_\alpha) = W(f_\beta)$.*

This theorem implies that there exist non-equivalent configurations having the same weight. These non-equivalent configurations can be referred to as *isomers*, if we apply the chemical term to the present case. The final target of this chapter is to obtain the number of such isomers (non-equivalent configurations) in the form of a generating fuction such as

$$\sum_{[\theta]} B_\theta X_1^{\theta_1} X_2^{\theta_2} \cdots X_m^{\theta_m}. \tag{13.27}$$

Note that the total number of configurations is represented by eq. 13.23 in Lemma 13.1.

Before we prove the Pólya-Redfield theorem in a general fashion, we re-examine the case shown in Table 13.2. When we sum up the generating functions in the rightmost column of this table, we have a generating function that affords the total number of fixed configurations. By the inspection of Table 13.1, we find that each orbit of configurations have 8 fixed configurations ($\mid \mathbf{P} \mid= 8$). This value is obtained alternatively by the application of eq. 13.4. Let $B_{(\theta_1\theta_2)}$ be the number of orbits of configurations that correspond to the weight $X^{\theta_1}Y^{\theta_2}$. Then, the total number of fixed configurations concerning the weight is calculated to be $8\Sigma_{[\theta]}\, B_{(\theta_1\theta_2)}X^{\theta_1}Y^{\theta_2}$. This value should be equal to the summation of the rightmost column of Table 13.2; therefore, we have

$$\sum_{[\theta]} B_{(\theta_1\theta_2)}X^{\theta_1}Y^{\theta_2} \;=\; \frac{1}{8}((X+Y)^4 + (X^2+Y^2)^2 + \cdots)$$

$$\;=\; X^4 + X^3Y + 2X^2Y^2 + XY^3 + Y^4. \tag{13.28}$$

This equation provides the number of isomers as the coefficient of each term.

This example can be extended easily to a general case. For simplicity's sake, we first define a *cycle index* as follows.

Definition 13.4 *Let* $\mathbf{P}$ *be a permutation group on* $\Delta = \{\delta_1, \delta_2, \cdots, \delta_n\}$. *Suppose that the cycle structure of* p ($\in \mathbf{P}$) *is represented by* $(1^{\nu_1}2^{\nu_2}\cdots n^{\nu_n})$, *where* $\Sigma_{i=1}^{n}\, i\nu_i = n$. *A* cycle index *of* $\mathbf{P}$ *is defined by*

$$CI(\mathbf{P}; s_d) = \frac{1}{\mid \mathbf{P} \mid}\sum_{p \in \mathbf{P}} s_1^{\nu_1}s_2^{\nu_2}\cdots s_n^{\nu_n}. \tag{13.29}$$

Then, we end up with the following theorem.

Theorem 13.8 (Pólya-Redfield) *Let* $\mathbf{P}$ *be a permutation group on* $\Delta = \{\delta_1, \delta_2, \cdots, \delta_n\}$. *Suppose that the cycle structure of* p ($\in \mathbf{P}$) *is represented by* $(1^{\nu_1}2^{\nu_2}\cdots n^{\nu_n})$, *where* $\Sigma_{i=1}^{n}\, i\nu_i = n$. *Each of the positions of* Δ *is occupied by a ligand that is selected from a codomain (or range),* $\mathbf{X} = \{X_1, X_2, \ldots, X_m\}$. *Consider configurations having* $\theta_1\, X_1, \theta_2\, X_2,\, \cdots,$ *and* $\theta_m\, X_m,$ *where*

$$[\theta]:\quad \theta_1 + \theta_2 + \cdots + \theta_m = n. \tag{13.30}$$

Let the symbol B_θ *denote the number of orbits of such configurations. A generating function for calculating* B_θ *is represented by*

$$G(X_1, X_2, \ldots, X_m) \;=\; \sum_{[\theta]} B_\theta X_1^{\theta_1}X_2^{\theta_2}\cdots X_m^{\theta_m} \tag{13.31}$$

$$\;=\; CI(\mathbf{P}; s_d), \tag{13.32}$$

where the summation is concerned with all of the partitions ([θ]) shown in eq. 13.30. The variables in the cycle index are ligand inventories expressed by

$$s_d = X_1^d + X_2^d + \cdots + X_m^d. \tag{13.33}$$

Proof. If the number $\sum_{[\theta]} A_\theta$ obtained in Lemma 13.1 is summed over all p ($\in \mathbf{P}$), we obtain the total number of fixed configurations as to be

$$\sum_{p \in \mathbf{P}} \sum_{[\theta]} A_\theta X_1^{\theta_1} X_2^{\theta_2} \cdots X_m^{\theta_m} = \sum_{p \in \mathbf{P}} s_1^{\nu_1} s_2^{\nu_2} \cdots s_n^{\nu_n}. \tag{13.34}$$

We use $\mathbf{F}$ and Π in the sense defined in Theorem 13.5. Let B_θ be the number of such orbits of $\mathbf{F}$ on the action of Π that have the weight $W(f_\alpha) = X_1^{\theta_1} X_2^{\theta_2} \cdots X_m^{\theta_m}$. According to eq. 13.4, an orbit Δ_α with the weight $W(f_\alpha)$ contains $|\Pi|$ of fixed configurations. Then, the orbits with the $W(f_\alpha)$ weight contains $B_\theta |\Pi|$. It follows that the total number of fixed configurations is represented by

$$\sum_{[\theta]} B_\theta \mid \Pi \mid X_1^{\theta_1} X_2^{\theta_2} \cdots X_m^{\theta_m}. \tag{13.35}$$

Obviously, eqs. 13.34 and 13.35 represent the same total number of fixed configurations in alternative manners. Since $\mathbf{P} \cong \Pi$ (Theorem 13.5) gives $|\mathbf{P}| = |\Pi|$, we obtain eq. 13.36.

$$\sum_{[\theta]} B_\theta X_1^{\theta_1} X_2^{\theta_2} \cdots X_m^{\theta_m} = \frac{1}{|\mathbf{P}|} \sum_{p \in \mathbf{P}} s_1^{\nu_1} s_2^{\nu_2} \cdots s_n^{\nu_n}. \tag{13.36}$$

This equation is equivalent to eq. 13.32, if the cycle index defined by Def. 13.4 is introduced.

Although this proof does not use the Cauchy-Frobenius lemma (Theorem 13.1) directly, it is based on the same idea that underlies the proof of Theorem 13.1. Theorem 13.1 will be employed in the next chapter in order to generalize the Pólya-Redfield theorem (Theorem 13.8).

Example **13.2** Let us re-examine the case of Table 13.2 by means of Theorem 13.8. We have the cycle index:

$$CI(\mathbf{P}; s_d) = \frac{1}{8}(s_1^4 + 3s_2^2 + 2s_1^2 s_2 + 2s_4),$$

to which we introduce the ligand inventory:

$$s_d = X^d + Y^d.$$

As a result, we have

$$G(X, Y) = CI(\mathbf{P}; X^d + Y^d) = X^4 + X^3 Y + 2X^2 Y^2 + XY^3 + Y^4.$$

This equation is identical with eq. 13.28.

Bibliography

[1] a) G. Pólya, *Acta Math.*, **68**, 145 (1937). b) G. Pólya, R. C. Read, *Combinatorial Enumeration of Groups, Graphs, and Chemical Compounds*, Springer, Berlin (1987).

[2] J. H. Redfield, *Am. J. Math.*, **49**, 433 (1927).

[3] C. L. Liu, *Introduction to Combinatorial Mathematics*, McGraw-Hill, New York (1969).

[4] F. Harary, *Graph Theory*, Addisson-Wesley, Reading (1971).

[5] F. Harary, E. M. Palmer, *Graphical Enumeration*, Academic, New York (1973).

[6] L. Lovász, *Combinatorial Problems and Exercises*, Akadeémiai Kiadó, Budapest (1979).

[7] G. Pólya, R. E. Tarjan, D. R. Woods, *Notes on Introductory Combinatorics*, Birkhäuser, Boston (1983).

[8] A. T. Balaban (Ed.), *Chemical Application of Graph Theory*, Academic, London (1976).

[9] N. Trinajstić, *Chemical Graph Theory*, Vol. II, CRC, Boca Raton (1983).

[10] J. Hinze (Ed.), *The Permutation Group in Physics and Chemistry (Lecture Notes in Chemistry, Vol. 12)*, Springer-Verlag, Berlin Heidelberg New York (1979)

[11] a) D. H. Rouvray, *Chem. Soc. Rev.*, **3**, 355 (1974). b) D. H. Rouvray, *Endeavour*, **34**, 28 (1975).

[12] K. Balasubramian, *Chem. Rev.*, **85**, 599 (1985).

[13] R. J. Hansen, P. C. Jurs, *J. Chem. Educ.*, **65**, 661 (1988).

[14] a) W. Hässelbarth, *Theor. Chem. Acta*, **66**, 91 (1984). b) W. Hässelbarth, *Theor. Chem. Acta*, **67**, 427 (1985).

[15] a) S. Fujita, *J. Chem. Inf. Comput. Sci.*, **26**, 212 (1986). b) S. Fujita, *J. Chem. Inf. Comput. Sci.*, **26**, 224 (1986). c) S. Fujita, *Bull. Chem. Soc. Jpn.*, **61**, 4189 (1989). d) S. Fujita, *J. Chem. Inf. Comput. Sci.*, **29**, 22 (1989).

Chapter 14

Obligatory Minimum Valencies[1]

In chemical enumerations, there are many cases in which we should take account of obligatory minimum valencies (OMVs). For example, consider an adamantane skeleton (1) of $\mathbf{T}_d$ symmetry, in which ten carbon atoms are replaced by other atoms.

The skeleton has four bridgehead positions and six bridge positions. Each of the bridgehead positons can take as a substituent an atom having three or more valency (*e.g.* carbon and nitrogen); however it cannot accept a mono-valent and a di-valent atom (*e.g.* hydrogen and oxygen). Hence, the OMV of the bridgehead positions is equal to 3. On the other hand, each of the bridge positions can take an atom possessing two or more valency. Mono-valent atoms are unsuitable as substituents. Thus, the OMV of the bridge positions is equal to 2.

A pioneering work to treat such OMV restriction has appeared in an enumeration of metal complexes[1] and in an enumeration of organic compounds.[2] The term OMV was used in a manual enumeration of organic reactions.[3] In the series of our work on imaginary transition structures,[4] we have reported enumerations of organic reactions.[5] During this work, we ourselves encountered such OMV restriction.[6] We have solved this problem by using two or more different weights that are assigned to the respective orbits of a domain.[6] Although these papers

---

[1]Reprinted in part with permission from S. Fujita, *Bull. Chem. Soc. Japan*, **61**, 4189–5206 (1988). ©(1988) The Chemical Society of Japan. See also S. Fujita, *J. Math. Chem.*, **5**, 99–120 (1990).

are restricted to the case of organic reactions, the resulting formulas are applicable to the enumeration of any kinds of chemical structures under the influence of the OMV. In this chapter, we extend our previous work to treat such orbits of a domain in a more explicit fashion.

14.1 Isomer Enumeration under the OMV Restriction

In Chapter 13, a compound has been formulated as an orbit of configurations. We also use the formulation in this section. Let f be a function from a domain Δ to a codomain $\mathbf{X}$, *i.e.*, $f : \Delta \to \mathbf{X}$, where

$$\Delta = \{\delta_1, \delta_2, \ldots, \delta_{|\Delta|}\} \quad \text{and} \quad \mathbf{X} = \{X_1, X_2, \ldots, X_{|\mathbf{X}|}\}$$

In chemical sense, the elements of the domain are the substitution positions of a parent skeleton; and they are called *positions*. The elements of the codomain are called *ligands* due to the terminology of chemistry. The function f is called a *configuration*, which corresponds to the chemical terms ("homomer", "isomer", *etc.*) according to their properties. Consider a set of such configurations:

$$\mathbf{F} = \{f_1, f_2, \ldots, f_{|\mathbf{F}|}\}.$$

If the domain (Δ) is subject to a perumtation group $\mathbf{P}$, the $\mathbf{F}$ set are divided into several orbits on the action of $\mathbf{P}$. Each of the resulting orbits corresponds to a compound.

The treatment described in Chapter 13 has taken no account of the orbits of Δ on the action of $\mathbf{P}$. In order to treat the OMV restriction, we examine cases in which the domain Δ has two or more orbits. For simplicity of discussion, we first deal with a domain having two orbits. Thus, suppose that the Δ domain is partitioned by $\mathbf{P}$ into two orbits: $\Delta_1 = \{\delta_{11}, \delta_{12}, \ldots, \delta_{1u}\}$ and $\Delta_2 = \{\delta_{21}, \delta_{22}, \ldots, \delta_{2v}\}$, where $u + v = |\Delta|$. The definition of an equivalence relation described in Chapter 13 (Definition 13.2 and Theorem 13.2) also holds for this case. The weight of a function is so redefined as to manipulate the two orbits.

Definition 14.1 (Weight of a function) *When $f(\delta)$ ($\delta \in \Delta_1$) is equal to X_ℓ ($\in$ $\mathbf{X}$), the weight $w_1(X_\ell)$ is assigned to this mapping. On the other hand, the weight $w_2(X_\ell)$ is assigned to the mapping in which $f(\delta)$ ($\delta \in \Delta_2$) is equal to X_ℓ. Then, we define a weight $W(f)$ for each function ($f : \Delta \to \mathbf{X}$) as follows:*

$$W(f) = \prod_{\delta \in \Delta_1} w_1(f(\delta)) \prod_{\delta \in \Delta_2} w_2(f(\delta)). \tag{14.1}$$

In this definition, the concept of weights is extended to meet such general cases that are influenced by OMVs.

Theorem 14.1 *Let $f_\alpha : \Delta \longrightarrow \mathbf{X}$ and $f_\beta : \Delta \longrightarrow \mathbf{X}$ be equivalent. Then.*

$$W(f_\alpha) = W(f_\beta). \tag{14.2}$$

Proof. Since $f_\alpha \sim f_\beta$, there exists an appropriate $p(\in \mathbf{P})$ that satisfies the following relationship:

$$f_\alpha(\delta) = f_\beta(p(\delta)) \quad \text{for} \quad \forall \delta \in \Delta.$$

Then,

$$
\begin{aligned}
W(f_\alpha) &= \prod_{\delta \in \Delta_1} w_1(f_\alpha(\delta)) \prod_{\delta \in \Delta_2} w_1(f_\alpha(\delta)) \\
&= \prod_{\delta \in \Delta_1} w_1(f_\beta(p(\delta))) \prod_{\delta \in \Delta_2} w_1(f_\beta(p(\delta))). \tag{14.3}
\end{aligned}
$$

Because of Def. 14.1,

$$W(f_\beta) = \prod_{\delta \in \Delta_1} w_1(f_\beta(\delta)) \prod_{\delta \in \Delta_2} w_1(f_\beta(\delta)). \tag{14.4}$$

Comparison of the right-hand side of eq. 14.3 with that of eq. 14.4 indicates their equality, since the set of $p(\delta)$'s is the same as the set of δ's except of the sequence. Therefore, we have $W(f_\alpha) = W(f_\beta)$.

Let $\mathbf{F}^{[\theta]}$ be a set of functions $(f : \Delta \to \mathbf{X})$, all of which have the same weight $W_\theta(f)$, *i.e.*,

$$\mathbf{F}^{[\theta]} = \{f_1^{[\theta]}, f_2^{[\theta]}, \ldots, f_\epsilon^{[\theta]}\}. \tag{14.5}$$

Let p be a permutation of $\mathbf{P}$ on Δ. This section shows that there is a permutation group on $\mathbf{F}^{[\theta]}$ onto which $\mathbf{P}$ is homomorphic. The following theorem corresponds to Theorem 13.5 (Chapter 13) except that the present one concerns homomorphism only.

Theorem 14.2 *A function $\pi^{[\theta]} \colon \mathbf{F}^{[\theta]} \to \mathbf{F}^{[\theta]}$ is defined in accord with p ($\in \mathbf{P}$) as follows:*

$$\pi^{[\theta]}(f(\delta)) = f(p(\delta)) \text{ for } \forall \delta \in \Delta \text{ and } \forall f \in \mathbf{F}^{[\theta]}.$$

(1) The function $\pi^{[\theta]}$ is a permutation on the set of $\mathbf{F}^{[\theta]}$.

(2) The mapping of $\mathbf{P}$ onto $\Pi^{[\theta]}$ defined above is homomorphic, wherein $\Pi^{[\theta]}$ is the set of all $\pi^{[\theta]}$.

Proof. (1) Suppose that both f_α and f_β of $\mathbf{F}^{[\theta]}$ ($f_\alpha \neq f_\beta$) are mapped by $\pi^{[\theta]}$ to the same function, *i.e.*,

$$\pi^{[\theta]}(f_\alpha(\delta)) = \pi^{[\theta]}(f_\beta(\delta)) \quad \text{for} \quad \forall \delta \in \Delta.$$

Because of the above assumption, we have

$$\pi^{[\theta]}(f_\alpha(\delta)) = f_\alpha(p(\delta)) \text{ and } \pi^{[\theta]}(f_\beta(\delta)) = f_\beta(p(\delta)).$$

As a result, we have $f_\alpha(p(\delta)) = f_\beta(p(\delta))$. This equation indicates that $f_\alpha = f_\beta$. This is contrary to the assumption. Therefore, $\forall f_\alpha \ (\in \mathbf{F}^{[\theta]})$ is transformed by $\pi^{[\theta]}$ into a function different from itself. This fact shows that $\pi^{[\theta]}$ is a permutation on the set of $\mathbf{F}^{[\theta]}$.

(2) Let $\pi_\gamma^{[\theta]}$ and $\pi_\epsilon^{[\theta]}$ be two permutations on $\mathbf{F}^{[\theta]}$. They correspond to p_γ and $p_\epsilon \ (\in \mathbf{P})$, respectively. Then the following equations are obtained in the light of the definition:

$$\pi_\gamma^{[\theta]}(f(\delta)) \;=\; f(p_\gamma(\delta)) \;\cdots\; (a)$$
$$\text{and } \pi_\epsilon^{[\theta]}(f(\delta)) \;=\; f(p_\epsilon(\delta)) \;\cdots\; (b).$$

Since $\mathbf{P}$ is a permutation group, we have

$$p_{\epsilon\gamma} = p_\epsilon p_\gamma \;\cdots\; (c).$$

Let $\pi_{\epsilon\gamma}^{[\theta]}$ correspond to $p_{\epsilon\gamma}(\in \mathbf{P})$. Then, we have

$$\pi_{\epsilon\gamma}^{[\theta]}(f(\delta)) \;=\; f(p_{\epsilon\gamma}(\delta)) \stackrel{(c)}{=} f(p_\epsilon p_\gamma(\delta))$$
$$=\; f(p_\epsilon(p_\gamma(\delta))) \stackrel{(b)}{=} \pi_\epsilon^{[\theta]}(f(p_\gamma(\delta))) \stackrel{(a)}{=} \pi_\epsilon^{[\theta]}\pi_\gamma^{[\theta]}(f(\delta)).$$

That is $\pi_{\epsilon\gamma}^{[\theta]} = \pi_\epsilon^{[\theta]}\pi_\gamma^{[\theta]}$, which indicates that the mapping of $\mathbf{P}$ onto $\pi^{[\theta]}$ is homomorphic.

Since the domain Δ is partitioned by $\mathbf{P}$ into two orbits Δ_1 and Δ_2, the cycle structure of each $p(\in \mathbf{P})$ is divided into two parts. Suppose that the p permutation has the following cycle structure:

$(\nu):$ $(1^{\nu_1} 2^{\nu_2} \cdots u^{\nu_u})$ for a permutation of Δ_1 part and

$(\lambda):$ $(1^{\lambda_1} 2^{\lambda_2} \cdots v^{\lambda_v})$ for a permutation of Δ_2 part,

where

$$1\nu_1 + 2\nu_2 + \cdots + u\nu_u = u \text{ and } 1\lambda_1 + 2\lambda_2 + \cdots + v\lambda_v = v.$$

Suppose that a function $f : \Delta \to \mathbf{X}$ substitutes $\theta_1 X_1, \theta_2 X_2, \ldots, \theta_{|\mathbf{X}|} X_{|\mathbf{X}|}$ for the positions of Δ_1 and $\eta_1 X_1, \eta_2 X_2, \ldots, \eta_{|\mathbf{X}|} X_{|\mathbf{X}|}$ for the positions of Δ_2, where

$$[\theta]: \;\; \theta_1 + \theta_2 + \ldots + \theta_{|\mathbf{X}|} = u \quad \text{and} \quad [\eta]: \;\; \eta_1 + \eta_2 + \ldots + \eta_{|\mathbf{X}|} = v.$$

The weight of the function f is represented by

$$W_{\theta;\eta}(f) = \prod_{i=1}^{|\mathbf{X}|} w_1(X_i)^{\theta_i} \prod_{i=1}^{|\mathbf{X}|} w_2(X_i)^{\eta_i}. \tag{14.6}$$

Let $A_{\theta;\eta}$ be the number of such functions (configurations) with the $W_{\theta;\eta}(f)$ weight that are invariant on the action of a permutation $(p \in \mathbf{P})$ having cycle structures $(\nu;\lambda)$. Let $B_{\theta;\eta}$ be the number of orbits of such configurations with the $W_{\theta;\eta}(f)$ weight that are invariant on the permutation (p). The following lemma is an extension of Lemma 13.1 (Chapter 13).

Lemma 14.1 *For each permuation p ($\in \mathbf{P}$) having cycle structures $(\nu;\lambda)$, a generating function for counting $A_{\theta;\eta}$ is represented by:*

$$\sum_{[\theta;\eta]} A_{\theta;\eta} \prod_{i=1}^{|\mathbf{X}|} w_1(X_i)^{\theta_i} \prod_{i=1}^{|\mathbf{X}|} w_2(X_i)^{\eta_i} = \prod_{d=1}^{u} s_d^{\nu_d} \prod_{d=1}^{v} t_d^{\lambda_d}, \tag{14.7}$$

wherein

$$s_d = \sum_{k=1}^{|\mathbf{X}|} w_1(X_k)^d \tag{14.8}$$

and

$$t_d = \sum_{k=1}^{|\mathbf{X}|} w_2(X_k)^d. \tag{14.9}$$

Proof. Consider such a cycle of length d that is contained in the permutation p ($\in \mathbf{P}$) and concerned with the Δ_1 part. In order for the p permutation to keep a function (configuration) invariant, the cycle also keeps the function invariant. This requires that the d positions of Δ_1 concering the cycle should have d ligands of the same kind, *i.e.* dX_1, dX_2, or $dX_{|\mathbf{X}|}$. Hence, the corresponding generating function (ligand inventory) is represented by

$$s_d \equiv \sum_{k=1}^{|\mathbf{X}|} w_1(X_k)^d. \tag{14.10}$$

Similarly, the Δ_2 part requires a ligand inventory,

$$t_d \equiv \sum_{k=1}^{|\mathbf{X}|} w_2(X_k)^d. \tag{14.11}$$

The same kinds of generating functions are obtained for all of the cycles involved in the permutation p. Hence, a generating function for counting $A_{\theta;\eta}$ is obtained as follows.

$$\sum_{[\theta;\eta]} A_{\theta;\eta} \prod_{i=1}^{|\mathbf{X}|} w_1(X_i)^{\theta_i} \prod_{i=1}^{|\mathbf{X}|} w_2(X_i)^{\eta_i} = \prod_{d=1}^{u} \Big(\sum_{k=1}^{|\mathbf{X}|} w_1(X_k)^d\Big)^{\nu_d} \prod_{d=1}^{v} \Big(\sum_{k=1}^{|\mathbf{X}|} w_2(X_k)^d\Big)^{\lambda_d}$$

$$= \prod_{d=1}^{u} s_d^{\nu_d} \prod_{d=1}^{v} t_d^{\lambda_d}, \tag{14.12}$$

which is in agreement with the cycle structure of p.

The cycles contained in the permutation p ($\in \mathbf{P}$) is divided into two parts. We assign a dummy variable s_d to a cycle of length d of Δ_1 part. Similarly, a variable t_d concerns a cycle of lengh d of Δ_2 part. Then, we obtain the following product for the p permutation.

$$\prod_{d=1}^{u} s_d^{\nu_d} \prod_{d=1}^{v} t_d^{\lambda_d} \tag{14.13}$$

Since the same kind of products can be obtained for all the permutations of $\mathbf{P}$, these products are summed up over $\mathbf{P}$. As a result, we arrive at a definition of a cycle index.

Definition 14.2 (Cycle index) *The following equation is defined as a cycle index for $\mathbf{P}$, the action of which provides a partition of the Δ domain.*

$$CI(\mathbf{P}; s_d, t_d) = \frac{1}{|\mathbf{P}|} \sum_{p \in \mathbf{P}} \prod_{d=1}^{u} s_d^{\nu_d} \prod_{d=1}^{v} t_d^{\lambda_d}. \tag{14.14}$$

Now we are ready for proving a theorem providing a generating function of $B_{\theta;\eta}$.

Theorem 14.3 *A generating function for the number $(B_{\theta;\eta})$ of orbits of configurations is represented by*

$$\sum_{\theta;\eta} B_{[\theta;\eta]} \prod_{i=1}^{|\mathbf{X}|} w_1(X_i)^{\theta_i} \prod_{i=1}^{|\mathbf{X}|} w_2(X_i)^{\eta_i} = CI(\mathbf{P}; \sum_{k=1}^{|\mathbf{X}|} w_1(X_k)^d, \sum_{k=1}^{|\mathbf{X}|} w_2(X_k)^d), \tag{14.15}$$

wherein the right-hand side indicates that the two terms denoted by the summations should be introduced to s_d and t_d of the $CI(\mathbf{P}; s_d, t_d)$ (Def. 14.2).

Proof. Let $\mathbf{F}^{[\theta,\eta]}$ be a set of functions that have the same weight represented by

$$W_{\theta;\eta} = \prod_{i=1}^{|\mathbf{X}|} w_1(X_i)^{\theta_i} \prod_{i=1}^{|\mathbf{X}|} w_2(X_i)^{\eta_i}. \tag{14.16}$$

Let $r_{\theta;\eta}$ be the number of orbits contained in $\mathbf{F}^{[\theta;\eta]}$. Then, a generating function for counting orbits is obviously as follows:

$$\sum_{[\theta;\eta]} r_{\theta;\eta} W_{\theta;\eta}.$$

Since the permutation group $(\Pi^{[\theta,\eta]})$ on $\mathbf{F}^{[\theta;\eta]}$ is homomorphic to $\mathbf{P}$ (Theorem 14.2), the number $(r_{\theta;\eta})$ is obtained by the Cauchy-Frobenius lemma (Theorem 13.4 in Chaper 13), *i.e.*,

$$r_{\theta;\eta} = \frac{1}{|\mathbf{P}|} \sum_{p \in \mathbf{P}} n_{p(\pi^{[\theta;\eta]})}, \tag{14.17}$$

wherein the symbol $p(\pi^{[\theta;\eta]})$ denotes a permutation corresponding to $\pi^{[\theta;\eta]}$ ($\in \Pi^{[\theta,\eta]}$) and the symbol $n_{p(\pi^{[\theta;\eta]})}$ designates the number of invariant configurations of $\mathbf{F}^{[\theta;\eta]}$ on the operation of $\pi^{[\theta;\eta]}$. Since this lemma is true for all p of $\mathbf{P}$, we have

$$\sum_{[\theta;\eta]} r_{\theta;\eta} W_{\theta;\eta} = \sum_{[\theta;\eta]} \left(\frac{1}{|\mathbf{P}|} \sum_{p \in \mathbf{P}} n_{p(\pi^{[\theta;\eta]})} \right) W_{\theta;\eta}$$

$$= \frac{1}{|\mathbf{P}|} \sum_{p \in \mathbf{P}} \sum_{[\theta;\eta]} n_{p(\pi^{[\theta;\eta]})} W_{\theta;\eta}. \tag{14.18}$$

The term

$$\sum_{[\theta;\eta]} n_{p(\pi^{[\theta;\eta]})} W_{\theta;\eta}$$

indicates the total number of functions invariant on the permutation p that corresponds to each $\pi^{[\theta,\eta]}$. The term is hence equal to the equation proved in Lemma 14.1. Therefore,

$$\sum_{[\theta;\eta]} n_{p(\pi^{[\theta;\eta]})} W_{\theta;\eta} = \sum_{[\theta;\eta]} A_{\theta;\eta} \prod_{i=1}^{|\mathbf{X}|} w_1(X_i)^{\theta_i} \prod_{i=1}^{|\mathbf{X}|} w_2(X_i)^{\eta_i}$$

$$= \prod_{d=1}^{u} s_d^{\nu_d} \prod_{d=1}^{v} t_d^{\lambda_d}, \tag{14.19}$$

wherein

$$s_d = \sum_{k=1}^{|\mathbf{X}|} w_1(X_k)^d \tag{14.20}$$

and

$$t_d = \sum_{k=1}^{|\mathbf{X}|} w_2(X_k)^d. \tag{14.21}$$

Since eq. 14.19 can be established for each p corresponding to $\pi^{[\theta,\eta]}$, the introduction of the equation into eq. 14.18 provides the following equation.

$$\sum_{[\theta;\eta]} r_{\theta;\eta} W_{\theta;\eta} = \sum_{[\theta;\eta]} \left(\frac{1}{|\mathbf{P}|} \sum_{p \in \mathbf{P}} n_{p(\pi^{[\theta;\eta]})} \right) W_{\theta;\eta}$$

$$= \frac{1}{|\mathbf{P}|} \sum_{p \in \mathbf{P}} \prod_{d=1}^{u} s_d^{\nu_d} \prod_{d=1}^{v} t_d^{\lambda_d}$$

$$= CI\left(\mathbf{P}; \sum_{k=1}^{|\mathbf{X}|} w_1(X_k)^d, \sum_{k=1}^{|\mathbf{X}|} w_2(X_k)^d \right), \tag{14.22}$$

wherein the latter relationship comes from Def. 14.2. Since the definition provides

$$\sum_{[\theta;\eta]} r_{\theta;\eta} W_{\theta;\eta} = \sum_{[\theta;\eta]} B_{\theta;\eta} \prod_{i=1}^{|\mathbf{X}|} w_1(X_i)^{\theta_i} \prod_{i=1}^{|\mathbf{X}|} w_2(X_i)^{\eta_i}, \tag{14.23}$$

we obtain the equation to be proved.

This theorem is an extension of Theorem 13.8 in two points: (1) the present theorem takes general weights into consideration; and (2) the present one treats a case in which a domain has two orbits.

Example **14.1** Let us examine the 8 carbon atoms of adamantane-2,6-dione (**2**) of $\mathbf{D}_{2d}$ symmetry, where the two carbonyl carbons are not taken into consideration.

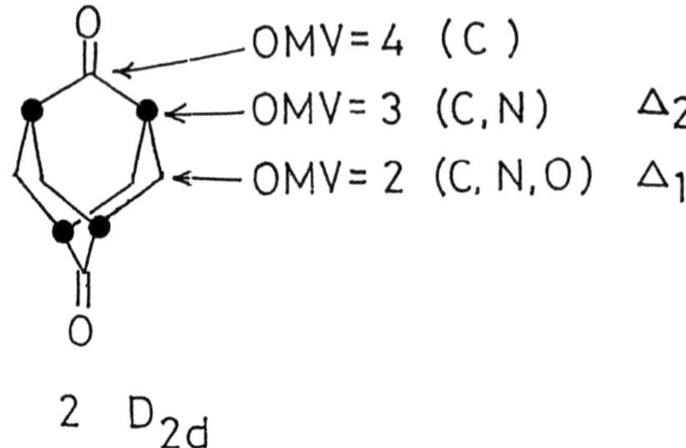

The 8 carbons are divided into two orbits, Δ_1 (4 bridges) and Δ_2 (4 bridge-heads), which are subject to the coset representations, $\mathbf{D}_{2d}(/\mathbf{C}_2')$ and $\mathbf{D}_{2d}(/\mathbf{C}_s)$, respectively. The obligatory minimum valency (OMV) of the bridge positions is equal to 2 and that of the bridgehead positions is equal to 3. Suppose that a set of 8 ligands are selected from the codomain, $\mathbf{X}=\{C, N, O\}$. Then, each bridge position can take C, N, O; but each bridgehead position does not take O. The weights are determined as follows.

$$w_1(C) = 1, w_1(N) = x, w_1(O) = y \text{ for } \Delta_1 \text{ and}$$
$$w_2(C) = 1, w_2(N) = x, w_2(O) = 0 \text{ for } \Delta_2$$

As a result, we obtain two ligand inventories.

$$s_d = 1 + x^d + y^d \text{ for } \Delta_1 \text{ and}$$
$$t_d = 1 + x^d \text{ for } \Delta_2.$$

The concrete form of a permutation group (**P**) on Δ_1 and dummy variables are shown in Table 14.1.   The counterpart for Δ_2 is obtained in the same line. Hence, the cycle index of this case is obtained to be

$$CI(\mathbf{P}; s_d, t_d) = \frac{1}{8}(s_1^4 t_1^4 + 2s_1^2 s_2 t_2^2 + s_2^2 t_2^2 + 2s_2^2 t_1^2 t_2 + 2s_4 t_4).$$

Table 14.1: Cycle structures and variables for 2

D_{2d}		P for Δ_1	cycle structure	dummy variable	P for Δ_2	cycle structure	dummy variable
I	$\sim$	$(1)(2)(3)(4)$	(1^4)	s_1^4	$(1)(2)(3)(4)$	(1^4)	t_1^4
$C_{2(1)}$	$\sim$	$(1)(2)(3\ 4)$	$(1^2 2)$	$s_1^2 s_2$	$(1\ 2)(3\ 4)$	(2^2)	t_2^2
$C_{2(2)}$	$\sim$	$(1\ 2)(3)(4)$	$(1^2 2)$	$s_1^2 s_2$	$(1\ 3)(2\ 4)$	(2^2)	t_2^2
$C_{2(3)}$	$\sim$	$(1\ 2)(3\ 4)$	(2^2)	s_2^2	$(1\ 4)(2\ 3)$	(2^2)	t_2^2
$\sigma_{d(1)}$	$\sim$	$(1\ 3)(2\ 4)$	(2^2)	s_2^2	$(1)(2\ 3)(4)$	$(1^2 2)$	$t_1^2 t_2$
S_4^3	$\sim$	$(1\ 4\ 2\ 3)$	(4)	s_4	$(1\ 3\ 4\ 2)$	(4)	t_4
S_4	$\sim$	$(1\ 3\ 2\ 4)$	(4)	s_4	$(1\ 2\ 4\ 3)$	(4)	t_4
$\sigma_{d(2)}$	$\sim$	$(1\ 4)(2\ 3)$	(2^2)	s_2^2	$(1\ 4)(2)(3)$	$(1^2 2)$	$t_1^2 t_2$

The above ligand inventories are introduced into this cycle index to afford a generating function.

$$
\begin{aligned}
G(x,y) \;=\; & CI(\mathbf{P}; 1 + x^d + y^d,\ 1 + x^d) \\
=\; & \frac{1}{8}((1 + x + y)^4(1 + x)^4 + 2(1 + x + y)^2(1 + x^2 + y^2)(1 + x^2)^2 \\
& + (1 + x^2 + y^2)^2(1 + x^2)^2 + 2(1 + x^2 + y^2)^2(1 + x)^2(1 + x^2) \\
& + 2(1 + x^4 + y^4)(1 + x^4)).
\end{aligned}
$$

When we introduce $x = y = 1$ into this equation, we find $G(1,1) = 213$, which is the total number of derivatives appearing in this enumeration.

14.2 Unit Cycle Indices

In order to manipulate the OMV restriction, we explicitly consider a partition of positions of a given skeleton. We then provide different sets of weights to different orbits of the skeleton. In the preceding section, we have solved enumeration problems in which a domain has two orbits. We here extend these results so as to be applicable to general cases in which such a domain has two or more orbits.

Suppose that $\Delta = \{\delta_1, \delta_2, \ldots, \delta_{|\Delta|}\}$ is a domain which contains vertices of the skeleton and $\mathbf{X} = \{X_1, X_2, \ldots, X_{|\mathbf{X}|}\}$ is a codomain which contains ligands. Let a finite group $\mathbf{G}$ act on Δ in the form of the permutation representation $\mathbf{P_G}$ on Δ. The action of $\mathbf{G}$ on Δ affords a partition to give $\Delta_{i\alpha}$ ($i = 1, 2, \ldots, s$ and $\alpha = 1, 2, \ldots, \alpha_i$), each of which is governed by the coset representation $(\mathbf{G}(/\mathbf{G}_i))$ according to the equation, $\mathbf{P_G} = \Sigma_{i=1}^{s} \alpha_i \mathbf{G}(/\mathbf{G}_i)$ (Theorem 5.7 in Chapter 5).

Note the relationship concerning $\mathbf{G}$, $\mathbf{P}$, and $\mathbf{P_G}$. As shown in the preceding sections, the $\mathbf{P}$ is a permutation group that is homomorphic to $\mathbf{G}$. The $\mathbf{P}$ is regarded as the set of permutations that are selected from the permutation representation $\mathbf{P_G}$ so as to avoid duplication. Similarly, we can consider a permutation group ($\mathbf{P}'$) that is associated with the $\Delta_{i\alpha}$ orbit. The $\mathbf{P}'$ group is a homomorphism of $\mathbf{G}$. Since, the coset representation (CR) $\mathbf{G}(/\mathbf{G}_i)$ governs the $\Delta_{i\alpha}$ orbit, the $\mathbf{P}'$ group is regarded as the set of permutations selected from the CR so as to avoid duplication. Although the results obtained in preceding sections are concerned with such permutation *groups*, they are also applicable to permutation *representations* or coset *representations* after trivial modification.

The symbol $w_{i\alpha}(X_k)$ (for $i = 1, 2, \ldots, s$, $\alpha = 1, 2, \ldots, \alpha_i$ and $k = 1, 2, \ldots, |\mathbf{X}|$) denotes a weight that is assigned to each orbit $\Delta_{i\alpha}$. We then define a weight of a function (configuration) as follows.

Definition 14.3 (Weight of function)

$$W(f) = \prod_{i=1}^{s} \prod_{\substack{\alpha=1 \\ \alpha_i \neq 0}}^{\alpha_i} \prod_{\delta \in \Delta_{i\alpha}} w_{i\alpha}(f(\delta)) \tag{14.24}$$

for a function $f : \Delta \to \mathbf{X}$.

This definition is an extension of Def. 14.1.

In the same line as Theorem 14.1, the weight $W(f_\gamma)$ can be proved to be equal to $W(f_\epsilon)$ if the two functions f_γ and f_ϵ: $\Delta \to \mathbf{X}$ are equivalent to each other.

Each CR, $\mathbf{G}(/\mathbf{G}_i)$, is a transitive permutation representation on $\Delta_{i\alpha}$. We can separately treat every orbit governed by such a CR as shown in the preceding section. Suppose that $G(/G_i)_g \in \mathbf{G}(/\mathbf{G}_i)$ corresponds to $g \in \mathbf{G}$. Suppose that the cycle structure of the $G(/G_i)_g$ permutation is represented by

$$(\nu) : (1^{\nu_1} 2^{\nu_2} \cdots m^{\nu_m}) \tag{14.25}$$

where $m = |\Delta_{i\alpha}| = |\mathbf{G}| / |\mathbf{G}_i|$ and

$$1\nu_1 + 2\nu_2 + \cdots + m\nu_m = m.$$

We then assign a variable s_d to each cycle of length d. Thereby, we define a *unit cycle index (UCI)* as follows.[2]

Definition 14.4 *A unit cycle index (UCI) for $G(/G_i)_g$ is represented by*

$$z_g^{(i\alpha)} = \left(\prod_{d=1}^{|\mathbf{X}|} s_d^{\nu_d} \right)_{(g)}^{(i\alpha)}, \tag{14.26}$$

where the superscript (iα) is concerned with the orbit $\Delta_{i\alpha}$ on which $\mathbf{G}(/\mathbf{G}_i)$ acts and the subscript (g) represents the correspondence to $g \in \mathbf{G}$.

Since every CR is determined algebraically through a coset decomposition of G by $\mathbf{G}_i$ (Chapter 5), we can precalculate UCIs for the elements of the CR. Table 14.2 lists all of the CRs for $\mathbf{C}_{2v}$ group, which are obtained by the corresponding

Table 14.2: Coset Representaions for $\mathbf{C}_{2v}$ group

Operation	$\mathbf{C}_{2v}(/\mathbf{C}_1)$	$\mathbf{C}_{2v}(/\mathbf{C}_2)$	$\mathbf{C}_{2v}(/\mathbf{C}_s)$	$\mathbf{C}_{2v}(/\mathbf{C}_s')$	$\mathbf{C}_{2v}(/\mathbf{C}_{2v})$
I	(1)(2)(3)(4)	(1)(2)	(1)(2)	(1)(2)	(1)
C_2	(1 2)(3 4)	(1)(2)	(1 2)	(1 2)	(1)
$\sigma_{v(1)}$	(1 3)(2 4)	(1 2)	(1)(2)	(1 2)	(1)
$\sigma_{v(2)}$	(1 4)(2 3)	(1 2)	(1 2)	(1)(2)	(1)

Table 14.3: Unit cycle indices for $\mathbf{C}_{2v}$ group

Operation	$\mathbf{C}_{2v}(/\mathbf{C}_1)$	$\mathbf{C}_{2v}(/\mathbf{C}_2)$	$\mathbf{C}_{2v}(/\mathbf{C}_s)$	$\mathbf{C}_{2v}(/\mathbf{C}_s')$	$\mathbf{C}_{2v}(/\mathbf{C}_{2v})$
I	s_1^4	s_1^2	s_1^2	s_1^2	s_1
C_2	s_2^2	s_1^2	s_2	s_2	s_1
$\sigma_{v(1)}$	s_2^2	s_2	s_1^2	s_2	s_1
$\sigma_{v(2)}$	s_2^2	s_2	s_2	s_1^2	s_1

coset decompositions. Table 14.3 collects unit cycle indices for the $\mathbf{C}_{2v}$ symmetry, which are easily obtained by the data of Table 14.2.

By using the UCIs ($z_g^{(i\alpha)}$), we obtain the definition of a cycle index.

Definition 14.5 (Cycle index based on UCIs)

$$CI(\mathbf{P_G}; s_d^{(i\alpha)}) = \frac{1}{|\mathbf{G}|} \sum_{g \in \mathbf{G}} \prod_{i=1}^{s} \prod_{\substack{\alpha=1 \\ \alpha_i \neq 0}}^{\alpha_i} z_g^{(i\alpha)}. \tag{14.27}$$

This definition is essentially equivalent to Pólya's cycle index except that it is based on the UCIs.

We can obtain the following theorem that is a generalization of Pólya's theorem. The proof reported elsewhere[7] is a simple extention of Theorem 14.3.

Theorem 14.4 *The number (B_θ) of configurations with a weight W_θ is obtained by means of a generating function:*

$$\sum_{[\theta]} B_\theta W_\theta = CI(\mathbf{P_G}; s_d^{(i\alpha)}), \tag{14.28}$$

where the corresponding ligand-inventories are represented by

$$s_d^{(i\alpha)} = \sum_{k=1}^{|\mathbf{X}|} w_{i\alpha}(X_k)^d. \tag{14.29}$$

---

[2]Do not confuse the term UCI with the term USCI (unit subduced cycle index). See Chapter 9.

A procedure of the enumeration in terms of Theorem 14.4 is summarized as follows:

1. determination of the symmetry **G** of a parent skeleton,

2. the counting of fixed points (marks) in the **G**-set of the skeleton on each operation of the symmetry **G** (Chapter 5),

3. determination of orbits of the **G**-set and assignment of the corresponding CRs (Chapter 5),

4. citation of unit cycle indices (UCIs) from a table of UCIs (*e.g.*, Table 14.3) and construction of a cycle index in accord with Definition 14.4, and

5. introduction of ligand-inventories to the variables of the cycle index (Theorem 14.4).

Advantages of the present procedure stem mainly from step 4, because of precalculated UCIs. This comes from the fact that they are independent of any particular **G**-set but dependent only upon the CRs. The full tabulation of UCIs for all point groups is a reasonable task and will be reported elsewhere.

The following example illustrates this procedure using precalculated UCIs.[7]

Example **14.2** The nine positions of noradamantane (**3**) are divided into four orbits, $\Delta_1 = \{4,5,6,7\}$, $\Delta_2 = \{2,3\}$, $\Delta_3 = \{8,9\}$, and $\Delta_4 = \{1\}$ (See Chapter 6).

The corresponding CRs are given by the equation,

$$\mathbf{P}_{\mathbf{C}_{2v}} = \mathbf{C}_{2v}(/\mathbf{C}_1) + \mathbf{C}_{2v}(/\mathbf{C}_s) + \mathbf{C}_{2v}(/\mathbf{C}'_s) + \mathbf{C}_{2v}(/\mathbf{C}_{2v}).$$

If Def. 14.5 is applied to this case, we obtain the following cycle index by using the precalculated UCIs (Table 14.2).

$$CI(\mathbf{P}_{\mathbf{C}_{2v}}; s_d^{(1)}, s_d^{(2)}, s_d^{(3)}, s_d^{(4)})$$

$$= (1/4)\Big((s_1^4)^{(1)}(s_1^2)^{(2)}(s_1^2)^{(3)}(s_1)^{(4)} + (s_2^2)^{(1)}(s_2)^{(2)}(s_2)^{(3)}(s_1)^{(4)}$$

$$+ (s_2^2)^{(1)}(s_1^2)^{(2)}(s_2)^{(3)}(s_1)^{(4)} + (s_1^4)^{(1)}(s_2)^{(2)}(s_1^2)^{(3)}(s_1)^{(4)}\Big). \qquad (14.30)$$

The superscripts (1) to (4) denote the correspondence to Δ_1 to Δ_4. When we select a codomain: $\mathbf{X} = \{C,N,O\}$, Δ_1 and Δ_4 can take C, N, and O because each OMV is equal to 2. On the other hand, Δ_2 and Δ_3 take C and N because of their OMVs $= 3$. Hence, we adopt the weights:

$$\begin{aligned}
w_1(C) &= 1, w_1(N) = x, w_1(O) = y, \quad \text{for } \Delta_1, \\
w_2(C) &= 1, w_2(N) = x, w_2(O) = 0, \quad \text{for } \Delta_2, \\
w_3(C) &= 1, w_3(N) = x, w_3(O) = 0, \quad \text{for } \Delta_3, \quad \text{and} \\
w_4(C) &= 1, w_4(N) = x, w_4(O) = y, \quad \text{for } \Delta_4.
\end{aligned}$$

Thereby, we determine ligand-inventories:

$$\begin{aligned}
s_d^{(1)} &= 1 + x^d + y^d \text{ for } \Delta_1, \\
s_d^{(2)} &= 1 + x^d \text{ for } \Delta_2 \\
s_d^{(3)} &= 1 + x^d \text{ for } \Delta_3, \quad \text{and} \\
s_d^{(4)} &= 1 + x^d + y^d \text{ for } \Delta_4.
\end{aligned}$$

These ligand-inventories are introduced into the cycle index (eq. 14.30). Theorem 14.4 provides a generating function:

$$\begin{aligned}
\sum_{[\theta]} B_\theta W_\theta &= CI(\mathbf{P_{C_{2v}}}; 1 + x^d + y^d, 1 + x^d, 1 + x^d, 1 + x^d + y^d) \\
&= (1/4)((1 + x + y)^5(1 + x)^4 + (1 + x^2 + y^2)(1 + x + y)(1 + x^2)^2 \\
&\quad + 2(1 + x^2 + y^2)^2(1 + x + y)(1 + x)^2(1 + x^2)^2 \\
&\quad + 2(1 + x^2 + y^2)^2(1 + x + y)(1 + x)^2(1 + x^2)^2 \\
&= 1 + 4x + 2y + 13x^2 + 11xy + 4y^2 + 27x^3 + 38x^2y + 21xy^2 + 4y^3 \\
&\quad + 39x^4 + 73x^3y + 59x^2y^2 + 17xy^3 + 2y^4 \\
&\quad + 39x^5 + 92x^4y + 96x^3y^2 + 42x^2y^3 + 8xy^4 + y^5 \\
&\quad + 27x^6 + 73x^5y + 96x^4y^2 + 54x^3y^3 + 15x^2y^4 + 2xy^5 \\
&\quad + 13x^7 + 38x^6y + 59x^5y^2 + 42x^4y^3 + 15x^3y^4 + 3x^2y^5 \\
&\quad + 4x^8 + 11x^7y + 21x^6y^2 + 17x^5y^3 + 8x^4y^4 + 2x^3y^5 \\
&\quad + x^9 + 2x^8y + 4x^7y^2 + 4x^6y^3 + 2x^5y^4 + x^4y^5.
\end{aligned} \tag{14.31}$$

The coefficient of $x^{\theta_1}y^{\theta_2}$ indicates the number of isomers with θ_1 N and θ_2 O. For example, the coefficient of x^2 is equal to 13, which shows that there are 13 derivatives with C_7N_2. On the other hand, there are only 4 derivatives with C_7O_2 in accord with the term $4y^2$. Figure 14.1 depicts these derivatives.

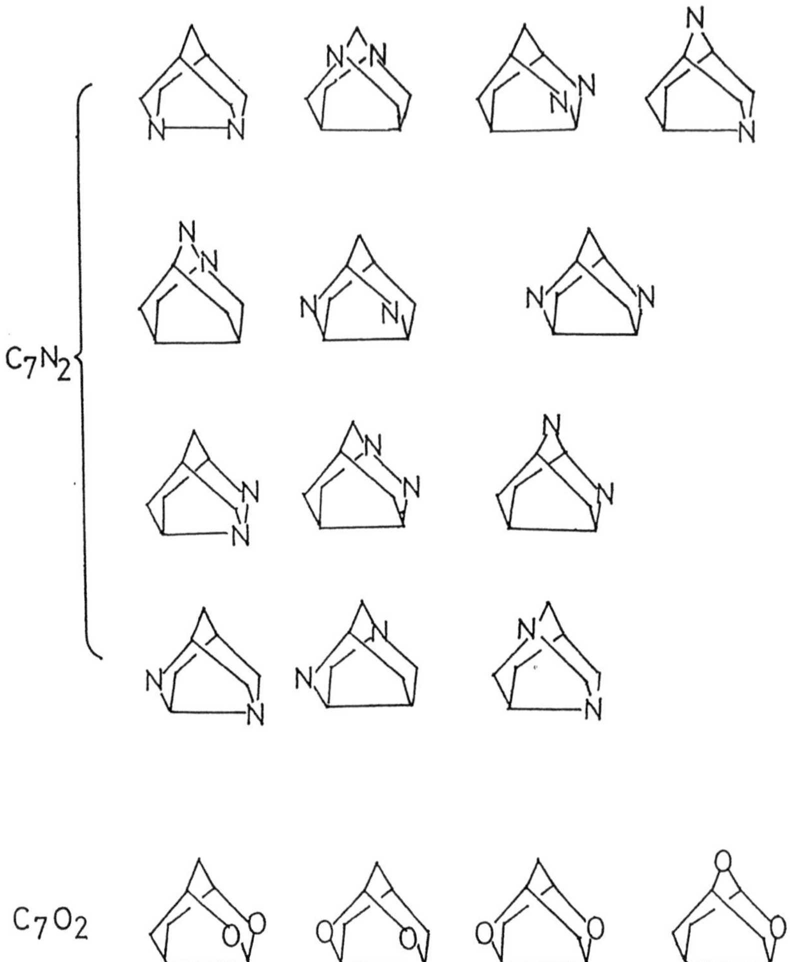

Figure 14.1: Derivatives with C_7N_2 and C_7O_2 based on **3**

Bibliography

[1] D. H. McDaniel, *Inorg. Chem.*, **11**, 2678 (1972).

[2] G. Pólya, R. C. Read, *Combinatorial Enumeration of Groups, Graphs, and Chemical Compounds*, Springer-Verlag, Berlin Heidelberg New York, pp. 125–126 (1987).

[3] J. B. Hendrickson, *Angew. Chem. Intern. Ed.*, **13**, 47 (1974).

[4] a) S. Fujita, *J. Chem. Inf. Comput. Sci.*, **26**, 212 (1986). b) S. Fujita, *J. Chem. Inf. Comput. Sci.*, **26**, 231 (1986). c) S. Fujita, *J. Chem. Inf. Comput. Sci.*, **26**, 238 (1986). d) S. Fujita, *J. Chem. Inf. Comput. Sci.*, **27**, 111 (1987). e) S. Fujita, *J. Chem. Inf. Comput. Sci.*, **27**, 115 (1987). f) S. Fujita, *J. Chem. Inf. Comput. Sci.*, **27**, 120 (1987). g) S. Fujita, *J. Chem. Inf. Comput. Sci.*, **28**, 1 (1988). h) S. Fujita, *J. Chem. Inf. Comput. Sci.*, **28**, 78 (1988). i) S. Fujita, *J. Chem. Inf. Comput. Sci.*, **28**, 128 (1988). j) S. Fujita, *J. Chem. Inf. Comput. Sci.*, **28**, 137 (1988). k) S. Fujita, *J. Chem. Soc. Perkin II*, 597 (1988). l) S. Fujita, *Pure Appl. Chem.*, **61**, 605 (1989). m) S. Fujita, *Yuki Gosei Kagaku Kyokaishi*, **47**, 396 (1989). n) S. Fujita, *J. Chem. Educ.*, **67**, 290 (1990).

[5] a) S. Fujita, *J. Chem. Inf. Comput. Sci.*, **26**, 212 (1986). b) S. Fujita, *J. Chem. Inf. Comput. Sci.*, **26**, 224 (1986). c) S. Fujita, *J. Chem. Inf. Comput. Sci.*, **27**, 99 (1987). d) S. Fujita, *J. Chem. Inf. Comput. Sci.*, **27**, 104 (1987).

[6] a) S. Fujita, *Bull. Chem. Soc. Jpn.*, **61**, 4189 (1988). b) S. Fujita, *J. Chem. Inf. Comput. Sci.*, **29**, 22 (1989). c) S. Fujita, *Bull. Chem. Soc. Jpn.*, **62**, 662 (1989).

[7] S. Fujita, *J. Math. Chem.*, **5**, 99 (1990).

Chapter 15

Compounds with Achiral Ligands Only [1]

Enumeration of compounds in chemistry is one of the most important fields to which the Pólya-Redfield theorem[1] has been applied. Ruch *et al.*[2] and later Brocas[3] have proved the concept of double coset to be very convenient for such compound-counting problems. Sheehan[4] has applied the concept of table of marks (see Chapter 5) to enumeration of graphs under an automorphic group. More recently, Hässelbarth[5] has developed a method that is also based on the concept of table of marks. This method is capable of enumerating compounds with respect to a given symmetry. Mead[6] has discussed an alternative method that is a combination of table of marks and double cosets. These methods, however, have taken no account of intransitivity of a domain explicitly; hence, they have disregarded the restriction due to OMVs (obligatory minimum valencies).[2]

The present chapter presents enumeration of compounds with a given symmetry and with a given weight, in which the transitivity of a domain is strictly taken into consideration. The first issue is to introduce a method that uses precalculated numbers of suborbits, which are derived by the subduction of coset representations (Chapter 9). The second issue is the counting of chemical structures with a given symmetry as well as a given weight. The method is also based on the concepts derived from subduction of coset representations (Chapter 9), especially on unit subduced cycle index (USCI) and a subduced cycle index (SCI).

15.1 Compounds with Given Symmetries

Let Δ be a domain that consists of $|\Delta|$ elements:

$$\Delta = \{\delta_1, \delta_2, \ldots, \delta_{|\Delta|}\}.$$

The elements of Δ are regarded as positions contained in a given skeleton of **G** symmetry. Let $\mathbf{P_G}$ be a permutation representation (PR) of **G** acting on Δ.

---

[1] This chapter is based on the article published in S. Fujita, *Theor. Chim. Acta*, **76**, 247–268 (1989).
[2] See Chapter 14.

Theorem 5.7 (Chapter 5) teaches us the the PR is represented by a sum of coset representations (CRs), *i.e.* $\mathbf{P_G} = \Sigma_{i=1}^{s} \alpha_i \mathbf{G}(/\mathbf{G}_i)$. As a result, the Δ domain is partitioned into the following orbits.

$$\Delta_{i1}, \Delta_{i2}, \ldots, \Delta_{i\alpha_i},$$

each of which is subject to $\mathbf{G}(/\mathbf{G}_i)$ $(i = 1, 2, \ldots, s)$.

Let $\mathbf{X}$ be a codomain (or range) that contains $|\mathbf{X}|$ elements called *ligands* (figures):

$$\mathbf{X} = \{X_1, X_2, \ldots, X_{|\mathbf{X}|}\}.$$

Let f be a function from the domain Δ to the codomain $(\mathbf{X})$, *i.e.*,

$$f : \Delta \to \mathbf{X} \tag{15.1}$$

wherein the functions are called *configurations*. Theorem 13.5 (Chapter 13) can be rewritten to meet the present case.

Theorem 15.1 *Consider a set of the configurations,*

$$\mathbf{F} = \{f_1, f_2, \ldots, f_\alpha, \ldots, f_\beta, \ldots, f_{|\mathbf{F}|}\}. \tag{15.2}$$

Then, any p_g belonging to $\mathbf{P_G}$ is associated with π_g represented by

$$\pi_g = \begin{pmatrix} f_1 & f_2 & \cdots & f_{|\mathbf{F}|} \\ f_1 p_g & f_2 p_g & \cdots & f_{|\mathbf{F}|} p_g \end{pmatrix}. \tag{15.3}$$

(1) This is a permutation on $\mathbf{F}$.

(2) Let $\Pi_\mathbf{G}$ be the set of all π_g $(p_g \in \mathbf{P_G})$. Then, $\Pi_\mathbf{G}$ is a homomorphism of $\mathbf{P_G}$.

(3) Moreover, $\mathbf{P_G} \cong \Pi_\mathbf{G}$ (i.e. isomorphic).

Proof. Abbreviated.

The $\Pi_\mathbf{G}$ is a permutation representation of $\mathbf{G}$. This formulation enables us to apply Theorem 5.7 (Chapter 5) to the present case.

Theorem 15.2 *Let a group $\mathbf{G}$ act on $\mathbf{F}$ via a permutation representation $\Pi_\mathbf{G}$. Then, we have*

$$\Pi_\mathbf{G} = \sum_{i=1}^{s} A_i \mathbf{G}(/\mathbf{G}_i), \tag{15.4}$$

wherein the multiplicity (A_i) of each coset representation is a non-negative integer and obtained by solving

$$\rho_j = \sum_{i=1}^{s} A_i m_{ij} \quad (for \ j = 1, 2, \ldots, s), \tag{15.5}$$

where ρ_j is the mark of $\mathbf{G}_j$ in $\Pi_\mathbf{G}$.

Although the marks (the fixed points) in the Theorem 5.7 (Chapter 5) are easily traceable because of the concrete nature of positions, the mark ρ_j is somewhat difficult to understand due to the abstract nature of functions or configurations. Equation 15.4 provides a partition of $\mathbf{F}$ by $\Pi_{\mathbf{G}}$ to create the corresponding orbits of $\mathbf{F}$. Suppose that one of the orbits governed by $\mathbf{G}(/\mathbf{G}_i)$ is represented by

$$\mathbf{F}_i = \{f_1^{(i)}, f_2^{(i)}, \ldots, f_m^{(i)}\}, \tag{15.6}$$

where $m = |\,\mathbf{G}\,|\,/\,|\,\mathbf{G}_i\,|$. The definition of $\mathbf{G}(/\mathbf{G}_i)$ indicates that the representation $\mathbf{G}(/\mathbf{G}_i)$ is associated with $\mathbf{F}_i$ in a transitive fashion. Since $\mathbf{G}(/\mathbf{G}_i)$ is a coset representation of $\mathbf{G}$ by $\mathbf{G}_i$, the subgroup $\mathbf{G}_i$ is a stabilizer on each function of $\mathbf{F}_i$. This fact indicates that there is an appropriate $f_k^{(i)}$ ($\in \mathbf{F}_i$) which is a fixed configuration on the action by $\mathbf{G}_i$. In other words, the fixed configuration $f_k^{(i)}$ has a symmetry of $\mathbf{G}_i$. It is to be noted that $f_k^{(i)}$ is invariant on the action of $\mathbf{G}_i$ (or one of its conjugate subgroups) but variant on that of $\mathbf{G}(/\mathbf{G}_i)$. Hence, all configurations of $\mathbf{F}_i$ are homomers that have the same symmetry $\mathbf{G}_i$; this is a chemical meaning corresponding to the fact that the $\mathbf{F}_i$ is an orbit of configurations. As a result, A_i also indicates the number of different (non-equivalent) configurations of symmetry $\mathbf{G}_i$. If we apply the discussions described in Chapter 7, The $\mathbf{G}_i$ is the local symmetry of each function in an abstract fashion. We summarize the discussion in the form of a corollary.

Corollary 15.1 *Let a group* $\mathbf{G}$ *act on* $\mathbf{F}$ *by acting on* Δ. *Then the number* (A_i) *of non-equivalent configurations of symmetry* $\mathbf{G}_i$ *($\leq \mathbf{G}$) constitutes the solution of the system of linear equations expressed by eq. 15.5.*

The above discussion implies that a set of such ρ_j's is considered as a fixed point vector (FPV), *i.e.*, FPV $= (\rho_1 \;\; \rho_2 \;\; \ldots \;\; \rho_s)$. This notation allows us to convert eq. 15.5 into matrix representaitons,

$$(\rho_1\, \rho_2\, \ldots\, \rho_s) \;=\; (A_1\, A_2\, \ldots\, A_s)M \tag{15.7}$$
$$\text{and}$$
$$(A_1\, A_2\, \ldots\, A_s) \;=\; (\rho_1\, \rho_2\, \ldots\, \rho_s)M^{-1}, \tag{15.8}$$

where M is the mark table and M^{-1} is its inverse.

The next important problem is the evaluation of ρ_j of eq. 15.5. The mark (ρ_j) of $\mathbf{G}_j$ in $\Pi_{\mathbf{G}}$ is the number of fixed functions (or fixed configurations) of $\mathbf{F}$ on the action of $\mathbf{G}_j$. In order for an appropriate $f^{(j)}$ ($\in \mathbf{F}$) to be a fixed configuration with respect to $\forall g \in \mathbf{G}_j$, the following equation is necessary to hold.

$$f^{(j)}(p_g(\delta)) = f^{(j)}(\delta) \quad (\text{for } \forall \delta \in \Delta \text{ and } \forall g \in \mathbf{G}_j) \tag{15.9}$$

In other words, $f^{(j)}$ has to be constant on the action of the $\mathbf{G}_j$ group. Because Δ is divided into orbits, $\Delta_{i1}, \Delta_{i2}, \ldots, \Delta_{i\alpha_i}$ ($i = 1, 2, \ldots, s$), by the action of $\mathbf{G}$ and because each of the orbits is subdivided by the action of $\mathbf{G}(/\mathbf{G}_i)\!\downarrow\!\mathbf{G}_j$ (Chapter 9), the number (β_{ij}) of suborbits $(\Delta_{jk\beta}^{(i\alpha)})$ for each $\Delta_{i\alpha}$ is given by Lemma 9.2 (Chapter 9). In order for $\exists f^{(j)}$ ($\in \mathbf{F}$) to be constant, all of the $\mid \Delta_{jk\beta}^{(i\alpha)} \mid$ positions of each suborbit have to take ligands of the same kind. This means that there are $\mid \mathbf{X}_{i\alpha} \mid$ ways of substitution for each suborbit subduced from $\Delta_{i\alpha}$, where

$$\mid \mathbf{X}_{i\alpha} \mid = \text{ no. of non-zero } w_{i\alpha}(X_k) \text{ for each } \Delta_{i\alpha}. \tag{15.10}$$

The $w_{i\alpha}(X_k)$ weight is determined so as to agree with the OMV of the $\Delta_{i\alpha}$ orbit (See Chapter 14). Since the number of the suborbits derived from $\Delta_{i\alpha}$ is the β_{ij} that appears in the subdivision by $\mathbf{G}(/\mathbf{G}_i)\!\downarrow\!\mathbf{G}_j$ (Lemma 9.2), the term,

$$\mid \mathbf{X}_{i\alpha} \mid^{\beta_{ij}}, \tag{15.11}$$

is the number of fixed configurations corresponding to $\Delta_{i\alpha}$. Finally, the total number of fixed configurations is obtained by collecting the terms (eq. 15.11) through all of the orbits, $\Delta_{i\alpha}$ ($\alpha = 1, 2, \ldots, \alpha_i;\ i = 1, 2, \ldots, s$).

Lemma 15.1 *The total number of fixed configurations on the operations of* $\mathbf{G}_j$ *is represented by*

$$\rho_j = \prod_{i=1}^{s} \prod_{\substack{\alpha=1 \\ \alpha_i \neq 0}}^{\alpha_i} \mid \mathbf{X}_{i\alpha} \mid^{\beta_{ij}} \quad (j = 1, 2, \ldots, s). \tag{15.12}$$

This lemma provides the FPV $= (\rho_1\ \rho_2\ \cdots\ \rho_s)$. This FPV is introduced into eq. 15.7 or eq. 15.8, producing the numbers of non-equivalent configurations as a row vector, $(A_1\ A_2\ \cdots\ A_s)$. Alternatively, this result is given as s equations. Thus, eqs. 15.5 and 15.12 lead to the following theorem.

Theorem 15.3 *Let a group* $\mathbf{G}$ *act on* $\mathbf{F}$ *by acting on* Δ. *Then, the number* (A_i) *of orbits (non-equivalent configurations) with symmetry* $\mathbf{G}_i$ *is obtained by the following equations.*

$$\prod_{i=1}^{s} \prod_{\substack{\alpha=1 \\ \alpha_i \neq 0}}^{\alpha_i} \mid \mathbf{X}_{i\alpha} \mid^{\beta_{ij}} = \sum_{i=1}^{s} A_i m_{ij} \quad (j = 1, 2, \ldots, s). \tag{15.13}$$

This theorem takes account of OMVs, because $\mid \mathbf{X}_{i\alpha} \mid$ can be selected in agreement with the OMV of each $\Delta_{i\alpha}$ orbit. The number (β_{ij}) of suborbits for each $\Delta_{i\alpha}$ given by Lemma 9.2 (Chapter 9) is equal to the sum of the powers appearing in the corresponding unit subduced cycle index (USCI). Hence, they can be preculculated from the data of USCI tables, which are found in Appendix C. For example, Table 15.1 is obtained for $\mathbf{D}_{2d}$ group from the USCI table (Table 9.3 of Chapter 9 and Appendix C).

In the preceding discussions, we have taken account of OMVs. There are, however, many cases in which the OMVs are unnecessary to consider. These cases

Table 15.1: The numbers of suborbits (β_{ij}'s) for $\mathbf{D}_{2d}$

	$\downarrow \mathbf{C}_1$	$\downarrow \mathbf{C}_2$	$\downarrow \mathbf{C}_2'$	$\downarrow \mathbf{C}_s$	$\downarrow \mathbf{S}_4$	$\downarrow \mathbf{C}_{2v}$	$\downarrow \mathbf{D}_2$	$\downarrow \mathbf{D}_{2d}$
$\mathbf{D}_{2d}(/\mathbf{C}_1)$	8	4	4	4	2	2	2	1
$\mathbf{D}_{2d}(/\mathbf{C}_2)$	4	4	2	2	2	2	2	1
$\mathbf{D}_{2d}(/\mathbf{C}_2')$	4	2	3	2	1	1	2	1
$\mathbf{D}_{2d}(/\mathbf{C}_s)$	4	2	2	3	1	2	1	1
$\mathbf{D}_{2d}(/\mathbf{S}_4)$	2	2	1	1	2	1	1	1
$\mathbf{D}_{2d}(/\mathbf{C}_{2v})$	2	2	1	2	1	2	1	1
$\mathbf{D}_{2d}(/\mathbf{D}_2)$	2	2	2	1	1	1	2	1
$\mathbf{D}_{2d}(/\mathbf{D}_{2d})$	1	1	1	1	1	1	1	1

are formulated here as special cases of the treatment described above. Suppose
that all of the weights are defined by the following equations due to this situation,

$$w_{i\alpha}(X_k) = X_k \tag{15.14}$$

for $i = 1, 2, \ldots, s$; $\alpha = 1, 2, \ldots, \alpha_i$; and $k = 1, 2, \ldots, |\mathbf{X}|$. It follows that

$$|\mathbf{X}_{i\alpha}| = |\mathbf{X}|. \tag{15.15}$$

Hence, the left-hand side of eq. 15.13 is transformed as follows.

$$\prod_{i=1}^{s} \prod_{\substack{\alpha=1 \\ \alpha_i \neq 0}}^{\alpha_i} |\mathbf{X}_{i\alpha}|^{\beta_{ij}} = \prod_{i=1}^{s} |\mathbf{X}|^{\alpha_i\beta_{ij}} = |\mathbf{X}|^{\sum_{i=1}^{s}\alpha_i\beta_{ij}}.$$

By introducing this into eq. 15.13, we end up with the following corollary.

Corollary 15.2 *Let a group of* $\mathbf{G}$ *act on* $\mathbf{F}$ *by acting on* Δ. *Then the number*
(A_i) of orbits with symmetry $\mathbf{G}_i$ *constitutes the solution of linear equations,*

$$|\mathbf{X}|^{\sum_{i=1}^{s}\alpha_i\beta_{ij}} = \sum_{i=1}^{s} A_i m_{ij} \quad (j = 1, 2, ..., s). \tag{15.16}$$

Equation 15.16 contains the number of suborbits as a concrete power, *i.e.*, $\sum_{i=1}^{s}\alpha_i\beta_{ij}$.
This can be calculated by means of a table such as Table 15.1, in which the β_{ij}
values are listed. It should be emphasized that the introduction of subduced representations has afforded a more detailed feature to the present approach.

The following example is an application of Theorems 15.2 and 15.3.

Example **15.1** We have already examine the allene skeleton (**1**) in Chapter 13 (Section 13.2). This skeleton has $\mathbf{D}_{2d}$ symmetry.

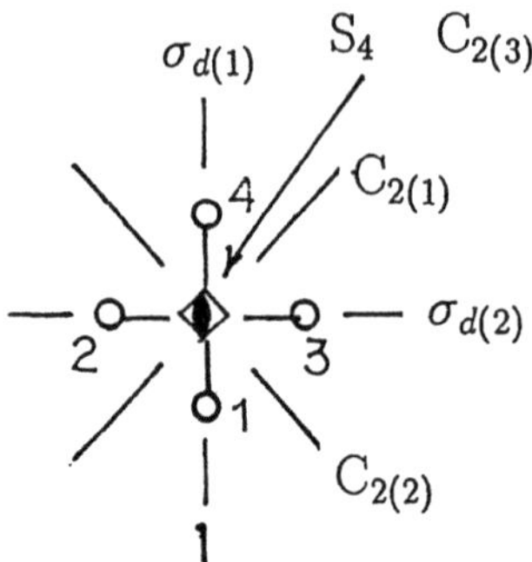

Suppose again that an appropriate set of ligands $(X^{\theta_1}Y^{\theta_2}; \theta_1 + \theta_2 = 4)$ are substituted for the four positions of **1**. All of the 16 configurations (f_1 to f_{16}) are depicted in Fig. 13.2. The set of the configurations is denoted as $\mathbf{F} = \{f_1, f_2, \ldots, f_{16}\}$. In the treatment of Chapter 13, we have counted fixed configurations with respect to each permutation, which are checked in Table 13.1. In the present treatment, we count fixed configurations with respect to each subgroups. Such subgroups are as follows:

$$\begin{aligned}
\mathbf{G}_1 &= \quad \mathbf{C}_1 = \{I\}, & \mathbf{G}_2 &= \mathbf{C}_2 = \{I, C_{2(3)}\}, \\
\mathbf{G}_3 &= \quad \mathbf{C}_2' = \{I, C_{2(1)}\}, & \mathbf{G}_4 &= \mathbf{C}_s = \{I, \sigma_{d(1)}\}, \\
\mathbf{G}_5 &= \quad \mathbf{S}_4 = \{I, C_{2(3)}, S_4^3, S_4\}, & \mathbf{G}_6 &= \mathbf{C}_{2v} = \{I, C_{2(3)}, \sigma_{d(1)}, \sigma_{d(2)}\}, \\
\mathbf{G}_7 &= \quad \mathbf{D}_2 = \{I, C_{2(1)}, C_{2(2)}C_{2(3)}\}, \\
&\quad \text{and} \\
\mathbf{G}_8 &= \quad \mathbf{D}_{2d} = \{I, C_{2(1)}, C_{2(2)}C_{2(3)}, \sigma_{d(1)}, S_4^3, S_4, \sigma_{d(2)}\}.
\end{aligned}$$

If we consider a domain $\mathbf{X} = \{X, Y\}$, we have $|\mathbf{X}| = 2$. Because the skeleton (**1**) has one orbit that is subject to $\mathbf{D}_{2d}(/\mathbf{C}_s)$, we use the values appearing the $\mathbf{D}_{2d}(/\mathbf{C}_s)$ row of Table 15.1 as the numbers of suborbits. Thereby, Lemma 15.1 afffords

$$\begin{aligned}
\rho_1 &= 2^4 = 16, \quad \rho_2 = 2^2 = 4, \quad \rho_3 = 2^2 = 4, \\
\rho_4 &= 2^3 = 8, \quad \rho_5 = 2^1 = 2, \quad \rho_6 = 2^2 = 4, \\
&\rho_7 = 2^1 = 2, \quad \text{and} \quad \rho_8 = 2^1 = 2,
\end{aligned}$$

where the power of each central term comes from the $\mathbf{D}_{2d}(/\mathbf{C}_s)$ row of Table 15.1. Hence, we have an FPV = (16 4 4 8 2 4 2 2), the elements of which are aligned in the order of $\mathbf{G}_1$ to $\mathbf{G}_8$.

Let us verify these values by examining the configurations depicted in Fig. 13.2 (Chapter 13). We calculate the number of fixed configurations on the basis of Table 13.1 (Chapter 13). Thus, we obtain Table 15.2, in which each configuration checked by $\sqrt{}$ is fixed with respect to each subgroup. The checked configurations

Table 15.2: Fixed configurations of F

$\mathbf{D}_{2d}$	f_1	f_2	f_3	f_4	f_5	f_6	f_7	f_8	f_9	f_{10}	f_{11}	f_{12}	f_{13}	f_{14}	f_{15}	f_{16}	FPs
											configuration						
$\mathbf{C}_1$	$\sqrt{}$	$\sqrt{}$	$\sqrt{}$	$\sqrt{}$	$\sqrt{}$	$\sqrt{}$	$\sqrt{}$	$\sqrt{}$	$\sqrt{}$	$\sqrt{}$	$\sqrt{}$	$\sqrt{}$	$\sqrt{}$	$\sqrt{}$	$\sqrt{}$	$\sqrt{}$	16
$\mathbf{C}_2$	$\sqrt{}$									$\sqrt{}$	$\sqrt{}$					$\sqrt{}$	4
$\mathbf{C}_2'$	$\sqrt{}$					$\sqrt{}$	$\sqrt{}$									$\sqrt{}$	4
$\mathbf{C}_s$	$\sqrt{}$	$\sqrt{}$	$\sqrt{}$							$\sqrt{}$	$\sqrt{}$			$\sqrt{}$	$\sqrt{}$	$\sqrt{}$	8
$\mathbf{S}_4$	$\sqrt{}$															$\sqrt{}$	2
$\mathbf{C}_{2v}$	$\sqrt{}$									$\sqrt{}$	$\sqrt{}$					$\sqrt{}$	4
$\mathbf{D}_2$	$\sqrt{}$															$\sqrt{}$	2
$\mathbf{D}_{2d}$	$\sqrt{}$															$\sqrt{}$	2
formula	X^4	X^3Y^1				X^2Y^2						X^1Y^3				Y^4	

are counted for each row and the resulting number is collected in the rightmost column of Table 15.2. These values are equal to those listed in the above FPV.

When the FPV is multiplied by the inverse of a mark table (M^{-1}) for $\mathbf{D}_{2d}$, we have

$$(16\ 4\ 4\ 8\ 2\ 4\ 2\ 2)M^{-1} = (0\ 0\ 1\ 2\ 0\ 1\ 0\ 2),$$

where the inverse M^{-1} is shown in Table 5.4 (Chapter 5) and in Appendix B. This is an embodiment of Theorem 15.3 in a matrix representation. The resulting vector indicates that there appear one $\mathbf{C}_2'$-, two $\mathbf{C}_s$-, one $\mathbf{C}_{2v}$-, and two $\mathbf{D}_{2d}$-derivatives. These derivatives obviously result in the following correspondences:

$$\begin{array}{lll}
\text{one } \mathbf{C}_2'\text{-derivative} & \longleftrightarrow & \{f_6,f_7,f_8,f_9\} \\
\text{two } \mathbf{C}_s\text{-derivatives} & \longleftrightarrow & \{f_2,f_3,f_4,f_5\} \text{ and } \{f_{12},f_{13},f_{14},f_{15}\} \\
\text{one } \mathbf{C}_{2v}\text{-derivative} & \longleftrightarrow & \{f_{10},f_{11}\} \\
\text{two } \mathbf{D}_{2d}\text{-derivatives} & \longleftrightarrow & \{f_1\} \text{ and } \{f_{16}\}.
\end{array}$$

These correspondences are verified by Fig. 13.2 (Chapter 13). Note that the set, $\{f_6,f_7,f_8,f_9\}$, is subject to the CR $\mathbf{D}_{2d}(/\mathbf{C}_2')$. The corresponding molecule belongs to the $\mathbf{C}_2'$ symmetry, which appears in the parentheses of the CR. Similar results are obtained for the other molecules. This example also exemplifies Corollary 15.2.

The next example illustrates a case in which two or more orbits take different weights.

Example **15.2** Let us revisit adamantane-1,6-dione (**2**), which has once been examined in Example 14.1 (Chapter 14).

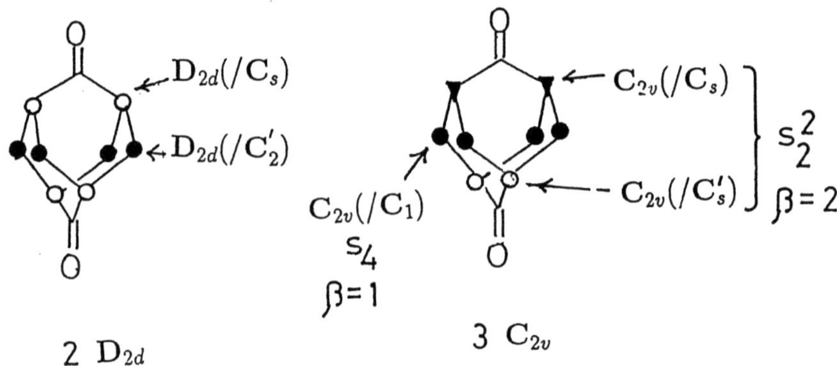

We take 8 positions of **2** into consideration; two carbonyl carbons are disregarded. The 8 positions are regarded as a domain, which is divided into two orbits that are subject to $\mathbf{D}_{2d}(/\mathbf{C}_2')$ (4 bridges: Δ_1) and $\mathbf{D}_{2d}(/\mathbf{C}_s)$ (4 bridgeheads: Δ_2). Let $\mathbf{X}=\{C,N,O\}$ be a codomain. According to the OMVs of the orbits, we have the following weights.

$$w_1(C) = 1, w_1(N) = x, w_1(O) = y \text{ for } \Delta_1 \text{ and}$$
$$w_2(C) = 1, w_2(N) = x, w_2(O) = 0 \text{ for } \Delta_2.$$

As a result, eq. 15.10 affords

$$|\mathbf{X}_1| = 3 \text{ for } \Delta_1 \text{ and}$$
$$|\mathbf{X}_2| = 2 \text{ for } \Delta_2.$$

Equation 15.12 is applied to this case, affording the following values of fixed configurations.

$$\rho_1 = 3^4 2^4 = 1296, \quad \rho_2 = 3^2 2^2 = 36, \quad \rho_3 = 3^3 2^2 = 108,$$
$$\rho_4 = 3^2 2^3 = 72, \quad \rho_5 = 3^1 2^1 = 6, \quad \rho_6 = 3^1 2^2 = 12,$$
$$\rho_7 = 3^2 2^1 = 18, \quad \text{and} \quad \rho_8 = 3^1 2^1 = 6,$$

where the powers of each central term come from the $\mathbf{D}_{2d}(/\mathbf{C}_2')$ and $\mathbf{D}_{2d}(/\mathbf{C}_s)$ rows of Table 15.1. The geometrical meaning of these powers can be understood intuitively. For example, the derivative (**3**) illustrates the powers appearing in ρ_6 for $\mathbf{C}_{2v}$, where $\mathbf{D}_{2d}(/\mathbf{C}_2')\downarrow\mathbf{C}_{2v}$ produces one suborbit ($\beta = 1$) and $\mathbf{D}_{2d}(/\mathbf{C}_s)\downarrow\mathbf{C}_{2v}$ produces two orbits ($\beta = 2$). When we collect these values, we have an FPV =

(1296 36 108 72 6 12 18 6). The FPV is multiplied by the inverse of a mark table (M^{-1}) for $\mathbf{D}_{2d}$ to afford

$$(1296\ 36\ 108\ 72\ 6\ 12\ 18\ 6)M^{-1} = (120\ 3\ 45\ 30\ 0\ 3\ 6\ 6),$$

where the inverse M^{-1} is shown in Table 5.4 (Chapter 5) and in Appendix B. The resulting vector indicates that there appear 120 $\mathbf{C}_1$-, three $\mathbf{C}_2$-, 45 $\mathbf{C}_2'$-, 30 $\mathbf{C}_s$-, three $\mathbf{C}_{2v}$-, 6 $\mathbf{D}_2$- and 6 $\mathbf{D}_{2d}$-derivatives. The total number of the derivatives is equal to 213. This value is identical with the value obtained in Example 14.1 (Chapter 14).

15.2 Enumeration of Compounds with Given Symmetry and Weight

In Chaper 14, we have enumerated compounds of a given weight under the OMV restriction. We have also obtained a solution to the question (the preceding section): Which is the number of compounds of a subsymmetry on the basis of a skeleton of a given symmetry under the same restriction? A further question is the combination of these: Which is the number of compounds with a given subsymmetry as well as a given weight under the OMV restriction?

This section uses the same formulation as described in the preceding section. Thus, we take account of transitivity of a domain in order to cover the restriction due to obligatory minimum valency (OMV). Suppose that the weight of a function' is given by Def. 14.3 (Chapter 14). In terms of the equivalence relation ($f_\gamma(\delta) = f_\epsilon(p_g(\delta))$) for $\forall \delta \in \Delta$), two equivalent functions are easily proved to have an equal weight (Theorem 14.1 in Chapter 14). Let $\mathbf{F}^{[\theta]}$ is a set of functions ($f : \Delta \to \mathbf{X}$) all of which have the same weight $W_\theta(f)$, $\mathbf{F}^{[\theta]} = \{f_1^{[\theta]}, f_2^{[\theta]}, \ldots, f_{|\mathbf{F}^{[\theta]}|}^{[\theta]}\}$. Suppose that p_g ($\in \mathbf{P_G}$) acts on f_γ ($\in \mathbf{F}^{[\theta]}$) by the equivalence relation. The resulting $f_\epsilon^{[\theta]}$ is also a member of $\mathbf{F}^{[\theta]}$ due to the definition. Hence, we obtain a permutation,

$$\lambda_g^{[\theta]} = \begin{pmatrix} f_1^{[\theta]} & f_2^{[\theta]} & \cdots & f_{|\mathbf{F}^{[\theta]}|}^{[\theta]} \\ f_1^{[\theta]} p_g & f_2^{[\theta]} p_g & \cdots & f_{|\mathbf{F}^{[\theta]}|}^{[\theta]} p_g \end{pmatrix}. \tag{15.17}$$

The set represented by

$$\Lambda_{\mathbf{G}}^{[\theta]} = \{\lambda_g^{[\theta]} \mid \forall g \in \mathbf{G}\}$$

can be proved to be a permutation representation on $\mathbf{F}^{[\theta]}$ in the same line as described for Theorem 14.2 (Chapter 14). This formulation permits us to apply the general theorem (Theorem 5.7 in Chapter 5) to this case.

Theorem 15.4 *Let $\Lambda_{\mathbf{G}}^{[\theta]}$ be a permutation representation on $\mathbf{F}^{[\theta]}$, each permutation $(\lambda_g^{[\theta]})$ of which is associated with $g(\in \mathbf{G})$ through $p_g(\in \mathbf{P_G})$. Then, we have*

$$\Lambda_{\mathbf{G}}^{[\theta]} = \sum_{i=1}^{s} A_{\theta i}\mathbf{G}(/\mathbf{G}_i), \tag{15.18}$$

where the multiplicity $(A_{\theta i})$ is a non-negative integer obtained by solving the equations,

$$\rho_{\theta j} = \sum_{i=1}^{s} A_{\theta i}m_{ij} \quad (j = 1, 2, \ldots, s). \tag{15.19}$$

The symbol $\rho_{\theta j}$ denotes the mark of $\mathbf{G}_j$ in $\Lambda_{\mathbf{G}}^{[\theta]}$. The subscript or superscript θ is concerned with all different values of weights W_θ. The multiplicity $A_{\theta i}$ is the number of non-equivalent cofigurations with the W_θ-weight and the $\mathbf{G}_i$-subsymmetry.

When the number of different weights is denoted as $\mid \theta \mid$, eq. 15.19 is written in a matrix expression.

$$\begin{pmatrix} \rho_{11} & \cdots & \rho_{1j} & \cdots & \rho_{1s} \\ \rho_{21} & \cdots & \rho_{2j} & \cdots & \rho_{2s} \\ \vdots & & \vdots & & \vdots \\ \rho_{\theta 1} & \cdots & \rho_{\theta j} & \cdots & \rho_{\theta s} \\ \vdots & & \vdots & & \vdots \\ \rho_{|\theta|1} & \cdots & \rho_{|\theta|j} & \cdots & \rho_{|\theta|s} \end{pmatrix} = \begin{pmatrix} A_{11} & \cdots & A_{1s} \\ A_{21} & \cdots & A_{2s} \\ \vdots & & \vdots \\ A_{|\theta|1} & \cdots & A_{|\theta|s} \end{pmatrix} \begin{pmatrix} m_{11} & \cdots & m_{1s} \\ m_{21} & \cdots & m_{2s} \\ \vdots & & \vdots \\ m_{s1} & \cdots & m_{ss} \end{pmatrix}. \tag{15.20}$$

We call the matrix $(\rho_{\theta j})$ a fixed-point matrix (FPM); and the matrix $(A_{\theta i})$ an isomer-counting matrix. The third matrix of eq. 15.20 is the matrix representation of a table of marks. Obviously, the following equations hold.

$$A_i = \sum_{[\theta]} A_{\theta i} \tag{15.21}$$

$$\rho_j = \sum_{[\theta]} \rho_{\theta j} \tag{15.22}$$

They have been alternatively obtained by Lemma 15.1 and Thorem 15.3. This point is discussed later in detail.

The next problem is the evaluation of $\rho_{\theta j}$. We will discuss a column of the FPM $(\rho_{\theta j})$. The series of elements:

$$\rho_{1j}, \rho_{2j}, \ldots, \rho_{\theta j}, \ldots, \rho_{|\theta|j}$$

of the jth column is concerned with configurations of symmetry $\mathbf{G}_j$. Suppose that $\mathbf{F}^{(j)}$ is a set of such configurations of symmetry $\mathbf{G}_j$. A subduced representation $\mathbf{G}/\mathbf{G}_i) \downarrow \mathbf{G}_j$ acts on $\Delta_{i\alpha}$ ($\alpha = 1, 2, \ldots, \alpha_i$) to produce the subdivision of each orbit $(\Delta_{i\alpha})$ into $\beta_k^{(ij)}$ suborbits of length d_{jk} ($\Delta_{jk\beta}^{i\alpha}$ for $\beta = 1, 2, \ldots, \beta_k^{(ij)}$), each of which is subject to the corresponding coset representation $(G_j(/G_k^{(j)}))$ (Chapter 9). In order for $f^{(j)} \in F^{(j)}$ to be constant, each suborbit of length d_{jk} has to take the same ligand, *i.e.*,

$$d_{jk}X_1, \text{ or } d_{jk}X_1, \ldots, \text{ or } d_{jk}X_{|\mathbf{X}|}.$$

Hence, for each suborbit of $\Delta_{i\alpha}$, the corresponding generating function (ligand inventory) is obtained, *i.e.*,

$$s_{d_{jk}} \equiv \sum_{\ell=1}^{|\mathbf{X}|} w_{i\alpha}(X_\ell)^{d_{jk}}. \tag{15.23}$$

Since this equation holds for all the suborbits of $\Delta_{i\alpha}$, the multiplication over all the suborbits of the orbit $\Delta_{i\alpha}$ (i.e., over all subgroups $G_k^{(j)}$) affords a generating function,

$$\prod_{k=1}^{v_j} \left(\sum_{\ell=1}^{|\mathbf{X}|} w_{i\alpha}(X_\ell)^{d_{jk}} \right)^{\beta_k^{(ij)}}, \tag{15.24}$$

wherein $\beta_k^{(ij)}$ is concerned with $G_j(/G_k^{(j)})$. Since eq. 15.24 is true for all the orbits of Δ, the multiplication over all α and all i provides a generating function,

$$\prod_{i=1}^{s} \prod_{\substack{\alpha=1 \\ \alpha_i \neq 0}}^{\alpha_i} \prod_{k=1}^{v_j} \left(\sum_{\ell=1}^{|\mathbf{X}|} w_{i\alpha}(X_\ell)^{d_{jk}} \right)^{\beta_k^{(ij)}} \quad (j = 1, 2, \ldots, s). \tag{15.25}$$

Expansion of eq. 15.25 affords a polynomial that contains monomials expressed concerning the weights, since the multiplication over k in eq. 15.25 contains monomials of total powers $(\sum_{k=1}^{v_j} d_{jk}\beta_k^{(ij)})$. We arrive at a generating function giving $\rho_{\theta j}$ as the coefficients of W_θ,

$$\sum_{[\theta]} \rho_{\theta j} W_\theta = \prod_{i=1}^{s} \prod_{\substack{\alpha=1 \\ \alpha_i \neq 0}}^{\alpha_i} \prod_{k=1}^{v_j} \left(\sum_{\ell=1}^{|\mathbf{X}|} w_{i\alpha}(X_\ell)^{d_{jk}} \right)^{\beta_k^{(ij)}} \quad (j = 1, 2, \ldots, s). \tag{15.26}$$

We have already defined unit subduced cycle indices (USCIs) in Chapter 9 (Def. 9.2), which is represented by

$$Z(G(/G_i) \downarrow G_j; s_{d_{jk}}) = \prod_{k=1}^{v_j} (s_{d_{jk}})^{\beta_k^{(ij)}}. \tag{15.27}$$

To formulate the above discussion, we define a subduced cycle index (SCI) by using the USCIs. Note that the last product of the right-hand side of eq. 15.26 (*i.e.* eq. 15.24) is alternatively obtained by the introduction of eq. 15.23 into eq. 15.27.

Definition 15.1 *The subduced cycle index (SCI) is defined to be*

$$ZI(G_j; s_{d_{jk}}^{(i\alpha)}) = \prod_{i=1}^{s} \prod_{\substack{\alpha=1 \\ \alpha_i \neq 0}}^{\alpha_i} Z(G(/G_i) \downarrow G_j; s_{d_{jk}}^{(i\alpha)})$$

$$= \prod_{\substack{i=1}}^{s} \prod_{\substack{\alpha=1 \\ \alpha_i \neq 0}}^{\alpha_i} \prod_{k=1}^{v_j} (s_{d_{jk}}^{(i\alpha)})^{\beta_k^{(ij)}}, \tag{15.28}$$

for each $\mathbf{G}_j$ *(j = 1, 2, \ldots, s). The superscript (iα) is associated with the corresponding orbit* $\Delta_{i\alpha}$.

The subduced cycle index (SCI) is composed of USCIs that has been precalculated from the data of subduction of coset representations (Chapter 9). We summarize the above results expressed by eq. 15.26 in the form of the following lemma.

Lemma 15.2 *If* $\mathbf{G}_j(\leq \mathbf{G})$ *acts on* Δ, *a generating function for marks* $\rho_{\theta j}$ *with a weight* W_θ *is represented by*

$$\sum_{[\theta]} \rho_{\theta j} W_\theta = ZI(\mathbf{G}_j; s_{d_{jk}}^{(i\alpha)}) \quad (j = 1, 2, \ldots, s), \tag{15.29}$$

where the right-hand side is substituted by ligand inventories,

$$s_{d_{jk}}^{(i\alpha)} = \sum_{\ell=1}^{|\mathbf{X}|} w_{i\alpha}(X_\ell)^{d_{jk}}. \tag{15.30}$$

The resulting generating functions are expanded and the coefficient of every term (corresponding to each weight W_θ) is listed to afford an FPM (eq. 15.20). The generating function for the $\mathbf{G}_j$ group affords the jth column of the FPM. Hence, we calculate the number of non-equivalent configurations (*i.e.*, the number of compounds) by introducing these values into the equations of Theorem 15.4.

When we take no account of OMVs, this lemma can be simplified as follows. Let the weight $w_{i\alpha}(X_\ell)$ be equal to X_ℓ. Then, the SCI (eq. 15.28) is transformed as follows.

$$\begin{aligned}
ZI(\mathbf{G}_j; s_{d_{jk}}^{(i\alpha)}) &= \prod_{i=1}^{s} \prod_{\substack{\alpha=1 \\ \alpha_i \neq 0}}^{\alpha_i} Z(\mathbf{G}(/\mathbf{G}_i) \downarrow \mathbf{G}_j; s_d^{(i\alpha)}) = \prod_{i=1}^{s} \prod_{\substack{\alpha=1 \\ \alpha_i \neq 0}}^{\alpha_i} \prod_{k=1}^{v_j} (s_{d_{jk}})^{\beta_k^{(ij)}} \\
&= \prod_{i=1}^{s} \prod_{k=1}^{v_j} \prod_{\substack{\alpha=1 \\ \alpha_i \neq 0}}^{\alpha_i} (s_{d_{jk}})^{\beta_k^{(ij)}} = \prod_{i=1}^{s} \prod_{k=1}^{v_j} (s_{d_{jk}})^{\alpha_i \beta_k^{(ij)}} = \prod_{k=1}^{v_j} \prod_{i=1}^{s} (s_{d_{jk}})^{\alpha_i \beta_k^{(ij)}} \\
&= \prod_{i=1}^{s} (s_{d_{jk}})^{\sum_{i=1}^{s} \alpha_i \beta_k^{(ij)}} \tag{15.31}
\end{aligned}$$

The superscript ($i\alpha$) should be omitted, since $s_{d_{jk}}$ is independent of $\Delta_{i\alpha}$. This provides a new form of subduced cycle index suitable to the special case.[3]

---

[3]Throughout this book, SCIs (and other related indices) with a prime (') are concerned with cases in which OMVs are not taken into consideration.

Definition 15.2 *When each ligand has an equal weight independent of the orbits of a domain, a subduced cycle index (SCI) is defined as*

$$ZI'(\mathbf{G}_j; s_{d_{jk}}) = \prod_{\substack{i=1 \\ \alpha_i \neq 0}}^{s} \prod_{\alpha=1}^{\alpha_i} Z(\mathbf{G}(/\mathbf{G}_i) \downarrow \mathbf{G}_j; s_{d_{jk}})$$

$$= \prod_{i=1}^{s} (s_{d_{jk}})^{\sum_{i=1}^{s} \alpha_i \beta_k^{(ij)}} \tag{15.32}$$

for $\mathbf{G}_j$ $(j = 1, 2, \ldots, s)$.

By the use of the definition, Lemma 15.2 can be converted into the following lemma to meet the present requirement.

Lemma 15.3 *If $\mathbf{G}_j(\leq \mathbf{G})$ acts on Δ, the generating function for marks $\rho_{\theta j}$ with a weight W_θ is represented by*

$$\sum_{[\theta]} \rho_{\theta j} W_\theta = ZI'(\mathbf{G}_j; s_{d_{jk}}) \quad (j = 1, 2, \ldots, s), \tag{15.33}$$

where the right-hand side is substituted by ligand-inventories,

$$s_{d_{jk}} = \sum_{\ell=1}^{|\mathbf{X}|} X_\ell^{d_{jk}}. \tag{15.34}$$

We can obtain the number of compounds by combining Lemma 15.3 and Theorem 15.4. Note that the power of the cycle index, $\sum_{i=1}^{s} \alpha_i \beta_k^{(ij)}$, comes from the subduced representation $\mathbf{G}(/\mathbf{G}_i) \downarrow \mathbf{G}_j$ in the present method.

Example **15.3** By the use of the theorems obtained in this section, we re-examine adamantane-1,6-dione (**2**), which has been examined in Example 15.2. We also take account of the two orbits that are subject to $\mathbf{D}_{2d}(/\mathbf{C}_2')$ (4 bridges: Δ_1) and $\mathbf{D}_{2d}(/\mathbf{C}_s)$ (4 bridgeheads: Δ_2) and of $\mathbf{X}=\{C,N,O\}$ as a codomain. According to the OMVs of the orbits, we have the following weights.

$$w_1(C) = 1, w_1(N) = x, w_1(O) = y \text{ for } \Delta_1 \text{ and}$$
$$w_2(C) = 1, w_2(N) = x, w_2(O) = 0 \text{ for } \Delta_2.$$

This OMV restriction affords the following ligand inventories.

$$s_d^{(1)} = 1 + x^d + y^d \text{ for } \Delta_1 \tag{15.35}$$
and
$$s_d^{(2)} = 1 + x^d \text{ for } \Delta_2 \tag{15.36}$$

In order to obtain an SCI for each subgroup, we use the $\mathbf{D}_{2d}(/\mathbf{C}_2')$ and $\mathbf{D}_{2d}(/\mathbf{C}_s)$ columns of the USCI table of $\mathbf{D}_{2d}$ group (Table 9.3 in Chapter 9 and Appendix

Table 15.3: Coefficients appearing in the generating functions

term	C_1	C_2	C_2'	C_s	S_4	C_{2v}	D_2	D_{2d}
$x^8,\ x^4y^4,\ y^4,\ 1$	1	1	1	1	1	1	1	1
$x^7y,\ y$	4	0	2	0	0	0	0	0
x^7,x	8	0	2	2	0	0	0	0
x^6y^2,y^2	6	2	2	2	0	0	2	0
$x^6y,\ xy$	28	0	2	0	0	0	0	0
x^6,x^2	28	4	4	4	0	2	2	0
x^5y^3,y^3	4	0	2	0	0	0	0	0
x^5y^2,xy^2	36	0	2	4	0	0	0	0
x^5y,x^2y	84	0	6	0	0	0	0	0
x^5,x^3	56	0	6	6	0	0	0	0
$x^4y^3,\ xy^3$	20	0	2	0	0	0	0	0
x^4y^2,x^2y^2	90	6	6	6	0	0	2	0
x^4y,x^3y	140	0	6	0	0	0	0	0
x^4	70	6	6	6	2	2	2	2
x^3y^4,xy^4	4	0	0	2	0	0	0	0
x^3y^3,x^2y^3	40	0	4	0	0	0	0	0
x^3y^2	120	0	4	8	0	0	0	0
x^2y^4	6	2	2	2	0	2	0	0

C). Equation 15.29 (Lemma 15.2) for the present case is calculated as follows.

$$C_1 \ : \ (s_1^4)^{(1)}(s_1^4)^{(2)} = (1 + x + y)^4(1 + x)^4,$$

$$C_2 \ : \ (s_2^2)^{(1)}(s_2^2)^{(2)} = (1 + x^2 + y^2)^2(1 + x^2)^2,$$

$$C_2' \ : \ (s_1^2 s_2)^{(1)}(s_2^2)^{(2)} = (1 + x + y)^2(1 + x^2 + y^2)(1 + x^2)^2,$$

$$C_s \ : \ (s_2^2)^{(1)}(s_1^2 s_2)^{(2)} = (1 + x^2 + y^2)^2(1 + x)^2(1 + x^2),$$

$$S_4 \ : \ (s_4)^{(1)}(s_4)^{(2)} = (1 + x^4 + y^4)(1 + x^4),$$

$$C_{2v} \ : \ (s_4)^{(1)}(s_2^2)^{(2)} = (1 + x^4 + y^4)(1 + x^2)^2,$$

$$D_2 \ : \ (s_2^2)^{(1)}(s_4)^{(2)} = (1 + x^2 + y^2)^2(1 + x^4),$$

and

$$D_{2d} \ : \ (s_4)^{(1)}(s_4)^{(2)} = (1 + x^4 + y^4)(1 + x^4),$$

in which the ligand inventories are introduced. The superscripts (1) and (2) correspond to the orbits Δ_1 and Δ_2. The resulting generating functions are expanded to afford a fixed-point matrix (FPM), which is shown in Table 15.3.

The FPM (Table 15.3) is multiplied by the inverse of a mark table for D_{2d} (Appendix B) to give the numbers of derivatives that are itemized with respect to

Table 15.4: Number of derivatives from adamantane-2,6-dione (2)

term	C_1	C_2	C_2'	C_s	S_4	C_{2v}	D_2	D_{2d}	total
x^8, x^4y^4, y^4, 1	0	0	0	0	0	0	0	1	1
x^7y,y	0	0	1	0	0	0	0	0	1
x^7,x	0	0	1	1	0	0	0	0	2
x^6y^2,y^2	0	0	0	1	0	0	1	0	2
x^6y, xy	3	0	1	0	0	0	0	0	4
x^6, x^2	2	0	1	1	0	1	1	0	6
x^5y^3,y^3	0	0	1	0	0	0	0	0	1
x^5y^2,xy^2	3	0	1	2	0	0	0	0	6
x^5y,x^2y	9	0	3	0	0	0	0	0	12
x^5,x^3	4	0	3	3	0	0	0	0	10
x^4y^3, xy^3	2	0	1	0	0	0	0	0	3
x^4y^2,x^2y^2	8	1	2	3	0	0	1	0	15
x^4y,x^3y	16	0	3	0	0	0	0	0	19
x^4	6	1	2	2	0	0	0	2	13
x^3y^4,xy^4	0	0	0	1	0	0	0	0	1
x^3y^3,x^2y^3	4	0	2	0	0	0	0	0	6
x^3y^2	12	0	2	4	0	0	0	0	18
x^2y^4	0	0	1	0	0	1	0	0	2
total	120	3	45	30	0	3	6	6	213

the weights (terms) and to their symmetries. The results are listed in Table 15.4. Figure 15.1 depicts derivatives with molecular formulas C_6N_2 (x^2), C_6O_2 (y^2), and C_6NO (xy). It should be noted that the number of the C_6N_2 isomers is different from that of the C_6O_2; this result stems from the OMV restriction.

When we sum up each row of Table 15.4, we obtain the total number of derivatives having the corresponding weight. The result is found in the rightmost column and identical with that of Example 14.1 (Chapter 14). The summation of each column provides the total number of the corresponding symmetry. The result (bottom) is identical with that of Example 15.2.

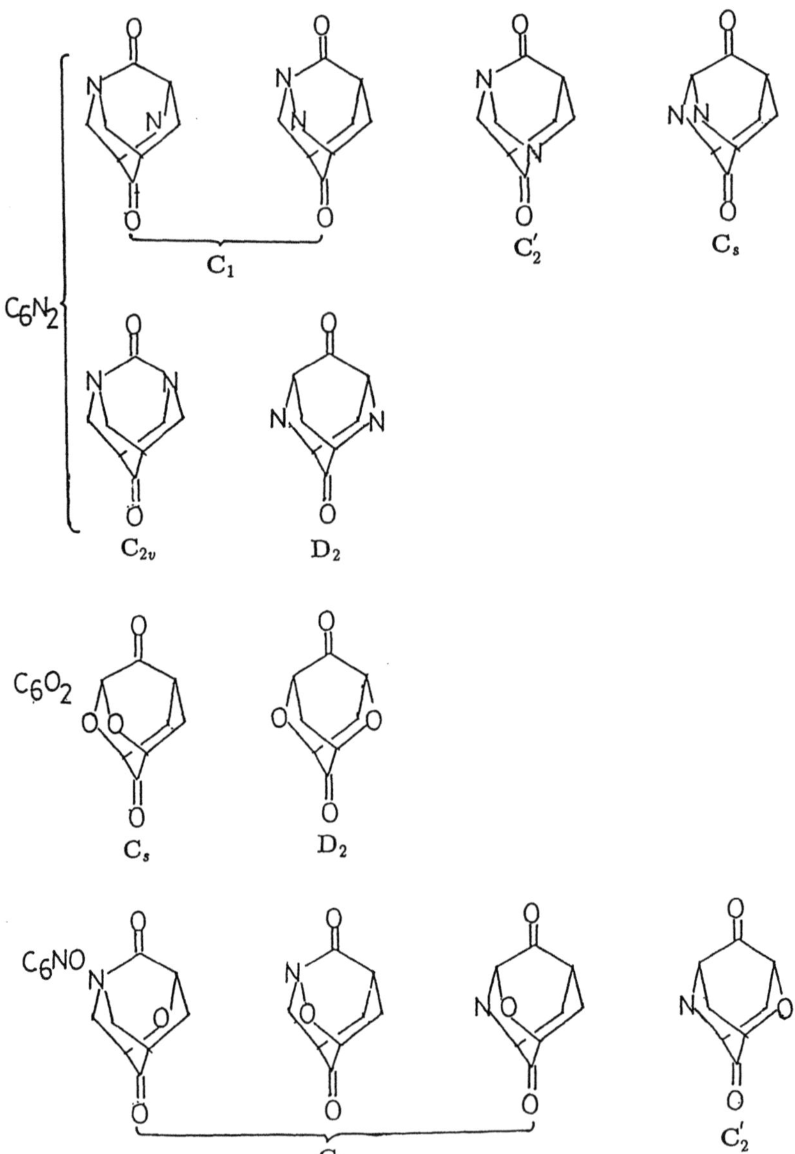

Figure 15.1: Derivatives from adamantane-2,6-dione (2)

Bibliography

[1] a) G. Pólya, *Acta Math.*, **68**, 145 (1937). b) G. Pólya, R. C. Read, *Combinatorial Enumeration of Groups, Graphs, and Chemical Compounds*, Springer, Berlin (1987).

[2] E. Ruch, W. Hässelbarth, B. Richter, *Theor. Chim. Acta*, **19**, 288 (1970).

[3] J. Brocas, *J. Am. Chem. Soc.*, **108**, 1135 (1986).

[4] J. Sheehan, *Canad. J. Math.*, **20**, 1068 (1968).

[5] W. Hässelbarth, *Theor. Chim. Acta*, **67**, 339 (1985).

[6] C. A. Mead, *J. Am. Chem. Soc.*, **109**, 2130 (1987).

Chapter 16

New Cycle Index [1]

In Chapter 15, we have discussed an application of unit subduced cycle indices (USCIs) to enumeration of compounds, which is based on a new type of generating functions. In a continuation of the work, the present chapter deals with the relationship between USCIs and Pólya's cycle indices.

16.1 New Cycle Indices Based On USCIs

By means of Theorem 15.4 (Chapter 15), the number of non-equivalent configurations (*i.e.* the number of compounds) with W_θ-weight and G_j-symmetry is represented by

$$A_{\theta i} = \sum_{j=1}^{s} \rho_{\theta j} \overline{m}_{ji}, \qquad (16.1)$$

where each $\overline{m}_{ji}$ is an element of the inverse of the mark table of G. When we sum up the number $A_{\theta i}$ over G_i, we obtain the total number of compounds with the weight W_θ as follows.

$$A_\theta = \sum_{i=1}^{s} A_{\theta i} = \sum_{i=1}^{s} \sum_{j=1}^{s} \rho_{\theta j} \overline{m}_{ji}. \qquad (16.2)$$

This equation leads to a generating function for A_θ, *i.e.*,

$$\sum_{[\theta]} A_\theta W_\theta = \sum_{[\theta]} \Big(\sum_{i=1}^{s} \sum_{j=1}^{s} \rho_{\theta j} \overline{m}_{ji} \Big) W_\theta = \sum_{j=1}^{s} \Big((\sum_{i=1}^{s} \overline{m}_{ji}) \sum_{[\theta]} \rho_{\theta j} W_\theta \Big). \qquad (16.3)$$

Obviously, the last term of eq. 16.3 $(\sum_{[\theta]} \rho_{\theta j} W_\theta)$ is a generating function for $\rho_{\theta j}$, which has already been evaluated in Chapter 15 (Lemma 15.2).

Definition 16.1 *A cycle index (CI) for the* G *group is defined as*

$$CI(G; s_{d_{jk}}^{(i\alpha)}) = \sum_{j=1}^{s} \Big((\sum_{i=1}^{s} \overline{m}_{ji}) ZI(G_j; s_{d_{jk}}^{(i\alpha)}) \Big) \qquad (16.4)$$

by using the subduced cycle indices (SCIs) defined in Def. 15.1 (Chapter 15).

---

[1]Reprinted in part with permission from S. Fujita, *J. Math. Chem.*, **5**, 99–120 (1990). ©(1990) J. C. Balzer AG.

The coefficient $\sum_{i=1}^{s} \overline{m}_{ji}$ for each $\mathbf{G}_j$ group is obtained by summing up the corresponding row of the inverse of the mark table of $\mathbf{G}$. This value will be proved to be positive if the $\mathbf{G}_j$ group is cyclic; otherwise, to be equal to zero. Such values for point groups are collected in each table of Appendix B and shown also in the bottom row of each USCI table (Appendix C). The superscript $(i\alpha)$ is added to meet the restriction of obligatory minimum valency (OMV).

It should be emphasized that the present definition (eq. 16.4) involves the SCIs that are in turn obtained as products of USCIs. Equations 16.3 and 16.4 afford a new generating function.

Theorem 16.1 *A generating function for the total number of orbits of configurations with the weight W_θ is represented by*

$$\sum_{[\theta]} A_\theta W_\theta = CI(\mathbf{G}; s_{d_{jk}}^{(i\alpha)}),\tag{16.5}$$

where the right-hand side is substituted by ligand inventories.

$$s_{d_{jk}}^{(i\alpha)} = \sum_{\ell=1}^{|\mathbf{X}|} w_{i\alpha}(X_\ell)^{d_{jk}}.\tag{16.6}$$

A procedure of the enumeration discussed in this section is composed of the following steps:

1. Assignment of coset representations (CRs) to the orbits of a given skeleton (See Chapters 5 and 6);

   - determination of the symmetry $\mathbf{G}$ of a parent skeleton,

   - counting of fixed points (marks) in the $\mathbf{G}$-set (a domain) of the skeleton on each operation of symmetry $\mathbf{G}$,

   - determination of coset representations (CR's) by the method described in Chapter 6,

2. Construction of a cycle index;

   - use of USCIs corrsponding to the CRs,

   - construction of an SCI for each subgroup

   - construction of a CI in the light of Def. 16.1, where the coefficient of each SCI is precalculated by summing up the elements of each row contained in the inverse of the mark table,

3. Introduction of ligand inventories to the CI; and

4. Expansion of the resulting generating function.

Example **16.1** Let us re-examine the 8 carbon atoms of adamantane-2,6-dione (**1**), which have been examined in Example 14.1 (Chapter 14) and Example 15.3 (Chapter 15). The domain (the 8 carbon atoms) have been shown to be divided into two orbits $(\mathbf{D}_{2d}(/\mathbf{C}_2')$ and $\mathbf{D}_{2d}(/\mathbf{C}_s))$. We adopt the corresponding USCIs from the USCI table of $\mathbf{D}_{2d}$, as shown in Example 15.3. In order to construct a cycle index, we use the data collected in the inverse of the mark table of $\mathbf{D}_{2d}$ group (the rightmost column of Table 5.4). Hence, eq. 16.4 is converted into

$$CI(\mathbf{D}_{2d}; s_d^{(1)}, s_d^{(2)}) = \frac{1}{8}(s_1^4)^{(1)}(s_1^4)^{(2)} + \frac{1}{8}(s_2^2)^{(1)}(s_2^2)^{(2)} + \frac{1}{4}(s_1^2 s_2)^{(1)}(s_2^2)^{(2)}$$
$$+\frac{1}{4}(s_2^2)^{(1)}(s_1^2 s_2)^{(2)} + \frac{1}{4}(s_4)^{(1)}(s_4)^{(2)}. \tag{16.7}$$

The superscripts (1) and (2) are concerned with the orbits of the domain (Δ_1 and Δ_2). This CI is identical with the CI obtained in Example 14.1, except that the notations of the dummy variables are different. Among the SCIs obtained in Example 15.3, those of $\mathbf{C}_1$, $\mathbf{C}_2$, $\mathbf{C}_2'$, $\mathbf{C}_s$, and $\mathbf{S}_4$ are effective in eq. 16.7. Note that these subgroups are all cyclic.

When each weight $w_{i\alpha}$ is independent of the orbits, Def. 16.1 is restated in a more simplified form.

Definition 16.2 *A cycle index (CI') for the* **G** *group is defined as*

$$CI'(\mathbf{G}; s_{d_{jk}}) = \sum_{j=1}^{s}((\sum_{i=1}^{s} \overline{m}_{ji})ZI'(\mathbf{G}_j; s_{d_{jk}})), \tag{16.8}$$

by using the subduced cycle indices (SCIs) defined in Def. 15.2 (Chapter 15).

Although the explicit form of the CI is different from that of Pólya's cycle index, they will be proved to be equivalent.

16.2 Correlation of New Cycle Indices to Pólya's Theorem

In Chapter 14 (Def. 14.5), we defined a cycle index based on unit cycle indices (UCIs). This definition is a generalization of Pólya's theorem so as to be applicable to problems concerning the OMV restriction. On the other hand, we have afforded an alternative definition of the cycle index that is constructed on the basis of SCIs derived from USCIs (Def. 16.1). These definitions are different in their explicit forms. However, since both the equations are generating functions for the same A_θ, their right-hand sides should be equal to each other. Then, we find

$$\frac{1}{|\mathbf{G}|}\sum_{g\in G}\prod_{i=1}^{s}\prod_{\substack{\alpha=1 \\ \alpha_i \neq 0}}^{\alpha_i}\Big(\prod_{d=1}^{|\mathbf{X}|} s_d^{\nu_d}\Big)_{(g)}^{(i\alpha)} = \sum_{j=1}^{s}\Big((\sum_{i=1}^{s}\overline{m}_{ji})\prod_{i=1}^{s}\prod_{\substack{\alpha=1 \\ \alpha_i \neq 0}}^{\alpha_i}\prod_{k=1}^{\nu_j}(s_{d_{jk}}^{(i\alpha)})^{\beta_k^{(ij)}}\Big), \tag{16.9}$$

where the UCI in the left-hand side and the SCI in the right-hand side are given in a concrete fashion. In order to characterize eq. 16.9, we should clarify the correspondence between g of the left-hand side and the subgroup $\mathbf{G}_j$ that is implied by the right-hand side. Let us prove the following Lemma in order to clarify the correspondence.

Lemma 16.1 *(1) Suppose that* $\mathbf{G}(/\mathbf{G}_i)$ *is a coset representaiton on* $\Delta_{i\alpha}$. *Let* $h = G(/G_i)_g$ *(for* $g \in \mathbf{G}$) *has a cycle structure:*

$$(\nu): \quad (1^{\nu_1} 2^{\nu_2} \cdots m^{\nu_m}), \tag{16.10}$$

where

$$\sum_{d=1}^{m} d\nu_d = m. \tag{16.11}$$

The permutation (h) can be represented by a product of cycles,

$$h = \prod_{d=1}^{m} \prod_{a=1}^{\nu_d} (a_1 \ a_2 \ \cdots a_d). \tag{16.12}$$

Let n *be the least common multiple of* d *(for* $\nu_d \neq 0$; $d = 1, 2, \ldots, m$*). Then the subgroup generated from the element* h,

$$\mathbf{H} = \{h, h^2, \ldots, h^{n-1}, h^n (= I)\}, \tag{16.13}$$

is a cyclic group. Moreover, the cyclic group $(\mathbf{H})$ *partitions* $\Delta_{i\alpha}$ *into* ν_d *orbits:* $\{a_1, a_2, \cdots, a_d\}$, *the length of which is equal to* d, *where* $a = 1, 2, \ldots, \nu_d$.

(2) If $q(1 \leq q \leq n)$ *is any divisor of* n, *an element* (h^q) *of* $\mathbf{H}$ *generates a cyclic subgroup,*

$$\mathbf{H}' = \{h^q, h^{2q}, \ldots, h^{wq}(= I)\}, \tag{16.14}$$

where $w = n/q$. *This provides a further partition of* $\Delta_{i\alpha}$ *in which the orbit* $\{a_1, a_2, \cdots, a_d\}$ *is subdivided, if* $1 \leq q \leq d$ *and* q *is a multiple of a divisor of* d.

(3) Consider the case in which $q'(1 \leq q' \leq n)$ *is no divisor of* n *and* q' *and* n *have at least one common divisor. Suppose that* q *is the greatest common divisor of* n *and* q'. *The element* $h^{q'}$ *generates a cyclic subgroup which is equal to* $\mathbf{H}'$.

(4) If q *and* n *are coprime, the element* (h^q) *has the same cycle structure as* h *and generates* $\mathbf{H}$ *that is equal to that generated by* h. *Note that* $h, h^q \notin \mathbf{H}'$.

We exemplify this lemma by the following example, before we begin the proof.

Example **16.2** Let us examine $\mathbf{D}_{2d}(/\mathbf{C}_1)$ whose elements are represented by

$$I \ \sim \ p_1 = (1)(2)(3)(4)(5)(6)(7)(8)$$
$$C_{2(1)} \ \sim \ p_2 = (1\ 2)(3\ 4)(5\ 7)(6\ 8)$$

$$
\begin{aligned}
C_{2(2)} &\sim p_3 = (1\ 3)(2\ 4)(5\ 6)(7\ 8) \\
C_{2(3)} &\sim p_4 = (1\ 4)(2\ 3)(5\ 8)(6\ 7) \\
\sigma_{d(1)} &\sim p_5 = (1\ 5)(2\ 6)(3\ 7)(4\ 8) \\
S_4^3 &\sim p_6 = (1\ 6\ 4\ 7)(2\ 5\ 3\ 8) \\
S_4 &\sim p_7 = (1\ 7\ 4\ 6)(2\ 8\ 3\ 5) \\
\sigma_{d(2)} &\sim p_8 = (1\ 8)(2\ 7)(3\ 6)(4\ 5).
\end{aligned}
$$

For understanding a geometrical meaning, see Fig. 7.2 (Chapter 7). First, we select $h = p_7$ corresponding to S_4. Then, we have

$$
\begin{aligned}
h &= p_7 = (1\ 7\ 4\ 6)(2\ 8\ 3\ 5) \\
h^2 &= p_4 = (1\ 4)(2\ 3)(5\ 8)(6\ 7) \\
h^3 &= p_6 = (1\ 6\ 4\ 7)(2\ 5\ 3\ 8) \\
h^4 &= p_1 = (1)(2)(3)(4)(5)(6)(7)(8),
\end{aligned}
$$

where $n = 4$ and the cycle structure (4^2). These permuations construct a cyclic group corresponding to $\mathbf{S_4}$ point group, *i.e.*, $\mathbf{H} = \{h, h^2, h^3, h^4(= I)\}$. By inspection, the permutations provide a partition, $\{1,7,4,6\}$ and $\{2,8,3,5\}$. This is an example of Lemma 16.1 (1).

Since $q = 2$ is a divisor of $n = 4$, we can construct $\mathbf{H}' = \{h^2, h^4(= I)\}$, the elements of which are represented by

$$
\begin{aligned}
h^2 &= p_4 = (1\ 4)(2\ 3)(5\ 8)(6\ 7) \\
h^4 &= p_1 = (1)(2)(3)(4)(5)(6)(7)(8).
\end{aligned}
$$

Note that $1 \leq 2(= q) \leq 4(= d)$ and that q is a multiple of a divisor of d. By inspection, the resulting orbits are $\{1,4\}$, $\{2,3\}$, $\{5,8\}$, and $\{6,7\}$. This example verifies Lemma 16.1 (2). There emerge no examples for (3) in this case.

Suppose that $q = 3$ and $n = 4$. Then we have $(h^3)^1 = h^3$, $(h^3)^2 = h^2$, $(h^3)^3 = h$, and $(h^3)^4 = I$. These construct a group equal to $\mathbf{H}$. This is an example of Lemma 16.1 (4).

Proof of Lemma 16.1. (1) Let $\mathbf{H}$ be a group generated by $h = G(/G_i)_g$ (for $g \in \mathbf{G}$),

$$
\mathbf{H} = \{h, h^2, \ldots, h^{n-1}, h^n(= I)\},
$$

The group $\mathbf{H}$ is obviously a cyclic group that is a subgroup of $\mathbf{G}$. Let us first consider the case in which h consists of a single cycle, *i.e.*,

$$
h = (a_1\ a_2\ \cdots\ a_n) = \begin{pmatrix} a_1 & a_2 & \cdots & a_{n-1} & a_n \\ a_2 & a_3 & \cdots & a_n & a_1 \end{pmatrix}.
$$

This permutation provides

$$
h^2 = \begin{pmatrix} a_1 & a_2 & \cdots & a_{n-2} & a_{n-1} & a_n \\ a_3 & a_4 & \cdots & a_n & a_1 & a_2 \end{pmatrix}
$$

$$
h^3 = \begin{pmatrix} a_1 & a_2 & \cdots & a_{n-2} & a_{n-1} & a_n \\ a_4 & a_5 & \cdots & a_1 & a_2 & a_3 \end{pmatrix}
$$

$$
\vdots
$$

$$
h^n = \begin{pmatrix} a_1 & a_2 & \cdots & a_{n-2} & a_{n-1} & a_n \\ a_1 & a_1 & \cdots & a_{n-2} & a_{n-1} & a_n \end{pmatrix} = I.
$$

Hence, each of these permutations is represented by a product of several cycles, all of which have length n or less. This means that the cyclic group $\mathbf{H}$ on Δ provides a single orbit of Δ, which has a length n.

(2) A cyclic group of finite order generally has only one subgroup whose order is equal to any divisor of the order of the cyclic group. The set of such subgroups is the non-redundant set of the subgroups of the cyclic group. Now, we consider h^q ($1 \le q < n$) to be a generator of such a cyclic group. When q ($1 \le q < n$) is a divisor of n, the generator h^q yields the following subgroup of $\mathbf{H}$,

$$
\mathbf{H}' = \{h^q, h^{2q}, \ldots, h^{wq}\},
$$

where $w = n/q$. The order of $\mathbf{H}'$ is equal to w. If we divide the permutation h^q into w parts of equal length q, then we can obtain

$$
h^q = \begin{pmatrix} \overbrace{a_1 \quad \cdots}^{q} & \overbrace{a_{q+1} \quad \cdots}^{q} & \overbrace{a_{2q+1} \quad \cdots}^{q} & \cdots & \overbrace{a_{(w-1)q+1} \quad \cdots}^{q} \\ a_{q+1} \quad \cdots & a_{2q+1} \quad \cdots & a_{3q+1} \quad \cdots & \cdots & a_1 \quad \cdots \end{pmatrix}
$$
$$
= (a_1 \, a_{q+1} \, \cdots \, a_{n-q+1})(a_2 \, \cdots \, a_{n-q+2}) \cdots (a_q \, \cdots \, a_n).
$$

This indicates that h^q is the product of q cycles of length w. Hence, $\mathbf{H}'$ on Δ provides q orbits, each of which has a length w.

(3) Consider the case in which q' ($1 \le q' < n$) is no divisor of n and in which n and q' have common divisors. Let q be the greatest common divisor of n and q', where $q' = aq'$. Because q is a divisor of n, we can assume $h^{wq} = I$ for $\exists w$. Then, we have $(h^{q'})^w = (h^{aq})^w = I$ and $(h^{q'})^{w+b} = (h^{q'})^w (h^{q'})^b = (h^{q'})^b$, where a and b are integers. These equations indicate that $h^{q'}$ generates $\mathbf{H}'$, which is equal to the cyclic group generated by h^q.

(4) If q' ($1 \le q' < n$) is not a muliple of any q (*i.e.*, if q' and n are coprime), the permutation $h^{q'}$ has the same cycle structure as h. Note that such a division as above is impossible. Therefore, $h^{q'}$ generates $\mathbf{H}$, which is also generated by h. In terms of the definition, $h^{q'}$ is not an element of $\mathbf{H}'$ defined above. In addition, h is not an element of any $\mathbf{H}'$ in this case.

These discussions can be easily extended to cases in which the permutation h consists of two or more cycles.

Lemma 16.1 indicates that the pemutation $h = G(/G_i)_g$ (for $g \in \mathbf{G}$) that has a cycle structure:

$$(\nu): \quad (1^{\nu_1} 2^{\nu_2} \cdots m^{\nu_m}) \tag{16.15}$$

creates a cyclic group $\mathbf{H}$; the $\mathbf{H}$ group produces suborbits according to the cycle structure (ν). In other words, the creation of suborbits can be expressed by a unit cycle index (UCI) of $s_1^{\nu_1} s_2^{\nu_2} \cdots s_m^{\nu_m} = \Pi_{d=1}^m s_d^{\nu_d}$. Such a UCI is obtained for each CR (*i.e.* for each i).

The above process is identical with the operation that selects the element of $\mathbf{H}$ from $\mathbf{G}(/\mathbf{G}_i)$. This is essentially equal to the subduction $\mathbf{G}(/\mathbf{G}_i) \downarrow \mathbf{H}$. As the result, Lemma 16.1 can be considered to show the $\mathbf{G}(/\mathbf{G}_i) \downarrow \mathbf{H}$, which provides ν_d orbits of length d. If we consider that $\mathbf{G}_j = \mathbf{H}$, Theorem 9.1 (Chapter 9) holds for this case. This fact combined with Lemma 16.1 permits us to recognize that $\mathbf{G}_j(/\mathbf{G}_k^{(j)})$ denotes a coset representation on the orbit, $\{a_1, a_2, \ldots, a_d\}$. The number of such orbits is ν_d. The length of the orbit is d, which is equal to the degree of $\mathbf{G}_j(/\mathbf{G}_k^{(j)})$, *i.e.*, $|\mathbf{G}_j| / |\mathbf{G}_k^{(j)}|$. It follows that

$$|\mathbf{G}_j| / |\mathbf{G}_k^{(j)}| = d \tag{16.16}$$

$$\text{and}$$

$$\beta_k^{(ij)} = \nu_d. \tag{16.17}$$

Equation 16.16 combined with Def. 9.2 (Chapter 9) affords

$$d_{jk} = d. \tag{16.18}$$

The USCI (Def. 9.2) is converted as follows in the light of eqs. 16.17 and 16.18.

$$Z(\mathbf{G}(/\mathbf{G}_i) \downarrow \mathbf{G}_j; s_{d_{jk}}^{(i\alpha)}) = \prod_{k=1}^{v_j} (s_{d_{jk}}^{(i\alpha)})^{\beta_k^{(ij)}}$$

$$= (\prod_{d=1}^m s_d^{\nu_d})_{(g)}^{(i\alpha)} = z_g^{(i\alpha)}. \tag{16.19}$$

where the product over k is converted into that over d. The superscript $(i\alpha)$ concerns the orbit $\Delta_{i\alpha}$. The subscript (g) represents that the term related to $G(/G_i)_g (= h)$. The subgroup $\mathbf{G}_j$ corresponds to g through $\mathbf{G}_j = \mathbf{H} = \{I, h, h^2, \ldots, h^{n-1}\}$. Note that eq. 16.19 is true only if $\mathbf{G}_j = \mathbf{H}$, *i.e.*, if $\mathbf{G}_j$ is a cyclic group.

If $g' \in \mathbf{G}$ is conjugate to $g \in \mathbf{G}$, or equivalently if the corresponding $G(/G_i)_{g'}$ $(= h')$ is conjugate to $G(/G_i)_g$ $(= h)$, the cycle structure of h is equal to that of h'. Hence, we have

$$z_g^{(i\alpha)} = z_{g'}^{(i\alpha)}. \tag{16.20}$$

The conjugate elements h and h' generate cyclic groups $\mathbf{H}$ and $\mathbf{H}'$ that are conjugate to each other.

If $g^d = g'$ ($d = 1, 2, ..., m$) for g and $g' \in \mathbf{G}$ and if $G(/G_i)_g$ ($= h$) and $G(/G_i)_{g'}$ ($= h'$) have the same cycle structure, the cyclic group $\mathbf{H}$ generated from h is obviously equal to that generated from h'.

Suppose that a cyclic group $\mathbf{G}_j$ has cyclic subgroups $\mathbf{G}_k^{(j)}$ ($k = 1, 2, ..., v_j - 1$). Since $| \mathbf{G}_k^{(j)} |$ is a divisor of $| \mathbf{G}_j |$, Lemma 16.1 (4) indicates that

(1) if $g \in \mathbf{G}_j$ corresponds to the generator $h = G(/G_i)_g$ that generates $\mathbf{G}_j$ ($= \mathbf{H}$), it is not an element of any $\mathbf{G}_k^{(j)}$. Hence,

$$g \in \mathbf{G}_j - \bigcup_{k=1}^{v_j-1} \mathbf{G}_k^{(j)},$$

and

(2) if $h' = G(/G_i)_{g'}$ has the same cycle structure as h, we have

$$g' \in \mathbf{G}_j - \bigcup_{k=1}^{v_j-1} \mathbf{G}_k^{(j)}, \quad \text{or} \quad g' \in \mathbf{G}_{j'} - \bigcup_{k=1}^{v_{j'}-1} \mathbf{G}_k^{(j')},$$

where $\mathbf{G}_{j'}$ is a conjugate group of $\mathbf{G}_j$.

Hence, if σ denotes the number of groups conjugate to $\mathbf{G}_j$ in $\mathbf{G}$, the number of elements that have the same cycle structure as h is represented by

$$N_{(h)} = \sigma \, | \, \mathbf{G}_j - \bigcup_{k=1}^{v_j-1} \mathbf{G}_k^{(j)} \, | \tag{16.21}$$

Let $\varphi(n)$ be Euler's function, which gives the number of integers from 1 to n that are co-prime to n. Then, Lemma 16.1 (4) shows that

$$| \, \mathbf{G}_j - \bigcup_{k=1}^{v_j-1} \mathbf{G}_k^{(j)} \, | = \varphi(| \, \mathbf{G}_j \, |). \tag{16.22}$$

Let $\mathbf{N_G}(\mathbf{G}_j)$ denote a normalizer of $\mathbf{G}_j$. The term σ has been proved[1] to be

$$\sigma = | \, \mathbf{G} : \mathbf{N_G}(\mathbf{G}_j) \, | = | \, \mathbf{G} \, | \, / \, | \, \mathbf{N_G}(\mathbf{G}_j) \, | \, . \tag{16.23}$$

By introducing eqs. 16.22 and 16.23 into eq. 16.21, the number $N_{(h)}$ is expressed as follows.

$$N_{(h)} = | \, \mathbf{G} \, | \, \varphi(| \, \mathbf{G}_j \, |) / | \, \mathbf{N_G}(\mathbf{G}_j) \, |, \tag{16.24}$$

If cyclic groups are generated from all $g \in \mathbf{G}$, they can be easily proved to construct a complete set of the cyclic subgroups of $\mathbf{G}$. This means that the summation over $g \in \mathbf{G}$ in the left-hand side of eq. 16.9 corresponds to that over all the cyclic subgroups of $\mathbf{G}$. Hence, in the right-hand side of eq. 16.9, the summation over j (or $\mathbf{G}_j$) is effective if and only if $\mathbf{G}_j$ is a cyclic group. These results can be summarized as follows.

Theorem 16.2 *In the inverse of the mark table of* **G**,

$$\sum_{i=1}^{s} \overline{m}_{ji} = \begin{cases} \varphi(|\,\mathbf{G}_j\,|)/\,|\,\mathbf{N_G}(\mathbf{G}_j)\,|, & \text{for any cyclic } \mathbf{G}_j \\ 0, & \text{for others.} \end{cases} \tag{16.25}$$

This equation is equivalent to the one derived alternatively by Kerber and Thürlings.[2] The following equations are easily proved.

$$\sum_{j=1}^{s} \overline{m}_{ji} = \begin{cases} 0 & (i \neq s) \\ 1 & (i = s) \end{cases} \tag{16.26}$$

16.3 Partial Cycle Indices

In the light of eq. 16.1, we have

$$\begin{aligned}
\sum_{[\theta]} A_{\theta i} W_\theta &= \sum_{[\theta]} \sum_{j=1}^{s} \rho_{\theta j} \overline{m}_{ji} W_\theta \\
&= \sum_{j=1}^{s} \overline{m}_{ji} \Big(\sum_{[\theta]} \rho_{\theta j} W_\theta \Big).
\end{aligned} \tag{16.27}$$

The last term $(\sum_{[\theta]} \rho_{\theta j} W_\theta)$ has been evaluated in Chapter 15 (Lemma 15.2). We here define a *partial cycle index (PCI)* by using the subduced cycle indices (SCIs) defined in Chapter 15 (Def. 15.1).

Definition 16.3 (Partial cycle index)

A partial cycle index (PCI) for $\mathbf{G}_i$ *group is represented by*

$$PCI(\mathbf{G}_i; s_{d_{jk}}^{(i\alpha)}) = \sum_{j=1}^{s} \overline{m}_{ji} ZI(\mathbf{G}_j; s_{d_{jk}}^{(i\alpha)}) \quad (i = 1, 2, \dots, s). \tag{16.28}$$

Thereby, eq. 16.27 is converted into the following theorem.

Theorem 16.3 *A generating function of* $A_{\theta i}$ *is represented by*

$$\sum_{[\theta]} A_{\theta i} W_\theta = PCI(\mathbf{G}_i; s_{d_{jk}}^{(i\alpha)}) \quad (i = 1, 2, \dots, s), \tag{16.29}$$

where the right-hand side is substituted by ligand inventories.

$$s_{d_{jk}}^{(i\alpha)} = \sum_{\ell=1}^{|\mathbf{X}|} w_{i\alpha}(X_\ell)^{d_{jk}}. \tag{16.30}$$

This is a generating-function version of Theorem 15.4 and Lemma 15.2 (Chapter 15). Thus, the derivation of Theorem 16.3 does not provide new matters beyond the previous propositions. However, it may be convenient if we focus our attention on a specific $\mathbf{G}_i$ (See Chapter 18). Kerber and Thürlings[2] have alternatively derived a similar equation, though their formulation lacks the concepts of USCI, SCI and PCI; and takes no account of OMVs.

If the weights are independent of the orbits, we are able to convert Def. 16.3 into the following expression by using the SCIs defined in Chapter 15 (Def. 15.2).

Definition 16.4 (Partial cycle index)

A partial cycle index (PCI) without OMV restriction is represented by

$$PCI'(\mathbf{G}_i; s_{d_{jk}}) = \sum_{j=1}^{s} \overline{m}_{ji} ZI'(\mathbf{G}_j; s_{d_{jk}}) \quad (i = 1, 2, \ldots, s). \tag{16.31}$$

Thereby, Theorem 16.3 is transformed into the following corollary.

Corollary 16.1 *A generating function of $A_{\theta i}$ is represented by*

$$\sum_{[\theta]} A_{\theta i} W_{\theta} = PCI'(\mathbf{G}_i; s_{d_{jk}}) \quad (i = 1, 2, \ldots, s), \tag{16.32}$$

where the right-hand side is substituted by ligand inventories,

$$s_{d_{jk}} = \sum_{\ell=1}^{|\mathbf{X}|} X_{\ell}^{d_{jk}}. \tag{16.33}$$

For an example of Theorem 16.3 (and Corollary 16.1), see Chapter 17.

By using the PCIs, we have alternative expressions of the cycle indices (Defs. 16.1 and 16.2) as follows.

Definition 16.5 (Cycle indices)

(1) For an alternative definition of the CI, we have

$$CI(\mathbf{G}_i; s_{d_{jk}}^{(i\alpha)}) = \sum_{i=1}^{s} PCI(\mathbf{G}_i; s_{d_{jk}}^{(i\alpha)}) \tag{16,34}$$

(2) For an alternative defintion of the CI', we have

$$CI'(\mathbf{G}; s_{d_{jk}}) = \sum_{i=1}^{s} PCI'(\mathbf{G}_i; s_{d_{jk}}) \tag{16.35}$$

Bibliography

[1] T. Ōyama, *Yugen Chikan-Gun (Theory of Permutation Groups of Finite Order)*, Syōkabō , Tokyo (1981).

[2] A. Kerber and K.-J. Thürlings, in *Lecture Notes in Mathematics*, Vol. 969, p. 161, Springer, New York (1982).

Chapter 17

Cage-Shaped Molecules with High Symmetries [1]

Chiral and achiral hydrocarbons of high symmetry have attracted continuous attention of synthetic and physical organic chemists because of their potential interesting properties.[1]–[5] Two strategies have been applied to the derivation of new compounds of given symmetries. The first is a vertex strategy in which substituents are placed on vertices of a parent skeleton.[6, 7] The second (edge strategy) consists of the insertion of methylene or other units into the edges (bonds) of a parent skeleton. In particular, the latter is versatile methodology used to guide synthetical studies concerning cage-shaped compounds of high symmetry. Thus, there have appeared many papers that reported successful syntheses of such compounds. For example, a selected list of achiral hydrocarbons reported contains prismane of D_{3h} symmetry,[8] iceane (tetracyclo[5.3.1.1^{2,6}.0^{4,9}]dodecane) of D_{3h} symmetry,[9]–[11], 1,2:3,4:5,6-tris(bicyclo[2.2.2]octa-2-eno)benzene of D_{3h} symmetry,[12] superphane ([2$_6$](1,2,3,4,5,6)cyclophane) of D_{6h} symmetry,[13] pentaprismane of D_{5h} symmetry,[14] a tetrahedrane of T_d symmetry,[15] cubane of O_h symmetry,[16] and dodecahedrane of I_h symmetry.[17] Chiral hydrocarbons of high symmetry have also been investigated synthetically and physicochemically, e.g., twistane of D_2 symmetry[18, 19] and D_3-trishomocubane of D_3 symmetry.[20]–[22] Although several theoretical studies have discussed the two methodologies,[23, 24] there have emerged no comprehensive studies concerning the following problem: what symmetries are realized and how many isomers are allowed on the basis of a skeleton of a given symmetry.

In the preceding chapters, we have developed new methods of enumerating isomers. The key concept is the subduction of coset representations, which is the foundation of various new concepts, e.g., unit subduced cycle index (USCI), subduced cycle index (SCI) and a new type of cycle index (CI). These methods have

---

[1]Reprinted with permission from S. Fujita, *Tetrahedron*, 46, 365–382 (1990). ©(1990) Pergamon Press PLC. See also S. Fujita, *Bull. Chem. Soc. Jpn.*, 62, 3771–3778 (1989).

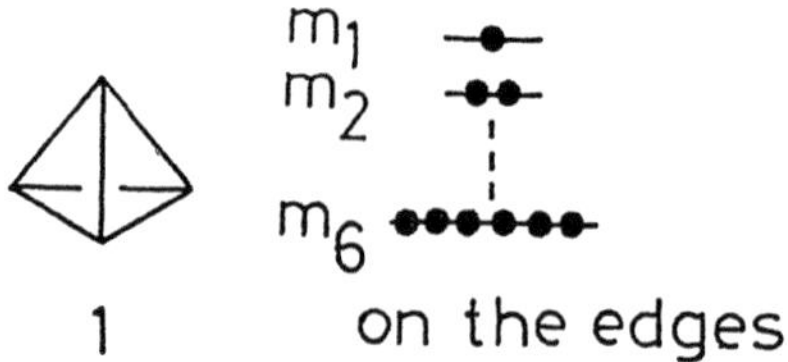

Figure 17.1: Substitution of polymethylenes on the edges of a tetrahedrane skeleton (1)

been applied to the enumeration of derivatives that are derived by the substitution of vertices of a given skeleton. These enumerations are concerned with the first methodology, *i.e.*, the vertex strategy for derivation of compounds. The present chapter is concerned with another mode of derivation of compounds, *i.e.* the edge strategy. This strategy is quite versatile to create cage-shaped molecules of high symmetries. We here discuss adamantane isomers and related molecules in detail.

17.1 Edge Strategy

Let us consider tetrahedrane (1) of $\mathbf{T}_d$ as a parent skeleton. The symmetrical properties of the vertices have been discussed in Chapter 6; they are represented by the SCR notation, $\mathbf{T}_d[/\mathbf{C}_{3v}(\mathbf{C}_4, \mathbf{H}_4)]$. The six edges (bonds) are also subject to the symmetry operations of $\mathbf{T}_d$ point group. By considering a set of the six edges as an orbit, we can discuss derivation by the edge strategy in the same line as the vertex strategy.

Tricyclodecanes and their homologs are derived by substitution of polymethylenes on the edges of the tetrahedrane skeleton (1). If we consider m_1 methylenes, m_2 ethylenes, $\cdots$ and m_6 hexamethylenes to be such substituents (Fig. 17.1), the resulting tricycloalkane has a molecular formula of $(\mathrm{CH})_4(\mathrm{CH}_2)_m$, where

$$m_1 + 2m_1 + \cdots + 6m_6 = m. \tag{17.1}$$

The integer m is equal to 6 when we consider the isomers of adamantane. It should be noted that this methodology takes account of their configurations, but not of their conformations.

The partition of the integer (m) is denoted as $(1^{m_1}2^{m_2}\cdots 6^{m_6})$, which is called a polymethylene index (PMI) in the present chapter. Since the PMIs indicate the modes of substitution, they are capable of characterizing the resulting isomers. For example, adamantane (2), which is derived by the substitution of 6 methylenes on

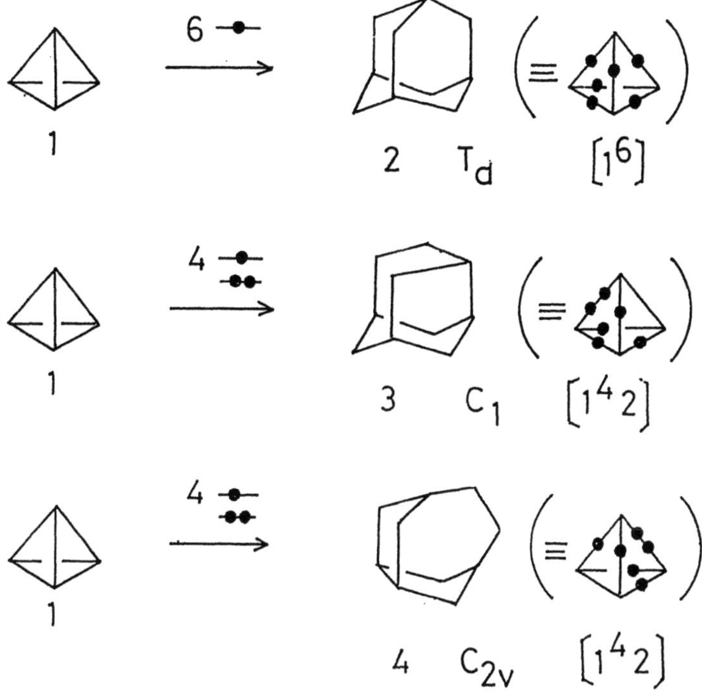

Figure 17.2: Polymethylene indices (PMI) and subsymmetries of T_d in adamantane isomers

the 6 edges of **1**, is chracterized by the PMI (1^6). Similarly, protoadamantane (**3**) and its isomer (**4**) have the same PMI ($1^4 2$).

Alternatively, the adamantane isomers can also be classified in terms of their symmetries (Fig. 17.2). Since the present skeleton has T_d symmetry, each of the isomers has one of the subsymmetries of T_d. Figure 17.2 shows that adamantane (**2**) of (1^6) has T_d symmetry, protoadamantane (**3**) has C_1 symmetry, and the isomer (**4**) has C_{2v} symmetry.

17.2 Tricyclodecanes with T_d and Its Subsymmetries

Adamantane. Adamantane rearrangements have long been investigated since Schleyer's discovery on the rearrangement of tetrahydrocyclopentadiene to adamantane.[25] Whitlock and Siefken have studied the mechanisms of such reactions by constructing a rearrangement graph, which contains various tricyclodecane intermediates.[26] Molecular mechanics has provided a method of calculating the stabilities of the intermediates.[27, 28] Various precursors and intermediates for adamantane have been reported and the relationship regarding them has been discussed in terms of "adamantaneland".[28] Recently, a renaissance on the adamantane chemistry has led to more detailed information on the isomeric intermediates and on the mechanisms of the rearrangements.[29, 30] In order to clarify the mechanisms, enumerations of potential intermediates are necessary. Computational and graph-theoretical investigations[28, 31] have enumerated isomers of adamantane. They are, however, concerned only with their constitutions but not with their configurations. In other words, the enumerations have regarded the isomers as two-dimentional objects or as chemical graphs. Moreover, the symmetries of the isomers have attracted little attention of chemists, because there exist no effective methods for enumerating molecular symmetries.

Enumeration of adamantane isomers. The above discussions indicate that our target is an enumeration with respect not only to PMIs but also to symmetries. The present method of enumerations contains the following steps.

1. Step 1 is classification of the edges of a parent skeleton into orbits and determination of coset representations. The six edges of the skeleton (**1**) construct an orbit that is subject to a coset representaion, $\mathbf{T}_d(/\mathbf{C}_{2v})$. There are two methods of accomplishing such assignment (Chapter 6).

   Method I. This method requires the mark table of $\mathbf{T}_d$ (Appendix A). When we examine the skeleton (**1**) and count fixed edges regarding every subgroup, we obtain a fixed-point vector, *i.e.*, FPV = (6 2 2 0 0 0 2 0 0 0 0), whose elements are the number of fixed edges. This vector is identical to the $\mathbf{T}_d(/\mathbf{C}_{2v})$ row of the mark table (Appendix A). Hence, we arrive at the assignment of the six edges.

   Method II. Alternatively, the FPV is multiplied by the inverse of the mark table for $\mathbf{T}_d$ (Appendix B). Then we obtain a vector, (0 0 0 0 0 0 1 0 0 0 0), which shows the appearance of $\mathbf{T}_d(/\mathbf{C}_{2v})$.

2. Step 2 is construction of a subduced cycle index (SCI) from unit subduced cycle indices (USCIs). We have precalculated USCIs for $\mathbf{T}_d$ symmetry as shown in Appendix C. Using the $\mathbf{T}_d(/\mathbf{C}_{2v})$ row of this table, we obtain an SCI for every subsymmetry. Note that, in this case, the SCI is equal to the USCI, because there emerges only one orbit.

3. Step 3 is introduction of a ligand inventory into the SCI. For the purpose of manipulating this case, we introduce a ligand inventory,

$$s_d = 1 + x_1^d + x_2^d + \cdots + x_6^d, \tag{17.2}$$

into the above SCIs. Thereby, we obtain fixed-point-counting polynomials. Since the variable x_r is concerned with the substitution of $(CH_2)_r$, the term $x_1^{m_1} x_2^{m_2} \cdots x_6^{m_6}$ corresponds to the PMI $(1^{m_1} 2^{m_2} \cdots 6^{m_6})$.

$$s_1^6 = (1 + x_1 + \cdots + x_6)^6 \quad \text{for} \quad C_1, \tag{17.3}$$

$$s_1^2 s_2^2 = (1 + x_1 + \cdots + x_6)^2 (1 + x_1^2 + \cdots + x_6^2)^2 \quad \text{for} \quad C_2, \tag{17.4}$$

$$s_1^2 s_2^2 = (1 + x_1 + \cdots + x_6)^2 (1 + x_1^2 + \cdots + x_6^2)^2 \quad \text{for} \quad C_s, \tag{17.5}$$

$$s_3^2 = (1 + x_1^3 + \cdots + x_6^3)^2 \quad \text{for} \quad C_3, \tag{17.6}$$

$$s_2 s_4 = (1 + x_1^2 + \cdots + x_6^2)(1 + x_1^4 + \cdots + x_6^4) \quad \text{for} \quad S_4, \tag{17.7}$$

$$s_2^3 = (1 + x_1^2 + \cdots + x_6^2)^3 \quad \text{for} \quad D_2, \tag{17.8}$$

$$s_1^2 s_4 = (1 + x_1 + \cdots + x_6)^2 (1 + x_1^4 + \cdots + x_6^4) \quad \text{for} \quad C_{2v}, \tag{17.9}$$

$$s_3^2 = (1 + x_1^3 + \cdots + x_6^3)^2 \quad \text{for} \quad C_{3v}, \tag{17.10}$$

$$s_2 s_4 = (1 + x_1^2 + \cdots + x_6^2)(1 + x_1^4 + \cdots + x_6^4) \text{for} \quad D_{2d}, \tag{17.11}$$

$$s_6 = 1 + x_1^6 + \cdots + x_6^6 \quad \text{for} \quad T, \tag{17.12}$$

and

$$s_6 = 1 + x_1^6 + \cdots + x_6^6 \quad \text{for} \quad T_d, \tag{17.13}$$

We expand these equations and then collect terms of the same power. Table 17.1 lists the resulting coefficients of the index terms $(x_1^{m_1} x_2^{m_2} \cdots x_6^{m_6})$, where $m_1 + 2m_2 + \cdots + 6m_6 = 6$. These partitions of the integer 6 correpond to adamantane isomers.

Table 17.1: Coefficients appearing in the generating functions

index term	PMI	coefficients for										
		C_1	C_2	C_s	C_3	S_4	D_2	C_{2v}	C_{3v}	D_{2d}	T	T_{2d}
x_1^6	(1^6)	1	1	1	1	1	1	1	1	1	1	1
$x_1^4 x_2$	$(1^4 2)$	30	2	2	0	0	0	2	0	0	0	0
$x_1^3 x_3$	$(1^3 3)$	60	4	4	0	0	0	0	0	0	0	0
$x_1^2 x_2^2$	$(1^2 2^2)$	90	6	6	0	0	6	0	0	0	0	0
$x_1^2 x_4$	$(1^2 4)$	60	4	4	0	0	0	0	0	0	0	0
$x_1 x_2 x_3$	(123)	120	0	0	0	0	0	0	0	0	0	0
$x_1 x_5$	(15)	30	2	2	0	0	0	2	0	0	0	0
x_2^3	(2^3)	20	4	4	2	0	0	0	2	0	0	0
$x_2 x_4$	(24)	30	2	2	0	0	0	2	0	0	0	0
x_3^2	(32)	15	3	3	0	1	3	1	0	1	0	0
x_6	(6)	6	2	2	0	0	0	2	0	0	0	0

4. In Step 4, Table 17.1 is regarded as a matrix, which is then multiplied by the

inverse of the mark table for $\mathbf{T}_d$ (Appendix B). The resulting matrix is found in Table 17.2, in which each row is concerned with an index term (or PMI) and each column corresponds to a subsymmetry; and their intersection is the number of adamantane isomers with the PMI and the subsymmetry.

Table 17.2: Number of adamantane isomers ($C_{10}H_{16}$)

index term	PMI	number of adamantane isomers of											total for PMI
		C_1	C_2	C_s	C_3	S_4	D_2	C_{2v}	C_{3v}	D_{2d}	T	T_{2d}	
x_1^6	(1^6)	0	0	0	0	0	0	0	0	0	0	1	1
$x_1^4 x_2$	$(1^4 2)$	1	0	0	0	0	0	1	0	0	0	0	2
$x_1^3 x_3$	$(1^3 3)$	1	1	2	0	0	0	0	0	0	0	0	4
$x_1^2 x_2^2$	$(1^2 2^2)$	2	0	3	0	0	1	0	0	0	0	0	6
$x_1^2 x_4$	$(1^2 4)$	1	1	2	0	0	0	0	0	0	0	0	4
$x_1 x_2 x_3$	(123)	5	0	0	0	0	0	0	0	0	0	0	5
$x_1 x_5$	(15)	1	0	0	0	0	0	1	0	0	0	0	2
x_2^3	(2^3)	0	1	0	0	0	0	0	2	0	0	0	3
$x_2 x_4$	(24)	1	0	0	0	0	0	1	0	0	0	0	2
x_3^2	(3^2)	0	0	1	0	0	0	0	0	1	0	0	2
x_6	(6)	0	0	0	0	0	0	1	0	0	0	0	1
total		12	3	8	0	0	1	4	2	1	0	1	32

Figure 17.3 depicts the 32 isomers enumerated in Table 17.2. Each of the isomers is accompanied with its PMI and its symmetry. As for each chiral molecule, an appropriate enantiomer is depicted. Note that this enumeration counts a pair of enantiomers once, according to the mathematical formulation. The total number (32) is equal to the value obtained by Balaban.[2] However, the present result provides more detailed information on the symmetry of the isomers, which has never been discussed. Thus, we are able to conclude that there emerge 16 chiral isomers, among which 12 isomers are asymmetric (C_1), three belong to C_2 and one is a D_2 isomer. Among the three C_2 molecules, it is easy to find the respective two-fold axes of **6** and **16** in terms of the perspectives in Fig. 17.3. However, the two-fold axis of the other C_2 molecule (**26**) is somewhat difficult to find. In order to do this task, it is helpful to refer to the desymmetrization procedure depicted in Fig. 17.2 or to construct a molecular model of this molecule.

There emerge no isomers belonging to C_3, S_4 or T symmetry. The non-existence of these symmetries is not accidental but inherent to the $\mathbf{T}_d(/C_{2v})$ orbit.

---

[2]Balaban's enumeration is concerned with constitution of isomers, because he considered the tetrahedrane skeleton to be a graph. See ref. [31].

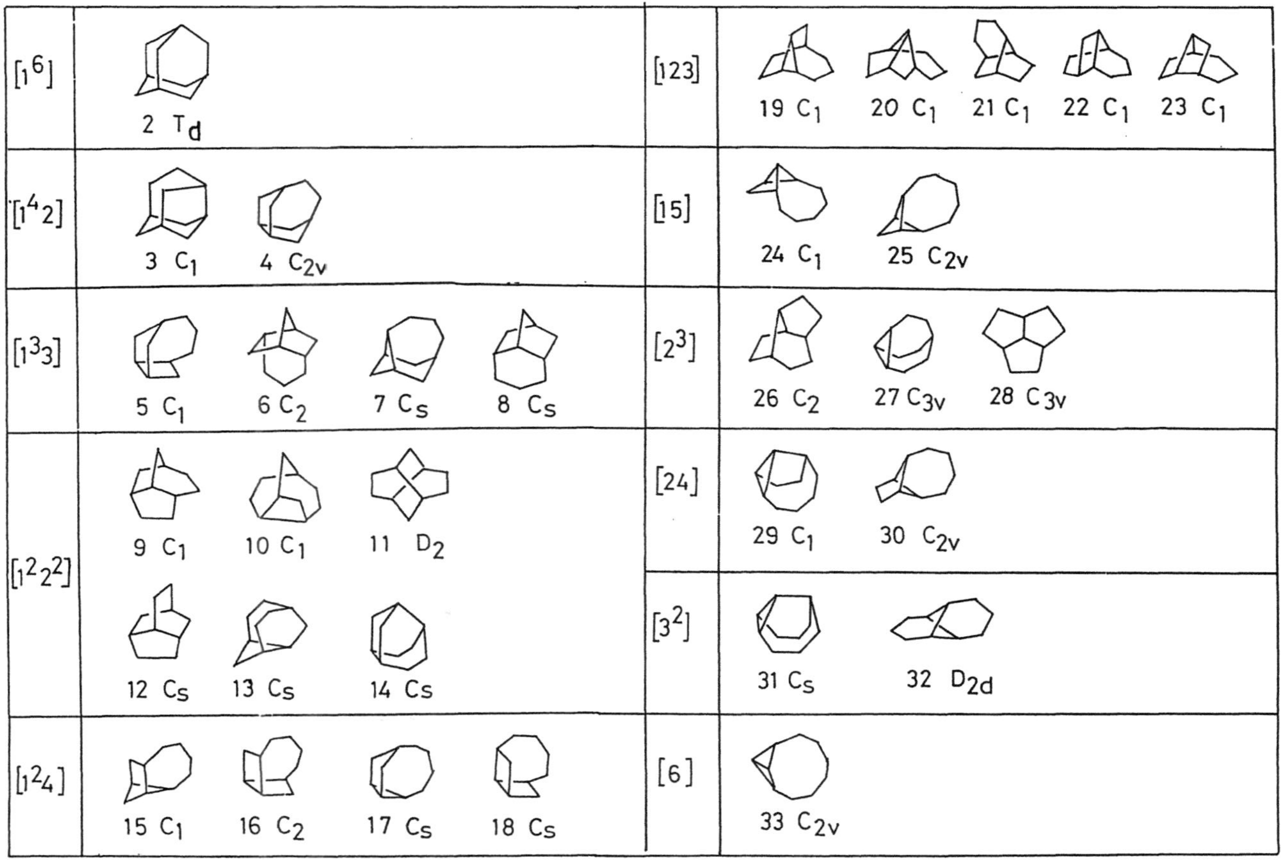

Figure 17.3: Adamantane Isomers derived from a tetrahedrane skeleton (1)

In other words, these symmetries are not realized even if we take account of substitution of any other units. This fact can be rationalized by means of a desymmetrization lattice, as discussed in Chapter 9. Thus, the C_3 molecule does not exist, since the USCI (s_3^2) of $T_d(/C_{2v}){\downarrow}\, C_3$ is equal to that (s_3^2) of $T_d(/C_{2v}){\downarrow}\, C_{3v}$. The non-existence of S_4 molecule is deduced by the fact that the USCI ($s_2 s_4$) of $T_d(/C_{2v}){\downarrow}\, S_4$ is equal to that ($s_2 s_4$) of $T_d(/C_{2v}){\downarrow}\, D_{2d}$. The T molecule can be concluded to be absent by comparing the USCI (s_6) of $T_d(/C_{2v}){\downarrow}T$ with that (s_6) of $T_d(/C_{2v}){\downarrow}\, T_d$.

When we compare a set of four ($1^3 3$) isomers with that of four ($1^2 4$) isomers, we find that they construct complemental series. This situation becomes clearer if the PMIs are rewritten in full forms *i.e.*, $0^2 1^3 3^1$ and $0^3 1^2 4^1$. Compare their powers and consider their meaning. A similar relationship is observed among ($1^4 2$), (15), and (24) series, since these PMIs can be converted into ($0^1 1^4 2^1$), ($0^4 1^1 5^1$), ($0^4 2^1 4^1$), respectively.

Enumeration of homologs. The enumeration of noradamantane isomers is accomplished in a similar way. In this case, we consider m_1 methylenes, m_2 ethylenes, $\cdots$, m_5 pentamethylenes as substituents for the edges of the tetrahedrane skeleton (1), where $m_1 + 2m_2 + \cdots + 5m_5 = 5$. The terms at issue are x_1^5, $x_1^3 x_2$, $x_1^2 x_3$, $x_1 x_2^2$, $x_1 x_4$, $x_2 x_3$, and x_5 in eqs. 17.3–17.13. The coefficients of the respective terms are calculated in the form of generating functions that are obtained in the process of expanding the equations. If our target is concerned only with noradamantane isomers, we can use another ligand inventory,

$$s_d = 1 + x_1^d + x_2^d + \cdots + x_5^d, \tag{17.14}$$

in place of eq. 17.2. Thereby, we can obtain the same results. The matrix containing the resulting coefficients is multiplied by the inverse of the mark table according to Step 4 above. The results are found in Table 17.3. There emerge 18 isomers in all. When we pay attention to two sets of isomers (four ($1^2 3$) and four (12^2) isomers), we find that they also construct complemental series, since the PMIs can be rewritten in full forms *i.e.*, ($0^2 1^2 3^1$) and ($0^2 1^1 2^2$). Their powers construct the same partition of the integer (5). A similar result is obtained for the set of four ($1^3 2$) isomers (full PMI: ($0^1 1^3 2^1$)); however, this series is not complemental with the two sets. Figure 17.4 illustrates these isomers.

The numbers of homoadamantane isomers are calculated in a similar way. The results have been reported in the orginal paper.[32] The edge strategy is versatile to create cage-shaped molecules by starting from achiral skeletons of high symmetries. In addition to the T_d skeleton above, the strategy has also been applied to several D_{3h} skeletons[33] and a dodecahedrane (I_h) skeleton.[34] Remaining skeletons to be examined will be cubane and octahedron of O_h symmetry. Finally, it is to be noted that each edge employed in the edge strategy should have two equivalent

$[_1 5]$	 34 C_{2v}		[5]	 51 C_{2v}
$[_1^3 2]$	 35 C_1	 36 C_2	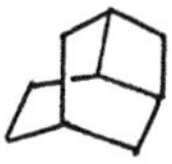 37 C_S	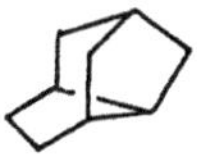 38 C_S
$[_1^2 3]$	 39 C_1	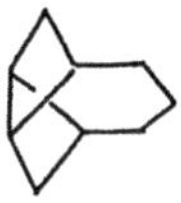 40 C_2	 41 C_S	 42 C_S
$[_1 2^2]$	 43 C_1	 44 C_2	 45 C_S	 46 C_S
$[_1 4]$	 47 C_1	 48 C_{2v}		
$[23]$	 49 C_1	 50 C_{2v}		

Figure 17.4: Noradamantane Isomers derived from a tetrahedrane skeleton (1)

Table 17.3: Number of noradamantane isomers (C_9H_{14})

index term	PMI	number of noradamantane isomers of											total for PMI
		C_1	C_2	C_s	C_3	S_4	D_2	C_{2v}	C_{3v}	D_{2d}	T	T_{2d}	
x_1^5	(1^5)	0	0	0	0	0	0	1	0	0	0	0	1
$x_1^3 x_2$	$(1^3 2)$	1	1	2	0	0	0	0	0	0	0	0	4
$x_1^2 x_3$	$(1^2 3)$	1	1	2	0	0	0	0	0	0	0	0	4
$x_1 x_2^2$	$(1 2^2)$	1	1	2	0	0	0	0	0	0	0	0	4
$x_1 x_4$	(14)	1	0	0	0	0	0	1	0	0	0	0	2
$x_2 x_3$	(23)	1	0	0	0	0	0	1	0	0	0	0	2
x_5	(5)	0	0	0	0	0	0	1	0	0	0	0	1
total		5	3	6	0	0	0	4	0	0	0	0	18

terminal vertices. Otherwise, subsymmetries produced may be complicated.

17.3 Use of Another Ligand-Inventory

Ligand-inventories can be selected in accord with our targets. If we use an alternative ligand-inventory,

$$s_d = 1 + x^d + x^{2d} + \cdots + x^{6d}, \tag{17.15}$$

in place of eq. 17.2, we obtain a summarized result. Note that the corresponding generating functions regarding eq. 17.15 are those in which the variables (x_r) of eqs. 17.3–17.13 are replaced by x^r. Table 17.4 collects the coefficients of the terms x^r ($r = 0$ to 8).

Table 17.4 regarded as a matrix is multiplied by the inverse of the mark table for T_d (Appendix B) to afford Table 17.5; the terms (x^6 and x^5) indicate the isomers of adamantane and noradamantane, respectively. Hence, the row for x^6 (adamantane isomers) is equal to the bottom of Table 17.2 that is obtained by summing up each column. Similarly, the row for x^5 (noradamantane isomers) in Table 17.5 is identical with the bottom row of Table 17.3.

17.4 New Type of Cycle Index

In Chaper 16, we have developed a new type of cycle index, which is derived from unit subduced cycle indices (USCIs) *via* a subduced cycle index (SCI). This type of cycle index is equivalent to Pólya's cycle index and is also effective to give a generating function for isomer enumeration.

Table 17.4: Coefficients derived by the ligand inventory (eq. 17.15)

index term[a]	Coefficients[b]										
	C_1	C_2	C_s	C_3	S_4	D_2	C_{2v}	C_{3v}	D_{2d}	T	T_{2d}
x^8	1251	51	51	0	3	15	11	0	3	0	0
x^7	786	38	38	0	0	0	10	0	0	0	0
x^6	462	30	30	3	2	10	10	3	2	1	1
x^5	252	20	20	0	0	0	8	0	0	0	0
x^4	126	14	14	0	2	6	6	0	2	0	0
x^3	56	8	8	2	0	0	4	2	0	0	0
x^2	21	5	5	0	1	3	3	0	1	0	0
x	6	2	2	0	0	0	2	0	0	0	0
1	1	1	1	1	1	1	1	1	1	1	1

[a] The index terms (x^7, x^6, and x^5) correspond to homoadamantane, adamantane, and noradamantane isomers.

[b] The coefficients of the term x^m are obtained by introducing the ligand inventory (eq. 17.15) into the SCIs (eqs. 17.3 to 17.13).

Table 17.5: Numbers of adamantane homologs of C_4 to C_{12}

index term[a]	Number of isomers											total number
	C_1	C_2	C_s	C_3	S_4	D_2	C_{2v}	C_{3v}	D_{2d}	T	T_{2d}	
x^8	37	7	20	0	0	1	4	0	3	0	0	72
x^7	21	7	14	0	0	0	5	0	0	0	0	47
x^6	12	3	8	0	0	1	4	2	1	0	1	32
x^5	5	3	6	0	0	0	4	0	0	0	0	18
x^4	2	1	4	0	0	0	2	0	2	0	0	11
x^3	1	1	0	0	0	0	2	2	0	0	0	6
x^2	0	0	1	0	0	0	1	0	1	0	0	3
x	0	0	0	0	0	0	1	0	0	0	0	1
1	0	0	0	0	0	0	0	0	0	0	1	1

[a] See the footnote of Table 17.4.

The total number of isomers with the index term (x^r) is calculated in the form of a generating function. We again use the $\mathbf{T}_d(/\mathbf{C}_{2v})$ row of the USCI table of $\mathbf{T}_d$ (Appendix C). Additionally, we utilize the factors collected at the bottom of the same table. These factors have been obtained by summing up the respective rows of the inverse of a mark table (Appendix B). We thereby obtain a cycle index (CI),

$$
\begin{aligned}
CI(\mathbf{T}_d; s_d) &= \frac{1}{24}s_1^6 + \frac{1}{8}s_1^2 s_2^2 + \frac{1}{4}s_1^2 s_2^2 + \frac{1}{3}s_3^2 + \frac{1}{4}s_2 s_4 \\
&= \frac{1}{24}(s_1^6 + 9 s_1^2 s_2^2 + 8 s_3^2 + 6 s_2 s_4).
\end{aligned}
\tag{17.16}
$$

This CI can be proved to be identical to that derived alternatively by Pólya's theorem (Chapter 16).

Introduction of eq. 17.15 into eq. 17.16 affords a generating function,

$$
\begin{aligned}
G(x) &=. \; CI(\mathbf{T}_d; 1 + x^d + x^{2d} + \cdots + x^{6d}) \\
&= (1 + x^{36}) + (x + x^{35}) + 3(x^2 + x^{34}) + 6(x^3 + x^{33}) + 11(x^4 + x^{32}) \\
&\quad + 18(x^5 + x^{31}) + 32(x^6 + x^{30}) + 47(x^7 + x^{29}) + 72(x^8 + x^{28}) \\
&\quad + 102(x^9 + x^{27}) + 140(x^{10} + x^{26}) + 182(x^{11} + x^{25}) + 235(x^{12} + x^{24}) \\
&\quad + 282(x^{13} + x^{23}) + 334(x^{14} + x^{22}) + 378(x^{15} + x^{21}) + 414(x^{16} + x^{20}) \\
&\quad + 434(x^{17} + x^{19}) + 447 x^{18}.
\end{aligned}
\tag{17.17}
$$

The coefficients of the terms x^r ($r = 0$ to 8) in eq. 17.17 are equal to the values of the rightmost column of Table 17.5, which are obtained by summing up the respective rows.

The number of isomers with a specific subsymmetry can be obtained by using Theorem 16.3 or Corollary 16.1 (Chapter 16). For example, we have a PCI for $\mathbf{C}_{3v}$ group,

$$
PCI'(\mathbf{C}_{3v}; s_d) = s_3^2 - s_6.
\tag{17.18}
$$

When we introduce the ligand-inventory (eq. 17.15) into this PCI, we have a generating function,

$$
\begin{aligned}
G'(x) &= PCI'(\mathbf{C}_{3v}; 1 + x^d + x^{2d} + \cdots + x^{6d}) \\
&= 2(x^3 + x^{33}) + 2(x^6 + x^{30}) + 4(x^9 + x^{27}) + \\
&\quad 4(x^{12} + x^{24}) + 6(x^{15} + x^{21}) + x^{18}.
\end{aligned}
\tag{17.19}
$$

The coefficients are identical with the $\mathbf{C}_{3v}$ column of Table 17.5.

Bibliography

[1] A. Nickon, E. F. Silversmith, *Organic Chemistry: The Name Game*, Pergamon, Oxford (1987).

[2] a) M. Nakazaki, K. Naemura, *Yuki Gosei Kagaku Kyokai Shi*, **35**, 883 (1977).
b) M. Nakazaki, *Top. Stereochem.*, **15**, 199 (1984).

[3] K. Naemura, *Yuki Gosei Kagaku Kyokai Shi*, **45**, 48 (1987).

[4] P. E. Eaton, *Tetrahedron*, **35**, 2189 (1979).

[5] L. A. Paquette, *Chem. Rev.*, **89**, 1051 (1989).

[6] V. Prelog, J. Thix, *Helv. Chim. Acta*, **65**, 2622 (1982).

[7] S. Fujita, *J. Chem. Educ.*, **63**, 744 (1986).

[8] T. J. Katz, N. Acton, *J. Am. Chem. Soc.*, **95**, 2738 (1973).

[9] C. A. Cupas, L Hodakowski, *J. Am. Chem. Soc.*, **96**, 4668 (1974).

[10] H. Tobler, R. O. Klaus, C. Ganter, *Helv. Chim. Acta*, **58**, 1455 (1975).

[11] D. P. G. Hamon, G. F. Taylor, *Aust. J. Chem.*, **29**, 1721 (1976).

[12] K. Komatsu, Y. Jinbu, G. R. Gillette, R. West, *Chem. Lett.*, 2029 (1988).

[13] Y. Sekine, V. Boekelheide, *J. Am. Chem. Soc.*, **103**, 1777 (1981).

[14] P. E. Eaton, Y. S. Or, S. J. Branca, *J. Am. Chem. Soc.*, **103**, 2134 (1981).

[15] G. Maier, S. Pfriem, U. Schäfer, K.-D. Malsch, R. Matusch, *Chem. Ber.*, **114**, 3965 (1981).

[16] P. E. Eaton, t. W. Cole, *J. Am. Chem. Soc.*, **86**, 962, 3157 (1964).

[17] L. A. Paquette, R. J. Ternansky, D. W. Balogh, G. Kentgen, *J. Am. Chem. Soc.*, **105**, 5446 (1983).

[18] K. Adachi, K. Naemura, M. Nakazaki, *Tetrahedron Lett.*, 5467 (1968).

[19] M. Tichy, J. Sicher, *Tetrahedron Lett.*, 4609 (1969).

[20] G. R. Underwood, B. Ramamoorthy, *Tethahedron Lett.*, 4125 (1970).

[21] G. Helmchen, G. Staiger, *Angew. Chem. Intern. Ed.*, **16** 117 (1977).

[22] M. Nakazaki, K. Naemura, N. Arashiba, *J. Org. Chem.*, **43**, 689 (1978).

[23] M. Farina, C. Morandi, *Tetrahedron*, **30**, 1819 (1974).

[24] W. Prelog, G. Helmchen, *Helv. Chim. Acta*, **55**, 2581 (1972).

[25] a) P. v. R. Schleyer, *J. Am. Chem. Soc.*, **79**, 3292 (1957). b) R. C. Fort, Jr, P. v. R. Schleyer, *Chem. Rev.*, **64**, 277 (1964).

[26] H. W. Whitlock, Jr., M. W. Siefken, *J. Am. Chem. Soc.*, **90**, 4929 (1968).

[27] E. M. Engler, M. Farsacin, A. Sevin, J. M. Cense, P. v. R. Schleyer, *J. Am. Chem. Soc.*, **95**, 5769 (1973).

[28] a) S. A. Godleski, P.v.Schleyer, E. Osawa, W. T. Wipke, *Prog. Phys. Org. Chem.*, **13**, 63 (1981). b) M. A. Mckervey, *Chem. Soc. Rev.*, **3**, 479 (1974).

[29] a) F. J. Jäggi and C. Ganter, *Helv. Chim. Acta*, **63**, 866 (1980). b) K. L. Ghatak, C. Ganter, *Helv. Chim. Acta*, **71**, 124 (1988).

[30] M. Farcasiu, E. W. Hagaman, e. Wenkert, P. v. R. Schleyer, *Tetrahedron Lett.*, **22**, 151 (1981).

[31] A. T. Balaban, *Rev. Roum. Chim.*, **31**, 795 (1986).

[32] S. Fujita, *Tetrahedron*, **46**, 365 (1990).

[33] S. Fujita, *Bull. Chem. Soc. Jpn.*, **63**, 1876 (1990).

[34] S. Fujita, *Bull. Chem. Soc. Jpn.*, **63**, 2759 (1990).

Chapter 18

Elementary Superposition[1]

Davidson[1] has pointed out that enumeration problems in chemistry can be solved by using the Redfield-Read superposition theorem, which was originally established by Redfield,[2] rediscovered independently by Read[3] and later refocussed by several mathematicians.[4, 5] Lloyd[6] has actually applied this theorem to chemical enumerations. All of these methods are capable of enumerating chemical objects with respect only to their weights (molecular formulas).

We ourselves[7] have reported a new enumeration method based on unit subduced cycle indices (USCIs) (Chapter 15). The USCIs can be transformed into a cycle index that has an explicitly different form but is essentially equivalent to Pólya's cycle index (Chapter 16).[8] The USCI approach has been applied to various enumeration problems.[9, 10, 11] Although this method is effective to chemical enumerations as well as to the related geometrical or graph-theoretical enumerations, there still remains an apparent disadvantage. Thus, the evaluation process of fixed points in this approach is based on the expansion of generating functions; this process should be simplified for some purposes. In a continuation of our work, we here propose the concept of *elementary superposition*, which provides a new methodology for solving the problem. From this concept, we are also able to derive the Redfield-Read theorem.

18.1   The USCI Approach

Since the USCI approach has been developed in Chapters 15 and 16, this section is devoted to giving a minimum set of equations for the present formulation. Any permutation representation ($\mathbf{P_G}$) acting on a domain (Δ) can be reduced into a sum of CRs in the light of

$$\mathbf{P_G} = \sum_{i=1}^{s} \alpha_i \mathbf{G}(/\mathbf{G}_i), \tag{18.1}$$

where α_i (≥ 0) is the multiplicity of $\mathbf{G}(/\mathbf{G}_i)$ (Theorem 5.7 in Chapter 5).

---

[1]Reprinted in part with permission from S. Fujita, *Bull. Chem. Soc. Jpn.*, **63**, 2770–2775 (1990). ©(1990) The Chemical Society of Japan.

We have defined a *unit subduced cycle index (USCI)*, a *subduced cycle index (SCI)*, a *partical cycle index (PCI)*, and a *cycle index (CI)*.[2]

Definition 18.1 (USCI, SCI, PCI and CI)
(1) A unit subduced cycle index (USCI) is defined as

$$Z(\mathbf{G}(/\mathbf{G}_i) \downarrow \mathbf{G}_j; s_{d_{jk}}) = \prod_{k=1}^{v_j} (s_{d_{jk}})^{\beta_k^{(ij)}}, \tag{18.2}$$

for $i = 1, 2, \ldots, s$ and $j = 1, 2, \ldots, s$, wherein

$$d_{jk} = \mid \mathbf{G}_j \mid / \mid \mathbf{G}_{jk} \mid. \tag{18.3}$$

See Def. 9.2 (Chapter 9).
(2) A subduced cycle index (SCI) is defined as

$$ZI'(\mathbf{G}_j; s_{d_{jk}}) = \prod_{\substack{i=1 \\ }}^{s} \prod_{\substack{\alpha=1 \\ (\alpha_i \neq 0)}}^{\alpha_i} Z(\mathbf{G}(/\mathbf{G}_i) \downarrow \mathbf{G}_j; s_{d_{jk}}) \tag{18.4}$$

$$= \prod_{k=1}^{v_j} (s_{d_{jk}})^{\sum_{i=1}^{s} \alpha_i \beta_k^{(ij)}} \tag{18.5}$$

$$\equiv \sum_{(\mu)} b_\mu s_1^{\mu_1} s_2^{\mu_2} \cdots s_q^{\mu_q}, \tag{18.6}$$

for $j = 1, 2, \ldots, s$, where the cycle structure (μ) is represented by

$$1\mu_1 + 2\mu_2 + \ldots + q\mu_q = q \quad (q = \mid \Delta \mid). \tag{18.7}$$

This cycle structure is dependent upon j; however, this dependence is not denoted for simplicity. See Def. 15.2 (Chapter 15).
(3) A partial cycle index (PCI) is defined as

$$PCI'(\mathbf{G}_i; s_{d_{jk}}) = \sum_{j=1}^{s} \overline{m}_{ji} ZI'(\mathbf{G}_j; s_{d_{jk}}) \ (i = 1, 2, \ldots, s), \tag{18.8}$$

where $\overline{m}_{ji}$ is the ji-element of the inverse of the mark table of $\mathbf{G}$. See Def. 16.4 (Chapter 16).
(4) A cycle index (CI) is defined as

$$CI'(\mathbf{G}; s_{d_{jk}}) = \sum_{j=1}^{s} (\sum_{i=1}^{s} \overline{m}_{ji}) ZI'(\mathbf{G}_j; s_{d_{jk}}). \tag{18.9}$$

See Def. 16.2 and Def. 16.5 (Chapter 16).

---

[2]We here take no account of OMV restriction.

Suppose that θ_1 of X_1, θ_2 of X_2, ..., and $\theta_{|\mathbf{X}|}$ of $X_{|\mathbf{X}|}$ are selected from a codomain, $\mathbf{X} = \{X_1, X_2, \ldots, X_{|\mathbf{X}|}\}$, wherein

$$[\theta] : \theta_1 + \theta_2 + \ldots + \theta_{|\mathbf{X}|} = \mid \Delta \mid . \tag{18.10}$$

The corresponding configuration, $f : \Delta \to \mathbf{X}$, has a weight (molecular formula) of

$$W_\theta = X_1^{\theta_1} X_2^{\theta_2} \cdots X_{|\mathbf{X}|}^{\theta_{|\mathbf{X}|}}. \tag{18.11}$$

The number $(\rho_{\theta j})$ of fixed configurations having W_θ and $\mathbf{G}_j$ is evaluated by the following lemma. This has been proved in Chapter 15 (Lemma 15.3).

Lemma 18.1 (Evaluation of $\rho_{\theta j}$)
The numbers ($\rho_{\theta j}$'s) of fixed configurations is given as the coefficients of generating functions,

$$\sum_{[\theta]} \rho_{\theta j} W_\theta = ZI'(\mathbf{G}_j; s_{d_{jk}}) \quad (j = 1, 2, \ldots, s), \tag{18.12}$$

where the variables in the right-hand side are replaced by ligand-inventories,

$$s_{d_{jk}} = \sum_{\ell=1}^{|\mathbf{X}|} X_\ell^{d_{jk}}. \tag{18.13}$$

Let the symmbol $A_{\theta i}$ be the number of orbits of configurations with W_θ and $\mathbf{G}_i$. If we consider a set containing all of the configurations with W_θ, the $A_{\theta i}$ is the number of orbits within this set of configurations. Hence, this is obtained by means of following theorem. The proof has been described in Chapter 15 (Theorem 15.4).

Theorem 18.1

$$A_{\theta i} = \sum_{j=1}^{s} \rho_{\theta j} \overline{m}_{ji} \quad (i = 1, 2, \ldots, s). \tag{18.14}$$

The following example illustrates an enumeration procedure based on the USCI approach. This approach can be applied to any combinatorial enumerations, although we select this example from a chemical field.

Example **18.1** (Derivatives of adamantane-2,6-dione (**1**))
Let us enumerate molecules derived from an adamantanedione (**1**). This parent molecule has $\mathbf{D}_{2d}$ symmetry. The twelve hydrogens of the molecule are considered to be substitution positions, which are replaced by X, Y or Z.

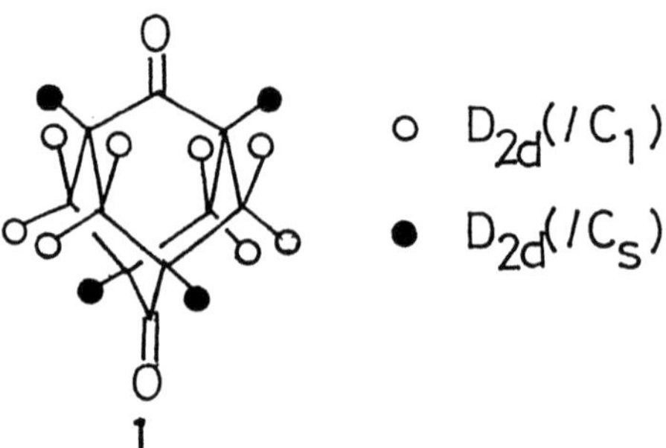

The twelve positions are divided into two orbits that are denoted by solid circles (4 bridgehead positions) and open circles (8 bridge positions). Assignment of CRs to these orbits is accomplished by examining a fixed-point vector (FPV). In this case, we obtain FPV = (12 0 0 2 0 0 0 0) by inspection of 1. The FPV is multiplied by the inverse of the mark table (Appendix B), giving a row vector, (1 0 0 1 0 0 0 0). It follows that

$$\mathbf{P_{D_{2d}}} = \mathbf{D}_{2d}(/\mathbf{C}_1) + \mathbf{D}_{2d}(/\mathbf{C}_s). \tag{18.15}$$

Obviously, the set of the 8 bridge positions is subject to $\mathbf{D}_{2d}(/\mathbf{C}_1)$ and that of the 4 bridgehead positions is governed by $\mathbf{D}_{2d}(/\mathbf{C}_s)$. Hence, we adopt USCIs appearing in the $\mathbf{D}_{2d}(/\mathbf{C}_1)$ and $\mathbf{D}_{2d}(/\mathbf{C}_s)$ rows of the USCI table (Appendix C). In accord with eq. 18.4, we construct an SCI for every subgroup:

$$\text{for } \mathbf{C}_1 \; : \; (s_1^8)(s_1^4) = (x + y + z)^{12} \tag{18.16}$$

$$\text{for } \mathbf{C}_2 \; : \; (s_2^4)(s_2^2) = (x^2 + y^2 + z^2)^6 \tag{18.17}$$

$$\text{for } \mathbf{C}_2' \; : \; (s_2^4)(s_2^2) = (x^2 + y^2 + z^2)^6 \tag{18.18}$$

$$\text{for } \mathbf{C}_s \; : \; (s_2^4)(s_1^2 s_2) = (x + y + z)^2(x^2 + y^2 + z^2)^5 \tag{18.19}$$

$$\text{for } \mathbf{S}_4 \; : \; (s_4^2)(s_4) = (x^4 + y^4 + z^4)^3 \tag{18.20}$$

$$\text{for } \mathbf{C}_{2v} \; : \; (s_4^2)(s_2^2) = (x^2 + y^2 + z^2)^2(x^4 + y^4 + z^4)^2 \tag{18.21}$$

$$\text{for } \mathbf{D}_2 \; : \; (s_4^2)(s_4) = (x^4 + y^4 + z^4)^3 \tag{18.22}$$

and

$$\text{for } \mathbf{D}_{2d} \; : \; (s_8)(s_4) = (x^4 + y^4 + z^4)(x^8 + y^8 + z^8) \tag{18.23}$$

The left-hand side of each equation denotes an SCI for each subgroup, in which the first set of parentheses contains the USCI of the $\mathbf{D}_{2d}(/\mathbf{C}_1)$ row and the second contains the USCI of $\mathbf{D}_{2d}(/\mathbf{C}_s)$. The right-hand side of each equation is obtained by introducing a ligand-inventory (for this case, $s_d = x^d + y^d + z^d$ in the light of Lemma 18.1).

The expansion of the right-hand sides of the equations affords generating functions, in which the $\rho_{\theta j}$ value with a partition $[\theta] = [\theta_1, \theta_2, \theta_3]$ ($\theta_1 + \theta_2 + \theta_3 = 12$) and with the G_j symmetry is given as the coefficient of the $x^{\theta_1} y^{\theta_2} z^{\theta_3}$ term. Obviously, the coefficients for $[\theta_1, \theta_2, \theta_3]$, $[\theta_2, \theta_3, \theta_1]$, $[\theta_3, \theta_1, \theta_2]$, $[\theta_1, \theta_3, \theta_2]$, $[\theta_2, \theta_1, \theta_3]$, and $[\theta_3, \theta_2, \theta_1]$ are equal to each other.

For illustrating a step of obtaining the numbers of isomers, we work out the case of $x^8 y^4$, which is alternatively represented by a ligand partition, [8,4,0]. We collect the coefficients of the $x^8 y^4$ term after the expansion of the above generating functions. We thereby obtain the FPV for the $x^8 y^4$ term, *i.e.*, FPV = (495 15 15 15 3 3 3 1). This FPV is multiplied by the inverse (Appendix B) to give a row vector, (54 2 6 6 1 1 1 1), which indicates the numbers in the order of the SSG. This means that, as $X_8 Y_4$-isomers, there emerge 54 C_1 (asymmetric) isomers, 2 C_2, 6 C_2', 6 C_s, isomers, one S_4, one C_{2v}, one D_2, and one D_{2d} isomer.

This type of calculation is repeated over all partitions $[\theta]$ to give Table 18.1, in which the intersection between $[\theta_1, \theta_2, \theta_3]$ and a subsymmetry indicates the number of isomers.

For the verification of the values in Table 18.1, Fig. 18.1 depicts isomers with [8,4,0] and symmetries other than C_1. The present enumeration clarifies both the molecular formula (or the ligand partition) and the symmetry of an isomer. Hence, we are able to draw a concrete structure of the isomer without difficulty.

18.2 Elementary Superposition

In the enumeration described in the previous section, the evaluation of the $\rho_{\theta j}$ value is accomplished by expanding generating functions (eq. 18.12). This method affords a full list of $\rho_{\theta j}$ over all of the $[\theta]$ partitions. However, there are many cases in which a specific case of $[\theta]$ is to be examined.

Suppose the domain Δ of G symmetry, which is originally divided in several orbits (eq. 18.1), is restricted within its subsymmetry G_j. This process results in the subduction of the domain,

$$\mathbf{P}_{\mathbf{G}_j} = \sum_{i=1}^{s} \alpha_i \mathbf{G}(/\mathbf{G}_i) \downarrow \mathbf{G}_j. \tag{18.24}$$

This subdivision of the domain is represented by the SCI (eq. 18.4 or eq. 18.5), which indicates a pattern of occupation of ligands. Although we consider a direct occupation of ligands on the subdivided domain in the previous treatment, we here introduce a slightly different formulation.

We consider a ligand partition $[\theta]$ represented by eq. 18.10. Let $\mathbf{S}^{[\theta_r]}$ be the symmetric group of degree θ_r associated with each ligand set ($r = 1, 2, \ldots, |\mathbf{X}|$).

Table 18.1: Number of Compounds Derived From The Skeleton (1)

Ligand partition	Number of isomers with								Total
	C_1	C_2	C_2'	C_s	S_4	C_{2v}	D_2	D_{2d}	
[12,0,0]	0	0	0	0	0	0	0	1	1
[11,1,0]	1	0	0	1	0	0	0	0	2
[10,2,0]	5	1	3	2	0	1	0	0	12
[10,1,1]	16	0	0	1	0	0	0	0	17
[9,3,0]	25	0	0	5	0	0	0	0	30
[9,2,1]	80	0	0	5	0	0	0	0	85
[8,4,0]	54	2	6	6	1	1	1	1	72
[8,3,1]	245	0	0	5	0	0	0	0	250
[8,2,2]	353	7	15	14	0	1	0	0	390
[7,5,0]	94	0	0	10	0	0	0	0	104
[7,4,1]	490	0	0	10	0	0	0	0	500
[7,3,2]	980	0	0	20	0	0	0	0	1000
[6,6,0]	104	4	10	8	0	2	0	0	128
[6,5,1]	688	0	0	10	0	0	0	0	698
[6,4,2]	1696	14	30	28	0	2	0	0	1770
[6,3,3]	2300	0	0	20	0	0	0	0	2320
[5,5,2]	2064	0	0	30	0	0	0	0	2094
[5,4,3]	3450	0	0	30	0	0	0	0	3480
[4,4,4]	4278	18	42	42	3	3	3	0	4389

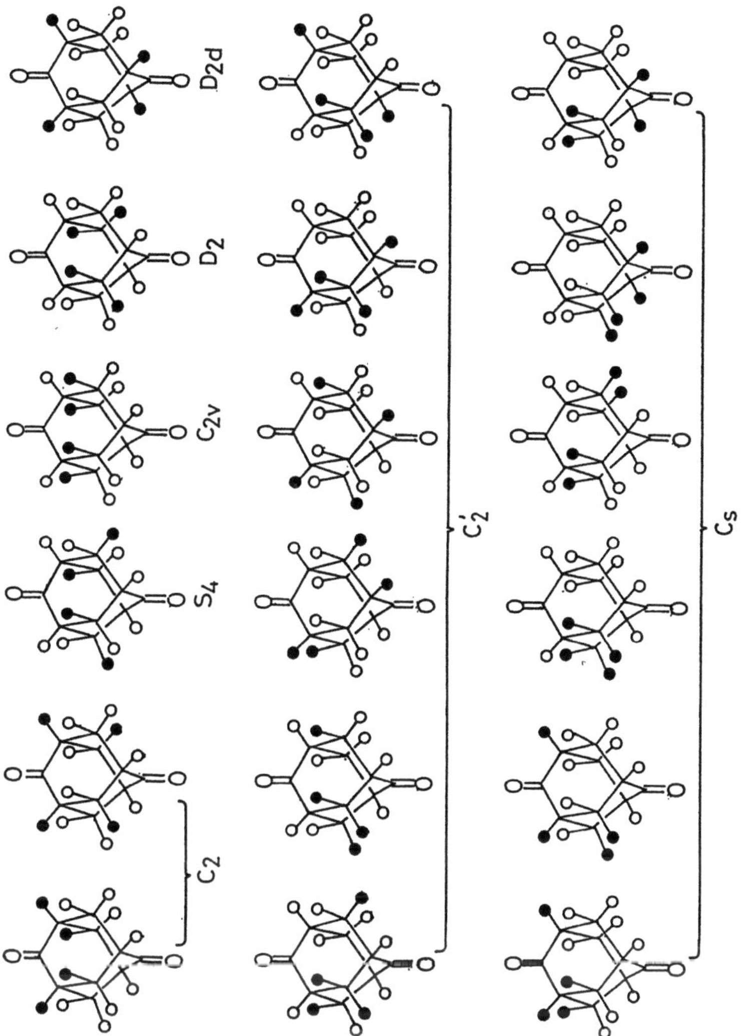

Figure 18.1: [8,4,0]-Molecules derived from adamantane-2,6-dione

We define a cycle index for the symmetric group of degree θ_r as being

$$CI(\mathbf{S}^{[\theta_r]}; s) = \frac{1}{\theta_r!} \sum_{(\nu(\theta_r))} n_{(\nu(\theta_r))} s_1^{\nu_1(\theta_r)} s_2^{\nu_2(\theta_r)} \cdots s_{\theta_r}^{\nu_{\theta_r}(\theta_r)}, \tag{18.25}$$

where the cycle structure $(\nu(\theta_r))$ is represented by

$$(\nu(\theta_r)) : 1\nu_1(\theta_r) + 2\nu_2(\theta_r) + \ldots + \theta_r \nu_{\theta_r}(\theta_r) = \theta_r, \tag{18.26}$$

and where

$$n_{(\nu(\theta_r))} = \frac{\theta_r!}{1^{\nu_1(\theta_r)}\nu_1(\theta_r)! 2^{\nu_2(\theta_r)}\nu_2(\theta_r)! \cdots \theta_r^{\nu_{\theta_r}(\theta_r)}\nu_{\theta_r}(\theta_r)!}. \tag{18.27}$$

We can construct a direct product of such symmetric groups, *i.e.*,

$$\mathbf{H} = \mathbf{S}^{[\theta_1]} \times \mathbf{S}^{[\theta_2]} \times \cdots \times \mathbf{S}^{[\theta_{|\mathbf{X}|}]}, \tag{18.28}$$

where $\mathbf{S}^{[\theta_r]}$ is effective only if θ_r is positive.

The cycle index of the direct product (eq. 18.28) is the product of the cycle indices of the factors. Hence, eq. 18.25 affords

$$CI(\mathbf{H}; s) = \prod_{r=1}^{|\mathbf{X}|} CI(\mathbf{S}^{[\theta_r]}; s) \tag{18.29}$$

$$\equiv \sum_{(\mu')} a_{\mu'} s_1^{\mu'_1} s_2^{\mu'_2} \cdots s_q^{\mu'_q}, \tag{18.30}$$

where the cycle structure (μ') is given by an equation similar to eq. 18.7.

Now, we consider an *elementary superposition* of $\mathbf{P}_{\mathbf{G}_j}$ (eq. 18.24) and $\mathbf{H}$ (eq. 18.28), both of which act on $|\Delta|$ elements. The term "elementary" stems from the fact that the SCI (eq. 18.5 corresponding to the partition represented by eq. 18.24) is a specific monomial but not a polynomial. Hence, the corresponding equivalent expression (eq. 18.6) indicates that $b_{\mu(\theta)} = 1$ for only one specific cycle structure $(\mu_{(\theta)})$ given by the SCI; otherwise $b_{\mu(\theta)} = 0$. The symbol $\mu_{(\theta)}$ is used to indicate that the cycle structure (eq. 18.7) is associated with the $[\theta]$ partition (eq. 18.10) through the condition of $(\mu) = (\mu')$. Although a strict proof is here abbreviated, we can obtain the following theorem.[3]

Theorem 18.2 (Elementary superposition)
The $\rho_{\theta j}$ value is evaluated by means of

$$\rho_{\theta j} = a_{\mu(\theta)}(1^{\mu_1(\theta)}\mu_1(\theta)! 2^{\mu_2(\theta)}\mu_2(\theta)! \cdots q^{\mu_q(\theta)}\mu_q(\theta)!) \tag{18.31}$$

for $j = 1, 2, \ldots, s$.

---

[3]The proof involves the application of the Cauchy-Frobenius lemma to the action of **H**. This will be reported elsewhere.

When we apply an operation $(*)$ introduced by Read[3] to this case, we are able to convert eq. 18.31 into

$$
\begin{aligned}
\rho_{\theta j} \;=\;& \sum_{(\mu)=(\mu')} a_{\mu'} b_\mu (1^{\mu_1}\mu_1! 2^{\mu_2}\mu_2! \cdots q^{\mu_q}\mu_q!) \\
\overset{\text{def.}}{=}\;& N\Big\{ \sum_{(\mu)=(\mu')} a_{\mu'} b_\mu (1^{\mu_1}\mu_1! 2^{\mu_2}\mu_2! \cdots q^{\mu_q}\mu_q!) s_1^{\mu_1} s_2^{\mu_2} \cdots s_q^{\mu_q} \Big\} \\
\overset{\text{def.}}{=}\;& N\{ CI(\mathbf{H}; s) * ZI'(\mathbf{G}_j; s_{d_{jk}}) \},
\end{aligned}
\tag{18.32}
$$

where the symbol $N\{\cdots\}$ generally denotes the sum of the coefficients appearing in the operand polynomial $(\cdots)$; in this case, however, the operand $(\cdots)$ is a monomial as discussed above. Equation 18.32 is an alternative expression of the elementary superposition (Theorem 18.2).

Now that we have arrived at such an alternative evaluation of $\rho_{\theta j}$, we introduce this equation to Theorem 18.1 in order to obtain the numbers $(A_{\theta i})$. For illustrating the present approach, we re-examine Example 18.1.

Example **18.2** (Enumeration based on the elementary superposition)
Let us obtain the number of $X_8 Y_4$-isomers with each subsymmetry which are derived from the parent skeleton (1). In this case, we consider

$$
\mathbf{P_{G_j}} = \mathbf{D}_{2d}(/\mathbf{C}_1) \downarrow \mathbf{G}_j + \mathbf{D}_{2d}(/\mathbf{C}_s) \downarrow \mathbf{G}_j
\tag{18.33}
$$

according to eq. 18.15, where $\mathbf{G}_j$ is selected from subgroups of $\mathbf{D}_{2d}$. This expression corresponds to each of the SCIs represented by eqs. 18.16-18.23.

In accord with the ligand partition [8,4,0] for this case, we adopt

$$
\mathbf{H} = \mathbf{S}^{[8]} \times \mathbf{S}^{[4]},
\tag{18.34}
$$

where $\mathbf{S}^{[8]}$ and $\mathbf{S}^{[4]}$ denote symmetric groups of degree 8 and of degree 4. Their cycle indices are calculated by means of eq. 18.25 to be

$$
\begin{aligned}
CI(\mathbf{S}^{[8]}; s) =\;& \\
& \frac{1}{8} s_8 + \frac{1}{7} s_1 s_7 + \frac{1}{12} s_2 s_6 + \frac{1}{12} s_1^2 s_6 + \frac{1}{15} s_3 s_5 + \frac{1}{10} s_1 s_2 s_5 + + \frac{1}{30} s_1^3 s_5 + \\
& \frac{1}{32} s_4^2 + \frac{1}{12} s_1 s_3 s_4 + \frac{1}{32} s_2^2 s_4 + \frac{1}{16} s_1^2 s_2 s_4 + \frac{1}{96} s_1^4 s_4 + \\
& \frac{1}{36} s_2 s_3^2 + \frac{1}{24} s_1 s_2^2 s_3 + \frac{1}{36} s_1^2 s_3^2 + \frac{1}{36} s_1^3 s_2 s_3 + \frac{1}{360} s_1^5 s_3 + \\
& \frac{1}{384} s_2^4 + \frac{1}{96} s_1^2 s_2^3 + \frac{1}{192} s_1^4 s_2^2 + \frac{1}{1440} s_1^6 s_2 + \frac{1}{40320} s_1^8
\end{aligned}
\tag{18.35}
$$

and

$$
CI(\mathbf{S}^{[4]}; s) = \frac{1}{4} s_4 + \frac{1}{3} s_1 s_3 + \frac{1}{8} s_2^2 + \frac{1}{4} s_1^2 s_2 + \frac{1}{24} s_1^4 .
\tag{18.36}
$$

The multiplication of these cycle indices affords $CI(\mathbf{H}; s)$ according to eq. 18.32. Thus, we obtain

$$
\begin{aligned}
CI(\mathbf{H}; s) \ = \ & CI(\mathbf{S}^{[8]}; s) \times CI(\mathbf{S}^{[4]}; s) \\
= \ & \ldots + \frac{1}{8} \times \frac{1}{4} s_8 s_4 + \ldots + (\frac{1}{4} \times \frac{1}{32} + \frac{1}{32} \times \frac{1}{8}) s_2^2 s_4^2 + \ldots + \\
& \frac{1}{32} \times \frac{1}{4} s_4^3 + \ldots + (\frac{1}{384} \times \frac{1}{4} + \frac{1}{96} \times \frac{1}{8}) s_1^2 s_2^5 + \ldots + \\
& \frac{1}{384} \times \frac{1}{8} s_2^6 + \ldots + \frac{1}{40320} \times \frac{1}{24} s_1^{12} + \ldots , \quad (18.37)
\end{aligned}
$$

where other non-effective terms are omitted.

We now consider the superposition of $\mathbf{P_{G_j}}$ (eq. 18.33) and $\mathbf{H}$ (eq. 18.34). For $\mathbf{C_1}$, the SCI is denoted as $(s_1^4)(s_1^8) = s_1^{12}$, as shown in eq. 18.16. The corresponding term of $CI(\mathbf{H}; s)$ is obtained by $(1/40320)s_1^8 \times (1/24)s_1^4$, in which the former comes from eq. 18.35 and the latter from eq. 18.36. Equation 18.31 is applied to this case to afford

$$
\rho_{\theta\mathbf{C_1}} = 1^{12} 12! \times \frac{1}{24} \times \frac{1}{40320} = 495. \quad (18.38)
$$

Similarly, other $\rho_{\theta j}$ values are obtained as follows:

$$
\rho_{\theta\mathbf{C_2}} \ = \ 2^6 6! \times \frac{1}{384} \times \frac{1}{8} = 15 \quad (18.39)
$$
$$
\text{for } \mathbf{C_2} \text{ of the SCI } (s_2^6),
$$

$$
\rho_{\theta\mathbf{C_2'}} \ = \ 2^6 6! \times \frac{1}{384} \times \frac{1}{8} = 15 \text{ for } \mathbf{C_2'} \text{ of the SCI } (s_2^6), \quad (18.40)
$$

$$
\rho_{\theta\mathbf{C_s}} \ = \ 1^2 2! 2^5 5! \times (\frac{1}{4} \times \frac{1}{384} + \frac{1}{8} \times \frac{1}{96}) = 15 \quad (18.41)
$$
$$
\text{for } \mathbf{C_s} \text{ of the SCI } (s_1^2 s_2^5),
$$

$$
\rho_{\theta\mathbf{S_4}} \ = \ 4^3 3! \times \frac{1}{4} \times \frac{1}{32} = 3 \text{ for } \mathbf{S_4} \text{ of the SCI } (s_4^3), \quad (18.42)
$$

$$
\rho_{\theta\mathbf{C_{2v}}} \ = \ 2^2 2! 4^2 2! \times (\frac{1}{4} \times \frac{1}{32} + \frac{1}{32} \times \frac{1}{8}) = 3 \quad (18.43)
$$
$$
\text{for } \mathbf{C_{2v}} \text{ of the SCI } (s_2^2 s_4^2),
$$

$$
\rho_{\theta\mathbf{D_2}} \ = \ 4^3 3! \times \frac{1}{4} \times \frac{1}{32} = 3 \text{ for } \mathbf{D_2} \text{ of the SCI } (s_4^3), \quad (18.44)
$$

and

$$
\rho_{\theta\mathbf{D_{2d}}} \ = \ 8^1 1! 4^1 1! \times \frac{1}{8} \times \frac{1}{4} = 1 \text{ for } \mathbf{D_{2d}} \text{ of the SCI } (s_8 s_4). \quad (18.45)
$$

These values are collected to afford an FPV = (495 15 15 15 3 3 3 1), which is multiplied by the inverse (Appendix B). The resulting row vector (54 2 6 6 1 1 1 1) is identical with the [8,4,0] row of Table 18.1.

18.3 Superposition for Other Indices

Let us apply the $*$ operation to the CI (eq. 18.30) and the partical cycle index (PCI$'$) (eq. 18.8) in a formal fashion. Thereby, we have

$$
\begin{aligned}
N\{CI(\mathbf{H}; s) &* PCI'(\mathbf{G}_i; s_{d_{jk}})\} \\
&= N\{CI(\mathbf{H}; s) * \sum_{j=1}^{s} \overline{m}_{ji} ZI'(\mathbf{G}_j; s_{d_{jk}})\} \\
&= \sum_{j=1}^{s} \overline{m}_{ji} N\{CI(\mathbf{H}; s) * ZI'(\mathbf{G}_j; s_{d_{jk}})\} \\
&= \sum_{j=1}^{s} \overline{m}_{ji} \rho_{\theta j} = A_{\theta i}
\end{aligned}
\tag{18.46}
$$

by referring to eqs. 18.32 and 18.14. Now we arrive at the following theorem.

Theorem 18.3 *The number of isomers with* W_θ *and* $\mathbf{G}_i$ *is represented by*

$$
A_{\theta i} = N\{CI(\mathbf{H}; s) * PCI'(\mathbf{G}_i; s_{d_{jk}})\}.
\tag{18.47}
$$

Example **18.3** As an alternative solution of Example 18.1, we examine a case with $[\theta] = [8, 4, 0]$ and $\mathbf{G}_i$ is equal to $\mathbf{C}_s$. The partial cycle index (PCI) of this case is represented by

$$
\begin{aligned}
PCI'(\mathbf{C}_s; s_d) &= \frac{1}{2}(s_2^4)(s_1^2 s_2) - \frac{1}{2}(s_4^2)(s_2^2) \\
&= \frac{1}{2} s_1^2 s_2^5 - \frac{1}{2} s_2^2 s_4^2,
\end{aligned}
$$

in the light of eq. 18.8. The $*$ operation between the PCI and eq. 18.37 affords

$$
\begin{aligned}
\frac{1}{2} \times 1^2 2! 2^5 5! \times (\frac{1}{4} \times \frac{1}{384} + \frac{1}{8} \times \frac{1}{96}) &- \frac{1}{2} \times 2^2 2! 4^2 2! \times (\frac{1}{4} \times \frac{1}{32} + \frac{1}{32} \times \frac{1}{8}) \\
&= \frac{1}{2} \times (15 - 3) = 6.
\end{aligned}
$$

This is identical with the value that appears in the intersection between the $[8, 4, 0]$-row and the $\mathbf{C}_s$-column of Table 18.1. We should find the multiplication of the FPV by the inverse (Example 18.2) is equivalent to the $*$ calculation in this example if we campare these processes carefully. For example, we find that $\rho_{\theta C_s} = 15$ and $\rho_{\theta C_{2v}} = 3$ (Example 18.2) appear in the present calculation.

Next, we apply the $*$ operation to the CI (eq. 18.30) and the cycle index CI$'$ (eq. 18.9) in a formal fashion. For simplicity's sake, we here employ an alternative definition of the CI$'$ (Def. 16.5 in Chapter 16). Thereby, we have

$$N\{CI(\mathbf{H}; s) * CI'(\mathbf{G}; s_{d_{jk}})\}$$

$$= N\{CI(\mathbf{H}; s) * \sum_{i=1}^{s} PCI'(\mathbf{G}_i; s_{d_{jk}})\}$$

$$= \sum_{i=1}^{s} N\{CI(\mathbf{H}; s) * PCI'(\mathbf{G}_i; s_{d_{jk}})\}$$

$$= \sum_{i=1}^{s} A_{\theta i} = A_{\theta} \qquad (18.48)$$

by using eq. 18.47. Hence, we end up with the following theorem.

Theorem 18.4 *The number of isomers with the W_θ is represented by*

$$A_\theta = N\{CI(\mathbf{H}; s) * CI'(\mathbf{G}; s_{d_{jk}})\}. \qquad (18.49)$$

This is equivalent to a two-component case of the Redfield-Read superposition theorem.[4]

Example **18.4** Let us re-examine Example 18.1 for illustrating Theorem 18.4. The cycle index (eq. 18.9) for this case is obtained to be

$$CI'(\mathbf{D}_{2d}; s_d)$$

$$= \frac{1}{8}(s_1^8)(s_1^4) + \frac{1}{8}(s_2^4)(s_2^2) + \frac{1}{4}(s_2^4)(s_2^2) + \frac{1}{4}(s_2^4)(s_1^2 s_2) + \frac{1}{4}(s_4^2)(s_4)$$

$$= \frac{1}{8}s_1^{12} + \frac{3}{8}s_2^6 + \frac{1}{4}s_1^2 s_2^5 + \frac{1}{4}s_4^3, \qquad (18.50)$$

where the USCIs of each monomial come from the $\mathbf{D}_{2d}(/\mathbf{C}_1)$ and $\mathbf{D}_{2d}(/\mathbf{C}_s)$ rows of the USCI table (Appendix C) and the coefficient of each monomial is taken up from the bottom of the same table. We now consider the superposition of eqs. 18.50 and 18.37. This operation affords the total number of isomers with the partition [8,4,0]. The $*$ operation leaves the terms s_1^{12}, s_2^6, $s_1^2 s_2^5$, and s_4^3. Their coefficients are added to be

$$1^{12}12! \times \frac{1}{40320} \times \frac{1}{24} \times \frac{1}{8} + 2^6 6! \times \frac{1}{384} \times \frac{1}{8} \times \frac{3}{8} +$$

$$1^2 2! 2^5 5! \times \left(\frac{1}{384} \times \frac{1}{4} + \frac{1}{96} \times \frac{1}{8}\right) \times \frac{1}{4} + 4^3 3! \times \frac{1}{32} \times \frac{1}{4} \times \frac{1}{4} = 72.$$

This value is identical with the total of the [8,4,0] row listed in Table 18.1.

---

[4]We are able to generalize Theorem 18.4 into a multi-component case, if we use the elementary superposition multiple times. This provides an alternative proof of the Redfield-Read theorem.

Bibliography

[1] R. A. Davidson, *J. Am. Chem. Soc.*, **103**, 312 (1981).

[2] J. H. Redfield, *Am. J. Math.* **49**, 433 (1927).

[3] R. C. Read, *J. London Math. Soc.*, **34**, 417 (1959).

[4] H. O. Foulkes, *Can. J. Math.*, **18**, 1060 (1966).

[5] F. Harary and E. Palmer, *Am. J. Math.*, **89**, 372 (1967).

[6] E. K. Lloyd, in *Studies in Physical and Theoretical Chemistry. Theory and Topology in Chemistry* , R. B. King and D. H. Rouvray (*eds.*), **51**, 537 (1987).

[7] S. Fujita, *Theor. Chim. Acta*, **76**, 247 (1989).

[8] S. Fujita, *J. Math. Chem.*, **5**, 99 (1990).

[9] S. Fujita, *Bull. Chem. Soc. Jpn.*, **62**, 3771 (1989).

[10] S. Fujita, *Bull. Chem. Soc. Jpn.*, **63**, 203 (1990).

[11] S. Fujita, *Tetrahedron*, **46**, 365 (1990).

Chapter 19

Compounds with Achiral and Chiral Ligands[1]

We have discussed subduction of coset representations (Chapter 9) and presented a systematic classification of molecular symmetry (Chapter 6). In addition, we have pointed out that several related concepts, *e.g.* unit subduced cycle indices (USCIs) and unit subduced cycle indices with chirality fittingness (USCI-CFs), are useful for qualitative discussions on molecular symmetry (Chapter 8). Chapters 8 and 9 have clarified their meanings, especially that of chirality fittingness. This chapter deals with a quantitative application of the USCI-CFs to enumeration problems.

19.1 Compounds with Given Symmetries

Orbit index. In order to make notations precise, we use the following formal expression containing achiral, neutral, and prochiral parts,

$$G_j(/G_k^{(j)}) = \chi_{ak}^{(j)} G_j^{(a)}(/G_k^{(j)}) + \chi_{bk}^{(j)} G_j^{(b)}(/G_k^{(j)}) + \chi_{ck}^{(j)} G_j^{(c)}(/G_k^{(j)}), \qquad (19.1)$$

where $\chi_{ak}^{(j)}$, $\chi_{bk}^{(j)}$, and $\chi_{ck}^{(j)}$ are all non-negative integers that satisfy the equation,

$$\chi_{ak}^{(j)} + \chi_{bk}^{(j)} + \chi_{ck}^{(j)} = 1. \qquad (19.2)$$

The superscripts and subscripts (a, b, and c) donote achiral, neutral, and prochiral parts. The right-hand side of eq. 19.1 indicates that only one of the three parts is effective. In the light of this definition, Theorem 9.1 (Chapter 9) is restated as follows.

$$G(/G_i) \downarrow G_j =$$
$$\sum_i^{v_j} \left[\chi_{ak}^{(j)} \beta_k^{(ij)} G_j^{(a)}(/G_k^{(j)}) + \chi_{bk}^{(j)} \beta_k^{(ij)} G_j^{(b)}(/G_k^{(j)}) + \chi_{ck}^{(j)} \beta_k^{(ij)} G_j^{(c)}(/G_k^{(j)}) \right], (19.3)$$

[1]Reprinted in part with permission form S. Fujita, *J. Math. Chem.*, **5**, 121–156 (1990). ©(1990) J. C. Baltzer AG.

for $i = 1, 2, \ldots, s$ and $j = 1, 2, \ldots, s$. The resulting suborbits are classified into three categories, *i.e.,,*

$$
\begin{cases}
\Delta_{k\beta a}^{(i\alpha)} & \text{that is subject to} \quad \mathbf{G}_j^{(a)}(/\mathbf{G}_k^{(j)}) \quad \text{(achiral part)} \\
\Delta_{k\beta b}^{(i\alpha)} & \text{that is subject to} \quad \mathbf{G}_j^{(b)}(/\mathbf{G}_k^{(j)}) \quad \text{(neutral part)} \\
\Delta_{k\beta c}^{(i\alpha)} & \text{that is subject to} \quad \mathbf{G}_j^{(c)}(/\mathbf{G}_k^{(j)}) \quad \text{(prochiral part).}
\end{cases}
$$

In the light of this notation, the unit subduced cycle index with chirality fittingness (USCI-CF) defined by Def. 9.3 (Chapter 9) is converted into an equivalent equation,

$$
ZC(\mathbf{G}(/\mathbf{G}_i) \downarrow \mathbf{G}_j; a_{d_{jk}}^{(i\alpha)}, b_{d_{jk}}^{(i\alpha)}, c_{d_{jk}}^{(i\alpha)}) = \prod_{k=1}^{v_j} \left((a_{d_{jk}}^{(i\alpha)})^{\chi_{ak}^{(j)}\beta_k^{(ij)}} (b_{d_{jk}}^{(i\alpha)})^{\chi_{bk}^{(j)}\beta_k^{(ij)}} (c_{d_{jk}}^{(i\alpha)})^{\chi_{ck}^{(j)}\beta_k^{(ij)}} \right).
$$

$$(19.4)$$

It should be noted that only one of the three terms in the product is effective in the light of eq. 19.2. The sum of the powers in each of the parts of eq. 19.4 is also useful to enumerate organic structures. The number of suborbits in each part is represented as follows.

$$
\beta_{ij}^{(a)} = \sum_{k=1}^{v_j} \chi_{ak}^{(j)}\beta_k^{(ij)},
\tag{19.5}
$$

$$
\beta_{ij}^{(b)} = \sum_{k=1}^{v_j} \chi_{bk}^{(j)}\beta_k^{(ij)},
\tag{19.6}
$$

and

$$
\beta_{ij}^{(c)} = \sum_{k=1}^{v_j} \chi_{ck}^{(j)}\beta_k^{(ij)}.
\tag{19.7}
$$

$$(19.8)$$

These are summarized in the form of

$$
(\beta_{ij}^{(a)}, \beta_{ij}^{(b)}, \beta_{ij}^{(c)}).
\tag{19.9}
$$

We call this term an *orbit index* for the subduction. The data of a table of USCI-CFs (Appendix D) afford the corresponding table of such orbit indices. Table 19.1 lists the orbit indices of $\mathbf{D}_{2d}$ group.

Enumeration. The formulation of this chapter is essentially the same as that of Chapter 15 except that the present formulation takes account of chiral ligands in addition to achiral ones. Let $\Delta = \{\delta_1, \delta_2, \ldots, \delta_{|\Delta|}\}$ be a domain which consists of $|\Delta|$ elements called positions. Let $\mathbf{X} = \{X_1, X_2, \ldots, X_{|\mathbf{X}|}\}$ be a codomain which contains $|\mathbf{X}|$ elements called figures. In chemistry, the figures may be ligands or atoms. Suppose that f is a function (called a configuration), *i.e.* $f : \Delta \to \mathbf{X}$. The mode of this mapping is restricted by the following weights: $w_{i\alpha}(X_\ell)$, for $i = 1, 2, \ldots, s$, $\alpha = 1, 2, \ldots, \alpha_i$, and $\ell = 1, 2, \ldots, |\mathbf{X}|$, which is assigned to each

Table 19.1: Orbit indices for D_{2d} group

	$\downarrow C_1$	$\downarrow C_2$	$\downarrow C_2'$	$\downarrow C_s$	$\downarrow S_4$	$\downarrow C_{2v}$	$\downarrow D_2$	$\downarrow D_{2d}$
$D_{2d}(/C_1)$	(0,8,0)	(0,4,0)	(0,4,0)	(0,0,4)	(0,0,2)	(0,0,2)	(0,2,0)	(0,0,1)
$D_{2d}(/C_2)$	(0,4,0)	(0,4,0)	(0,2,0)	(0,0,2)	(0,0,2)	(0,0,2)	(0,2,0)	(0,0,1)
$D_{2d}(/C_2')$	(0,4,0)	(0,2,0)	(0,3,0)	(0,0,2)	(0,0,1)	(0,0,1)	(0,2,0)	(0,0,1)
$D_{2d}(/C_s)$	(0,4,0)	(0,2,0)	(0,2,0)	(2,0,1)	(0,0,1)	(2,0,0)	(0,1,0)	(1,0,0)
$D_{2d}(/S_4)$	(0,2,0)	(0,2,0)	(0,1,0)	(0,0,1)	(2,0,0)	(0,0,1)	(0,1,0)	(1,0,0)
$D_{2d}(/C_{2v})$	(0,2,0)	(0,2,0)	(0,1,0)	(2,0,0)	(0,0,1)	(2,0,0)	(0,1,0)	(1,0,0)
$D_{2d}(/D_2)$	(0,2,0)	(0,2,0)	(0,2,0)	(0,0,1)	(0,0,1)	(0,0,1)	(0,2,0)	(0,0,1)
$D_{2d}(/D_{2d})$	(0,1,0)	(0,1,0)	(0,1,0)	(1,0,0)	(1,0,0)	(1,0,0)	(0,1,0)	(1,0,0)

element X_ℓ of the codomain X in agreement with the hehavior of each orbit $\Delta_{i\alpha}$. We then define a weight $W(f)$ for each function f in order to meet the present requirement (*cf.* Def. 14.3 in Chapter 14).

Definition 19.1 *A weight of the function (f) is represented by*

$$W(f) = \prod_{\substack{i=1 \\ }}^{s} \prod_{\substack{\alpha=1 \\ \alpha_i \neq 0}}^{\alpha_i} \prod_{\delta \in \Delta_{i\alpha}} w_{i\alpha}(f(\delta)) \tag{19.10}$$

$$= \prod_{\substack{i=1 \\ }}^{s} \prod_{\substack{\alpha=1 \\ \alpha_i \neq 0}}^{\alpha_i} \left(\prod_{\delta \in \Delta_{k\beta a}^{(i\alpha)}} w_{i\alpha}(f(\delta)) \prod_{\delta \in \Delta_{k\beta b}^{(i\alpha)}} w_{i\alpha}(f(\delta)) \prod_{\delta \in \Delta_{k\beta c}^{(i\alpha)}} w_{i\alpha}(f(\delta)) \right). \tag{19.11}$$

The products in the parentheses are monomials of total powers of $d_{jk}\beta_{ij}^{(a)}, d_{jk}\beta_{ij}^{(b)}$ and $d_{jk}\beta_{ij}^{(c)}$, respectively. A set of all the functions $(f : \Delta \to X)$ is defined as

$$F = \{f_1, f_2, \ldots, f_\gamma, \ldots, f_\epsilon, \ldots, f_{|F|}\}.$$

Suppose that a group G acts on the domain (Δ) in the form of the corresponding permutation representation P_G on Δ and that the group G acts simultaneously on the codomain X *via* a permutation group Q_G on X.

Definition 19.2 *For $p_g \in P_G$ and $q_g \in Q_G$, a binary relation between $f_\gamma(\in F)$ and $f_\epsilon(\in F)$ is defined as*

$$q_g f_\gamma(\delta) = f_\epsilon(p_g(\delta)) \quad for \quad \forall \delta \in \Delta, \tag{19.12}$$

which holds for $\exists g \in G$.

This binary relation is an equivalence relation. Hence, this affords a partition of the set (F) into equivalence classes.[2] In order to simplify our discussions, $q_g(X_\ell)$

---

[2]This type of action was discussed in W. Hässelbarth, *Theor. Chim. Acta*, **67**, 339 (1985).

is an operation that keeps X_ℓ invariant for a proper rotation $g \in \mathbf{G}$ but gives its antipode $(\overline{X}_\ell)$ for an improper rotation $g \in \mathbf{G}$. Compare this definition with Def. 13.2 (Chapter 13).

Let $\lambda_g : f_\epsilon(\delta) \to f_\gamma(\delta)$ (for $\forall \delta \in \Delta$) be a mapping corresponding to $g \in \mathbf{G}$ and let $\Lambda_{\mathbf{G}}$ be a set containing all λ_g. Then, $\Lambda_{\mathbf{G}}$ is proved to be a permutation representation of $\mathbf{G}$.[3] Hence, this formulation enables us to apply Theorem 5.7 (Chapter 5) to the present case.

Theorem 19.1 *Suppose that the group* $\mathbf{G}$ *acts on* $\mathbf{F}$ *by the simultaneous action of* $\mathbf{G}$ *on* Δ *and* $\mathbf{X}$*. The action constructs a permutation representation* $\Lambda_{\mathbf{G}}$ *on* $\mathbf{F}$*. The multiplicity of each transitive coset representation* $\mathbf{G}\,(/\mathbf{G}_i)$ *in* $\Lambda_{\mathbf{G}}$ *is determined by*

$$\Lambda_{\mathbf{G}} = \sum_{i=1}^{s} B_i \mathbf{G}(/\mathbf{G}_i), \tag{19.13}$$

wherein B_i*'s are non-negative integers. The multiplicitis* B_i*'s are obtained by solving the following equations.*

$$\rho_j = \sum_{i=1}^{s} B_i m_{ij} \quad (for\ j = 1, 2, \ldots, s), \tag{19.14}$$

where ρ_j *is the mark of of* $\mathbf{G}_j$ *in* $\Lambda_{\mathbf{G}}$*.*

This theorem corresponds to Theorem 15.2 (Chapter 15).

Each orbit corresponding to a transitive $\mathbf{G}(/\mathbf{G}_i)$ contains functions (configurations) of symmetry $\mathbf{G}_i$. Hence, B_i is the number of different configurations of symmetry $\mathbf{G}_i$, which can be regarded as the number of compounds of $\mathbf{G}_i$ symmetry.

The mark ρ_j is the number of fixed functions (configurations) of $\mathbf{F}$ with respect to $\mathbf{G}_j$. Suppose that an appropriate configuration $f^{(j)} \in \mathbf{F}$ is fixed to all the elements $g \in \mathbf{G}_j$. This requires

$$q_g f^{(j)}(\delta) = f^{(j)}(p_g(\delta)) \ (for\ \forall \delta \in \Delta\ and\ \forall g \in \mathbf{G}_j). \tag{19.15}$$

Let us now go back to the division to orbits and further subdivision to suborbits. Then, in order for $f^{(j)}(\in \mathbf{F})$ to be constant with respect to eq 19.15, all the positions of each suborbit have to take the same figure (or ligand) of suitable chirality. According to the chirality fittingness defined in Chapter 8, we examine the following three cases.

1. If the suborbit is homospheric (*i.e.* if the corresponding coset representation (CR) is an achiral part), there are $|\mathbf{X}_{i\alpha}^{(a)}|$ ways of substitution for each suborbit $\Delta_{k\beta a}^{(i\alpha)}$, where $|\mathbf{X}_{i\alpha}^{(a)}|$ is the number of non-zero $w_{i\alpha}(X_\ell^{(a)})$ with achiral $X_\ell^{(a)}$ for each suborbit $\Delta_{k\beta a}^{(i\alpha)}$. Since the number of the suborbits is $\beta_{ij}^{(a)}$, the number

---

[3]For the proof, see S. Fujita, *J. Math. Chem.*, **5**, 99, 121 (1990).

of fixed configurations for achiral parts contained in each $\Delta_{i\alpha}$ is represented
by

$$\mid \mathbf{X}_{i\alpha}^{(a)} \mid^{\beta_{ij}^{(a)}} . \tag{19.16}$$

2. Similarly, the number of fixed configurations for a neutral part (corresponding
   a hemispheric suborbit) is obtained as follows.

$$\mid \mathbf{X}_{i\alpha}^{(b)} \mid^{\beta_{ij}^{(b)}}, \tag{19.17}$$

where $\mid \mathbf{X}_{i\alpha}^{(b)} \mid$ is the number of non-zero $w_{i\alpha}(X_\ell)$ with achiral and chiral X_ℓ
for each suborbit $\Delta_{k\beta b}^{(i\alpha)}$. This value is equal to $\mid \mathbf{X}_{i\alpha} \mid$.

3. For counting the number of fixed configurations in a prochiral part (corresponding to an enantiospheric suborbit), a saturation of each orbit with
   chiral ligands is represented by either of the following two manners due to its
   chirality fillingness (See Chapter 8).

Q	Q	$\cdots$	Q	Q
Q′	Q′	$\cdots$	Q′	Q′

or

Q′	Q′	$\cdots$	Q′	Q′
Q	Q	$\cdots$	Q	Q

,

where Q is equal to one of chiral X_ℓ's and Q′ is equal to one of $\overline{X}_\ell$'s. These
possibilities have been referred to as a compensated chiral packing in Chapter
8. Thus, we obtain

$$\mid \mathbf{X}_{i\alpha}^{(c)} \mid^{\beta_{ij}^{(c)}}, \tag{19.18}$$

where

$$
\begin{aligned}
\mid \mathbf{X}_{i\alpha}^{(c)} \mid \; &= \text{the number of non-zero } w_{i\alpha}(X_\ell) \text{ with achiral } X_\ell \text{ plus twice} \\
&\quad \text{the number of non-zero } w_{i\alpha}(X_\ell) \text{ with chiral } X_\ell \text{ (one of the} \\
&\quad \text{antipodes) for each suborbit } \Delta_{k\beta c}. \\
&= \text{the number of non-zero } w_{i\alpha}(X_\ell) \text{ with achiral } X_\ell \text{ and chiral } X_\ell \\
&= \mid \mathbf{X}_{i\alpha} \mid.
\end{aligned}
$$

The product of eqs. 19.16, 19.17, and 19.18 provides the number of fixed
configurations for each orbit $\Delta_{i\alpha}$. A further multiplication of the products over all
α and all i is equal to ρ_j.

Lemma 19.1 *The mark ρ_j is calculated by the equation,*

$$\rho_j = \prod_{i=1}^{s} \prod_{\substack{\alpha=1 \\ \alpha_i \neq 0}}^{\alpha_i} \left(\mid \mathbf{X}_{i\alpha}^{(a)} \mid^{\beta_{ij}^{(a)}} \mid \mathbf{X}_{i\alpha}^{(b)} \mid^{\beta_{ij}^{(b)}} \mid \mathbf{X}_{i\alpha}^{(c)} \mid^{\beta_{ij}^{(c)}} \right) \quad (j = 1, 2, \ldots, s). \tag{19.19}$$

This lemma corresponds to Lemma 15.1 (Chapter 15).

Lemma 19.1 is introduced into eq. 19.14 to afford the following theorem.

Theorem 19.2 *The number (B_i) of non-equivalent configurations with symmetry* $\mathbf{G}_i$ *is obtained by solving the following equations.*

$$\prod_{\substack{i=1 \\ }}^{s} \prod_{\substack{\alpha=1 \\ \alpha_i \neq 0}}^{\alpha_i} \left(\mid \mathbf{X}_{i\alpha}^{(a)} \mid^{\beta_{ij}^{(a)}} \mid \mathbf{X}_{i\alpha}^{(b)} \mid^{\beta_{ij}^{(b)}} \mid \mathbf{X}_{i\alpha}^{(c)} \mid^{\beta_{ij}^{(c)}} \right) = \sum_{i=1}^{s} B_i m_{ij} \qquad (19.20)$$

$$(for\ j = 1, 2, \ldots, s).$$

Theorem 19.2 takes account of chiral ligands as well as achiral ligands. This theorem corresponds to Theorem 15.3 (Chapter 15) that dealt with achiral ligands only. Since $\mid \mathbf{X}_{i\alpha}^{(b)} \mid = \mid \mathbf{X}_{i\alpha}^{(c)} \mid = \mid \mathbf{X}_{i\alpha} \mid$, a simpler expression is derived to be

$$\prod_{\substack{i=1 \\ }}^{s} \prod_{\substack{\alpha=1 \\ \alpha_i \neq 0}}^{\alpha_i} \left(\mid \mathbf{X}_{i\alpha}^{(a)} \mid^{\beta_{ij}^{(a)}} \mid \mathbf{X}_{i\alpha} \mid^{\beta_{ij}^{(b)} + \beta_{ij}^{(c)}} \right) = \sum_{i=1}^{s} B_i m_{ij} \qquad (19.21)$$

$$(for\ j = 1, 2, \ldots, s),$$

Example **19.1** We have already examined the allene skeleton (**1**) of $\mathbf{D}_{2d}$ symmetry in Chapter 15 (Example 15.1). The four terminal positions of **1** are considered to be a domain. This domain consists of one orbit. Let $\mathbf{X} = \{\text{A,B,p},\overline{\text{p}}\}$ be a codomain, where A and B are achiral ligands and p and $\overline{\text{p}}$ are chiral and antipodal to each other. We presume the following weights:

$$w(\text{A}) = A,\ w(\text{B}) = B,\ w(\text{p}) = p,\ \text{and}\ w(\overline{\text{p}}) = \overline{p}.$$

By using these weights, the number of allowed ligands are obtained as follows.

$$\mid \mathbf{X}^{(a)} \mid = 2, \mid \mathbf{X}^{(b)} \mid = 4, \mid \mathbf{X}^{(c)} \mid = 4$$

From Table 19.1, we take the $\mathbf{D}_{2d}(/\mathbf{C}_s)$ row. For example, the orbit index (0, 4, 0) for $\mathbf{D}_{2d}(/\mathbf{C}_s)\!\downarrow \mathbf{C}_1$ affords $\rho_{\mathbf{C}_1} = 2^0 4^4 4^0 = 256$ by using Lemma 19.1. Similarly, we obtain

$$\begin{aligned}
\rho_{\mathbf{C}_2} &= 2^0 4^2 4^0 = 16 \ \ \text{for} \ \ \mathbf{C}_2, \ \rho_{\mathbf{C}_2'} = 2^0 4^2 4^0 = 16 \ \ \text{for} \ \ \mathbf{C}_2', \\
\rho_{\mathbf{C}_s} &= 2^2 4^0 4^1 = 16 \ \ \text{for} \ \ \mathbf{C}_s, \rho_{\mathbf{S}_4} = 2^0 4^0 4^1 = 4 \ \ \text{for} \ \ \mathbf{S}_4, \\
\rho_{\mathbf{C}_{2v}} &= 2^2 4^0 4^0 = 4 \ \ \text{for} \ \ \mathbf{C}_{2v}, \ \rho_{\mathbf{D}_2} = 2^0 4^1 4^0 = 4 \ \ \text{for} \ \ \mathbf{D}_2, \\
&\text{and} \\
\rho_{\mathbf{D}_{2d}} &= 2^1 4^0 4^0 = 2 \ \ \text{for} \ \ \mathbf{D}_{2d}.
\end{aligned}$$

Thereby, we have a fixed-point vector, FPV = (256 16 16 16 4 4 4 2). When the FPV is multiplied by the inverse of mark table (M^{-1}) for $\mathbf{D}_{2d}$ (Appendix B and Table 5.4 in Chapter 5), we have

$$(256\ 16\ 16\ 16\ 4\ 4\ 4\ 2)M^{-1} = (24\ 2\ 6\ 6\ 1\ 1\ 1\ 2)$$

Figure 19.1 lists all of these derivatives.

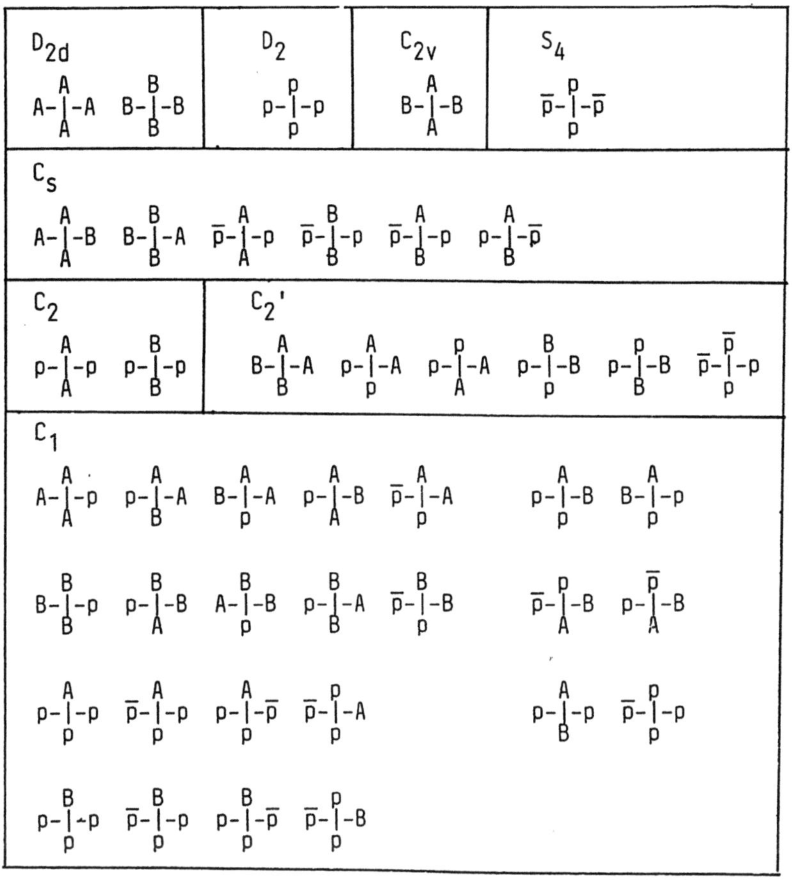

D$_{2d}$

A–|–A (A above, A below)   B–|–B (B above, B below)

D$_2$

p–|–p (p above, p below)

C$_{2v}$

B–|–B (A above, A below)

S$_4$

$\bar{p}$–|–$\bar{p}$ (p above, p below)

C$_s$

A–|–B (A above, A below)   B–|–A (B above, B below) $\bar{p}$–|–p (A above, p below) $\bar{p}$–|–p (B above, p below) $\bar{p}$–|–p (A above, B below)   p–|–$\bar{p}$ (A above, B below)

C$_2$

p–|–p (A above, A below)   p–|–p (B above, B below)

C$_2{}'$

B–|–A (A above, B below)   p–|–A (A above, p below)   p–|–A (p above, A below)   p–|–B (B above, p below)   p–|–B (p above, B below) $\bar{p}$–|–p ($\bar{p}$ above, p below)

C$_1$

A–|–p (A above, A below)   p–|–A (A above, B below)   B–|–A (A above, p below)   p–|–B (A above, A below) $\bar{p}$–|–A (A above, p below)   p–|–B (A above, p below)   B–|–p (A above, p below)

B–|–p (B above, B below)   p–|–B (B above, A below)   A–|–B (B above, p below)   p–|–A (B above, B below) $\bar{p}$–|–B (B above, p below) $\bar{p}$–|–B (p above, A below)   p–|–B ($\bar{p}$ above, A below)

p–|–p (A above, p below) $\bar{p}$–|–p (A above, p below)   p–|–$\bar{p}$ (A above, p below) $\bar{p}$–|–A (p above, p below)   p–|–p (A above, B below) $\bar{p}$–|–p (p above, p below)

p–|–p (B above, p below) $\bar{p}$–|–p (B above, p below)   p–|–$\bar{p}$ (B above, p below) $\bar{p}$–|–B (p above, p below)

Figure 19.1: Derivatives based on the allene skeleton

19.2 Compounds with Given Symmetries and Weights

This section deals with the most general case that takes account of OMVs. Thus, we enumerate the number of configurations with a given symmetry as well as a given weight on the basis of a given skeleton. If $f_\gamma : \Delta \to \mathbf{X}$ and $f_\epsilon : \Delta \to \mathbf{X}$ are equivalent, their weight can be proved to be equal to each other, $i.e.$ $W(f_\gamma) = W(f_\epsilon)$, where the weights are given by eq. 19.11.[4]

Let $\mathbf{F}^{[\theta]}$ be a set of functions $(f : \Delta \to \mathbf{X})$, all of which have the same weight $W_\theta(f)$: $\mathbf{F}^{[\theta]} = \{f_1^{[\theta]}, f_2^{[\theta]}, \ldots, f_\gamma^{[\theta]}, \ldots, f_\epsilon^{[\theta]}, \ldots, f_\phi^{[\theta]}\}$, where $\phi = \mid \mathbf{F}^{[\theta]} \mid$. Then, we obtain a permutation:

$$
\lambda_g^{[\theta]} = \begin{pmatrix} q_g f_1^{[\theta]} p_g^{-1} & \cdots & q_g f_\gamma^{[\theta]} p_g^{-1} & \cdots & q_g f_\phi^{[\theta]} p_g^{-1} \\ f_1^{[\theta]} & \cdots & f_\gamma^{[\theta]} & \cdots & f_\phi^{[\theta]} \end{pmatrix}
$$

$$
= \begin{pmatrix} f_1^{[\theta]} & \cdots & f_\gamma^{[\theta]} & \cdots & f_\phi^{[\theta]} \\ q_g^{-1} f_1^{[\theta]} p_g & \cdots & q_g^{-1} f_\gamma^{[\theta]} p_g & \cdots & q_g^{-1} f_\phi^{[\theta]} p_g \end{pmatrix}. \tag{19.22}
$$

Theorem 19.3 *Let the symbol $\Lambda_{\mathbf{G}}^{[\theta]}$ denote the set of λ_g for $\forall g \in \mathbf{G}$. Then, the $\Lambda_{\mathbf{G}}^{[\theta]}$ is a permutation representation of $\mathbf{G}$.*

This theorem corresponds to Theorem 14.2 (Chapter 14) and to the similar result described in Chapter 15.

Proof. This is a proof for $\Lambda_{\mathbf{G}}^{[\theta]}$ being homomorphic to $\mathbf{G}$.

$$
\lambda_g^{[\theta]} = \begin{pmatrix} f_\gamma^{[\theta]}(\delta) \\ q_g^{-1} f_\gamma^{[\theta]}(p_g(\delta)) \end{pmatrix}
$$

$$
\lambda_{g'}^{[\theta]} = \begin{pmatrix} f_\gamma^{[\theta]}(\delta) \\ q_{g'}^{-1} f_\gamma^{[\theta]}(p_{g'}(\delta)) \end{pmatrix} = \begin{pmatrix} f_\gamma^{[\theta]}(p_g(\delta)) \\ q_{g'}^{-1} f_\gamma^{[\theta]}(p_{g'}p_g(\delta)) \end{pmatrix}
$$

$$
= \begin{pmatrix} q_g^{-1} f_\gamma^{[\theta]}(p_g(\delta)) \\ q_g^{-1} q_{g'}^{-1} f_\gamma^{[\theta]}(p_{g'}p_g(\delta)) \end{pmatrix} = \begin{pmatrix} q_g^{-1} f_\gamma^{[\theta]}(p_g(\delta)) \\ q_{g'g}^{-1} f_\gamma^{[\theta]}(p_{g'}p_g(\delta)) \end{pmatrix},
$$

because $p_{g'}p_g = p_{g'g}$ and $q_{g'}q_g = q_{g'g}$. Thereby, we obtain

$$
\lambda_{g'}^{[\theta]} \lambda_g^{[\theta]} = \begin{pmatrix} f_\gamma^{[\theta]}(\delta) \\ q_{g'g}^{-1} f_\gamma^{[\theta]}(p_{g'}p_g(\delta)) \end{pmatrix} = \lambda_{g'g}^{[\theta]}. \tag{19.23}
$$

This equation indicates that the mapping of $\mathbf{G}$ (or $\mathbf{P_G}$) onto $\Lambda_{\mathbf{G}}^{[\theta]}$ is homomorphic. In other words, the $\Lambda_{\mathbf{G}}^{[\theta]}$ is a permutation representation of $\mathbf{G}$.

This result allows us to apply Theorem 5.7 (Chapter 5) to the present case. Thereby, we end up with the following theorem.

---

[4]For the proof, see S. Fujita, *J. Math. Chem.*, 5,121 (1990), Appendix H.

Theorem 19.4

$$\Lambda_{\mathbf{G}}^{[\theta]} = \sum_{i=1}^{s} B_{\theta i} \mathbf{G}(/\mathbf{G}_i),$$

(19.24)

where the $B_{\theta i}$ value is obtained by solving

$$\rho_{\theta j} = \sum_{i=1}^{s} B_{\theta i} m_{ij} \quad (j = 1, 2, \ldots, s).$$

(19.25)

The symbol $(B_{\theta i})$, which originally denotes the multiplicity of a transitive coset representation $(\mathbf{G}(/\mathbf{G}_i))$, also indicates the number of non-equivalent configurations with $\mathbf{G}_i$ symmetry as well as a weight W_θ. This theorem corresponds to Theorem 15.4 (Chapter 15). The latter theorem is concerned with cases that take account of achiral ligands only.

The next task is the evaluation of $\rho_{\theta j}$. Let us now define a subduced cycle index with chirality fittingness (SCI-CF) on the basis of USCI-CFs (eq. 19.4)

Definition 19.3 *A subduced cycle index with chirality fittingness (SCI-CF) is represented by*

$$
\begin{aligned}
ZIC&(\mathbf{G}_j; a_{d_{jk}}^{(i\alpha)}, b_{d_{jk}}^{(i\alpha)}, c_{d_{jk}}^{(i\alpha)}) \\
&= \prod_{\substack{i=1 \\ \alpha_i \neq 0}}^{s} \prod_{\alpha=1}^{\alpha_i} ZC(\mathbf{G}(/\mathbf{G}_i) \downarrow \mathbf{G}_j; a_{d_{jk}}^{(i\alpha)}, b_{d_{jk}}^{(i\alpha)}, c_{d_{jk}}^{(i\alpha)}) \\
&= \prod_{\substack{i=1 \\ \alpha_i \neq 0}}^{s} \prod_{\alpha=1}^{\alpha_i} \prod_{k=1}^{v_j} \left[(a_{d_{jk}}^{(i\alpha)})^{\chi_{ak}^{(j)}\beta_k^{(ij)}} (b_{d_{jk}}^{(i\alpha)})^{\chi_{bk}^{(j)}\beta_k^{(ij)}} (c_{d_{jk}}^{(i\alpha)})^{\chi_{ck}^{(j)}\beta_k^{(ij)}} \right]
\end{aligned}
$$

(19.26)

for $j = 1, 2, \ldots, s$.

This definition corresponds to Def. 15.1, which is concerned with an SCI without chirality fittingness. Now we arrive at Lemma 19.2.

Lemma 19.2 *The marks $(\rho_{\theta j}$'s) with the weights W_θ's are given by the following generating functions,*

$$\sum_{[\theta]} \rho_{\theta j} W_\theta = ZIC(\mathbf{G}_j; a_{d_{jk}}^{(i\alpha)}, b_{d_{jk}}^{(i\alpha)}, c_{d_{jk}}^{(i\alpha)}) \quad (j = 1, 2, \ldots, s),$$

(19.27)

wherein the variables in the right-hand side are substituted by

$$a_{d_{jk}}^{(i\alpha)} = \sum_{\ell=1}^{|\mathbf{X}|} w_{i\alpha}(X_\ell^{(a)})^{d_{jk}},$$

(19.28)

$$b_{d_{jk}}^{(i\alpha)} = \sum_{\ell=1}^{|\mathbf{X}|} w_{i\alpha}(X_\ell)^{d_{jk}},$$

(19.29)

$$c_{d_{jk}}^{(i\alpha)} = \sum_{\ell=1}^{|\mathbf{X}|} w_{i\alpha}(X_\ell^{(a)})^{d_{jk}} + 2 \sum_{\ell=1}^{|\mathbf{X}|} \left(w_{i\alpha}(X_\ell^{(c)}) w_{i\alpha}(\overline{X}_\ell^{(c)}) \right)^{d_{jk}/2}$$

(19.30)

Proof. Consider a series of $\rho_{\theta j}$'s for an appropiate $\mathbf{G}_j$ (the θ runs over all weights). These elements are the numbers of fixed configurations of symmetry $\mathbf{G}_j$.

1. A homospheric suborbit (an achiral part) of length d_{jk} can take achiral ligands of the same kind (Chapter 8). Hence, it takes $d_{jk}X_\ell^{(a)}$ for all achiral ligands. This indicates a ligand inventory,

$$a_{d_{jk}}^{(i\alpha)} = \sum_{\ell=1}^{|\mathbf{X}|} w_{i\alpha}(X_\ell^{(a)})^{d_{jk}}, \tag{19.31}$$

where $X_\ell^{(a)}$ denotes an achiral ligand.

2. A hemispheric suborbit (a neutral part) of length d_{jk} can take achiral and chiral ligands of the same kind (Chapter 8). Hence,

$$\begin{aligned}
b_{d_{jk}}^{(i\alpha)} &= \sum_{\ell=1}^{|\mathbf{X}|} w_{i\alpha}(X_\ell)^{d_{jk}} \\
&= \sum_{\ell=1}^{|\mathbf{X}|} w_{i\alpha}(X_\ell^{(a)})^{d_{jk}} + \sum_{\ell=1}^{|\mathbf{X}|} w_{i\alpha}(X_\ell^{(c)})^{d_{jk}} + \sum_{\ell=1}^{|\mathbf{X}|} w_{i\alpha}(\overline{X}_\ell^{(c)})^{d_{jk}}, \tag{19.32}
\end{aligned}$$

where X_ℓ denotes any type of ligands.

3. An enantiospheric suborbit takes two modes of packing (Chapter 8). Achiral ligands take a paired achiral packing. This packing gives rise to the same effect as that of the homospheric suborbit and affords the same type of ligand inventory. Chiral ligands occupy the enantiospheric suborbit in the manner of compensated chiral packing, which accomodates $d_{jk}/2\ X_\ell^{(c)}$ and the same number of antipodes $(d_{jk}/2\ \overline{X}_\ell^{(c)})$. There are two ways of such packing, as shown in the preceding section. Hence, we obtain the following ligand inventory.

$$c_{d_{jk}}^{(i\alpha)} = \sum_{\ell=1}^{|\mathbf{X}|} w_{i\alpha}(X_\ell^{(a)})^{d_{jk}} + 2\sum_{\ell=1}^{|\mathbf{X}|} \left(w_{i\alpha}(X_\ell^{(c)})w_{i\alpha}(\overline{X}_\ell^{(c)})\right)^{d_{jk}/2}, \tag{19.33}$$

where $X_\ell^{(c)}$ and $\overline{X}_\ell^{(c)}$ denote a pair of antipodal ligands. The first term of the right-hand side is concerned with the paired achiral packing and the second with the compensated chiral packing.

Since these equations are true for each orbit $(\Delta_{i\alpha})$, the product over all of the suborbits of $\Delta_{i\alpha}$ (*i.e.* over all of the subgroups $\mathbf{G}_k^{(j)}$) yields a generating function:

$$\prod_{k=1}^{v_j} \left(\sum_{\ell=1}^{|\mathbf{X}|} w_{i\alpha}(X_\ell^{(a)})^{d_{jk}}\right)^{\chi_{ak}^{(j)}\beta_k^{(ij)}} \left(\sum_{\ell=1}^{|\mathbf{X}|} w_{i\alpha}(X_\ell)^{d_{jk}}\right)^{\chi_{bk}^{(j)}\beta_k^{(ij)}}$$

$$\times \left(\sum_{\ell=1}^{|\mathbf{X}|} w_{i\alpha}(X_\ell^{(a)})^{d_{jk}} + 2\sum_{\ell=1}^{|\mathbf{X}|} \left(w_{i\alpha}(X_\ell^{(c)})w_{i\alpha}(\overline{X}_\ell^{(c)})\right)^{d_{jk}/2}\right)^{\chi_{ck}^{(j)}\beta_k^{(ij)}}$$

This equation is alternatively obtained by the introduction of eqs. 19.31 to 19.33 into the SCI (eq. 19.26). Since this equation holds for all of the orbits of Δ, the product over all α and all i provides a generating function of $\rho_{\theta j}$. Examination of the concrete form of the generating function shows that it is equal to the equation which is given in the present lemma.

Lemma 19.2 gives generating functions for calculating marks $(\rho_{\theta j})$, which are in turn introduced into the equations in Theorem 19.2 to afford the number $(B_{\theta i})$ of non-equivalent configurations of symmetry $\mathbf{G}_i$. In the evaluation of the $\rho_{\theta j}$ by Lemma 19.2, a term containing $w_{i\alpha}(X_\ell^{(c)})^m w_{i\alpha}(\overline{X}_\ell^{(c)})^n$ (for each ℓ) repsesents a molecule that is enantiomeric to a molecule represented by a term containing $w_{i\alpha}(X_\ell^{(c)})^n w_{i\alpha}(\overline{X}_\ell^{(c)})^m$. The coefficients of the enantiomeric pair should be summed up, since the present treatment regards such enantiomers as being equivalent.[5]

It should be noted that we are able to obtain any SCI of Def. 19.3 by an appropriate multiplication of such USCIs as collected in Appendix C.

19.3 Compounds with Given Weights

In Chapter 16, we have discussed enumeration of compounds with given weights, where we take acount of achiral ligands only. In this section, we extend the treatment of Chapter 16 so as to be applicable to cases that contain chiral ligands in addition to achiral ones.

Equation 19.25 is converted into

$$B_{\theta i} = \sum_{j=1}^{s} \rho_{\theta j} \overline{m}_{ji} \quad (i = 1, 2, \ldots, s), \tag{19.34}$$

where each $\overline{m}_{ji}$ is an element of the inverse of the mark table of $\mathbf{G}$. The total number (B_θ) of non-equivalent configurations (*i.e.* compounds) with the weight W_θ is obtained by summing up eq. 19.34 over all $\mathbf{G}_i$'s. Thus, we have

$$B_\theta = \sum_{i=1}^{s} B_{\theta i} = \sum_{i=1}^{s} \sum_{j=1}^{s} \rho_{\theta j} \overline{m}_{ji}. \tag{19.35}$$

It follows that

$$\sum_{[\theta]} B_\theta W_\theta = \sum_{[\theta]} \Big(\sum_{i=1}^{s} \sum_{j=1}^{s} \rho_{\theta j} \overline{m}_{ji} \Big) W_\theta = \sum_{j=1}^{s} \Big((\sum_{i=1}^{s} \overline{m}_{ji}) \sum_{[\theta]} \rho_{\theta j} W_\theta \Big). \tag{19.36}$$

The last term have been given in Lemma 19.2 (eq. 19.27). For simplicity's sake, we define a cycle index with chirality fittingness as follows.

---

[5] For illustrative examples, see S. Fujita, *Bull. Chem. Soc. Jpn.*, **63**, 203, 2759 (1990).

Definition 19.4 *A cycle index with chirality fittingness (CI-CF) for* **G** *group is represented by*

$$CIC(\mathbf{G}; a_{d_{jk}}^{(i\alpha)}, b_{d_{jk}}^{(i\alpha)}, c_{d_{jk}}^{(i\alpha)}) = \sum_{j=1}^{s}\left((\sum_{i=1}^{s}\overline{m}_{ji})ZIC(\mathbf{G}_j; a_{d_{jk}}^{(i\alpha)}, b_{d_{jk}}^{(i\alpha)}, c_{d_{jk}}^{(i\alpha)})\right) \qquad (19.37)$$

in terms of the subduced cycle index with chirality fittingness (SCI-CF).

Equations 19.36 and 19.37 afford a theorem,

Theorem 19.5 *The total number (B_θ) of non-equivalent W_θ-configurations (compounds) with chiral and achiral ligands is given by the following generating function,*

$$\sum_{[\theta]} B_\theta W_\theta = CIC(\mathbf{G}_j; a_{d_{jk}}^{(i\alpha)}, b_{d_{jk}}^{(i\alpha)}, c_{d_{jk}}^{(i\alpha)}), \qquad (19.38)$$

wherein the right-hand side is substituted by

$$a_{d_{jk}}^{(i\alpha)} = \sum_{\ell=1}^{|\mathbf{X}|} w_{i\alpha}(X_\ell^{(a)})^{d_{jk}}, \qquad (19.39)$$

$$b_{d_{jk}}^{(i\alpha)} = \sum_{\ell=1}^{|\mathbf{X}|} w_{i\alpha}(X_\ell)^{d_{jk}}, \qquad (19.40)$$

$$c_{d_{jk}}^{(i\alpha)} = \sum_{\ell=1}^{|\mathbf{X}|} w_{i\alpha}(X_\ell^{(a)})^{d_{jk}} + 2\sum_{\ell=1}^{|\mathbf{X}|}\left(w_{i\alpha}(X_\ell^{(c)})w_{i\alpha}(\overline{X}_\ell^{(c)})\right)^{d_{jk}/2} \qquad (19.41)$$

This theorem is an extension of Theorem 16.1 (Chapter 16).

19.4 Special Cases

Achiral and chiral ligands without OMVs. This section deals with the derivation of a special case in which OMV is not considered. For this purpose, the weights $(w(f)$'s) are independent of participant orbits, Thereby, the weight of a function (configuration) is found to be

$$W(f) = \prod_{\delta \in \Delta} w(f(\delta)). \qquad (19.42)$$

Suppose that $|\mathbf{X}_{i\alpha}^{(a)}| = |\mathbf{X}^{(a)}|$ and $|\mathbf{X}_{i\alpha}^{(b)}| = |\mathbf{X}_{i\alpha}^{(c)}| = |\mathbf{X}|$. Then, Theorem 19.2 is converted into the following corollary in order to meet the special case.

Corollary 19.1 *Without taking OMV into consideration, the number (B_i) of non-equivalent configurations with symmetry $\mathbf{G}_i$ is obtained by solving the following equations.*

$$\left(|\mathbf{X}^{(a)}|^{\sum_{i=1}^{s}\alpha_i\beta_{ij}^{(a)}}\right)\left(|\mathbf{X}|^{\sum_{i=1}^{s}\alpha_i(\beta_{ij}^{(b)}+\beta_{ij}^{(c)})}\right) = \sum_{i=1}^{s} B_i m_{ij} \quad (for\ j=1,2,\ldots,s) \qquad (19.43)$$

The definitions of this paragraph indicate that variables $a_{d_{jk}}^{(i\alpha)}, b_{d_{jk}}^{(i\alpha)}$, and $c_{d_{jk}}^{(i\alpha)}$, are independent of their orbits. Thus, we can omit the superscript $(i\alpha)$, *i.e.*

$$a_{d_{jk}}^{(i\alpha)} = a_{d_{jk}}, \quad b_{d_{jk}}^{(i\alpha)} = b_{d_{jk}}, \quad \text{and } c_{d_{jk}}^{(i\alpha)} = c_{d_{jk}}.$$

By using these variables, we transform the right-hand side of Def. 19.3 into a simpler form that is suitable to the special case. Hence, we find a new definition for a subduced cycle index (SCI).

Definition 19.5 *A subduced cycle index (SCI) without consideration of OMVs is defined as*

$$ZIC'(\mathbf{G}_j; a_{d_{jk}}, b_{d_{jk}}, c_{d_{jk}}) = \prod_{k=1}^{v_j} \left((a_{d_{jk}})^{q_{ak}^{(j)}} (b_{d_{jk}})^{q_{bk}^{(j)}} (c_{d_{jk}})^{q_{ck}^{(j)}} \right), \tag{19.44}$$

where the powers of the respective terms are represented by

$$q_{ak}^{(j)} = \sum_{i=1}^{s} \alpha_i \chi_{ak}^{(j)} \beta_k^{(ij)}, \tag{19.45}$$

$$q_{bk}^{(j)} = \sum_{i=1}^{s} \alpha_i \chi_{bk}^{(j)} \beta_k^{(ij)}, \tag{19.46}$$

and

$$q_{ck}^{(j)} = \sum_{i=1}^{s} \alpha_i \chi_{ck}^{(j)} \beta_k^{(ij)}. \tag{19.47}$$

Using the subduced cycle index, we transform Lemma 19.2 into Lemma 19.3 that is suitable for the special case of this section.

Lemma 19.3 *If $\mathbf{G}_j$ ($\leq \mathbf{G}$) acts on Δ and no OMVs are considered, a generating function for marks $\rho_{\theta j}$ with a weight W_θ is as follows.*

$$\sum_{[\theta]} \rho_{\theta j} W_\theta = ZIC'(\mathbf{G}_j; a_{d_{jk}}, b_{d_{jk}}, c_{d_{jk}}), \tag{19.48}$$

where

$$a_{d_{jk}} = \sum_{\ell=1}^{|\mathbf{X}|} (X_\ell^{(a)})^{d_{jk}}, \tag{19.49}$$

$$b_{d_{jk}} = \sum_{\ell=1}^{|\mathbf{X}|} (X_\ell)^{d_{jk}} = \sum_{\ell=1}^{|\mathbf{X}|} (X_\ell^{(a)})^{d_{jk}} + \sum_{\ell=1}^{|\mathbf{X}|} (X_\ell^{(c)})^{d_{jk}} + \sum_{\ell=1}^{|\mathbf{X}|} (\overline{X}_\ell^{(c)})^{d_{jk}}, \tag{19.50}$$

and

$$c_{d_{jk}} = \sum_{\ell=1}^{|\mathbf{X}|} (X_\ell^{(a)})^{d_{jk}} + 2 \sum_{\ell=1}^{|\mathbf{X}|} \left(X_\ell^{(c)} \overline{X}_\ell^{(c)} \right)^{d_{jk}/2}. \tag{19.51}$$

Lemma 19.3 affords a set of marks $\rho_{\theta j}$ that is necessary to enumerate isomers with a subsymmetry $\mathbf{G}_j$ and a weight W_θ based on a parent skeleton of symmetry $\mathbf{G}$.

The values of $\rho_{\theta j}$'s are introduced to Theorem 19.4 in order to obtain the number $(B_{\theta i})$ of isomers with W_θ and $\mathbf{G}_i$.

Achiral ligands with OMVs. This special case has been discussed in Chapter 15. Suppose that β_{ij} is the number of suborbits concerned with $\mathbf{G}(/\mathbf{G}_i)\downarrow \mathbf{G}_j$. Then, we have

$$
\begin{aligned}
\beta_{ij} &= \beta_{ij}^{(a)} + \beta_{ij}^{(b)} + \beta_{ij}^{(c)} \\
&= \sum_{k=1}^{v_j}(\chi_{ak}^{(j)} + \chi_{bk}^{(j)} + \chi_{ck}^{(j)})\beta_k^{(ij)} = \sum_{l=1}^{v_j} \beta_k^{(ij)}.
\end{aligned}
\tag{19.52}
$$

By introducing $| \mathbf{X}_{i\alpha}^{(a)} | = | \mathbf{X}_{i\alpha}^{(b)} | = | \mathbf{X}_{i\alpha}^{(c)} | = | \mathbf{X}_{i\alpha} |$ into Theorem 19.2, we obtain an equation for the special case.

$$
\prod_{\substack{i=1}}^{s} \prod_{\substack{\alpha=1 \\ \alpha_i \neq 0}}^{\alpha_i} | \mathbf{X}_{i\alpha} |^{\beta_{ij}} = \sum_{i=1}^{s} B_i m_{ij}.
\tag{19.53}
$$

This equation is identical with that of Theorem 15.3 (Chapter 15).

Under the conditions of this paragraph, we suppose that

$$
a_{d_{jk}} = b_{d_{jk}} = c_{d_{jk}} = s_{d_{jk}}.
$$

Hence, the definition of SCI (Def. 19.3) is converted into

$$
ZI(\mathbf{G}_j; s_{d_{jk}}^{(i\alpha)}) = \prod_{\substack{i=1}}^{s} \prod_{\substack{\alpha=1 \\ \alpha_i \neq 0}}^{\alpha_i} \prod_{k=1}^{v_j}(s_{d_{jk}}^{(i\alpha)})^{\beta_k^{(ij)}}.
\tag{19.54}
$$

This equation is identical with that of Def. 15.1 (Chapter 15). This fact indicates that the present derivation affords an alternative proof of Lemma 15.2 (Chaper 15).

19.5　Other Indices

In the same line as Section 16.3 (Chapter 16), we define partial cycle indices with chirality fittingness (PCI-CF) as follows.

Definition 19.6 *A partial cycle index with chirality fittingness (PCI-CF) is represented by*

$$
PCIC(\mathbf{G}_i; a_{d_{jk}}^{(i\alpha)}, b_{d_{jk}}^{(i\alpha)}, c_{d_{jk}}^{(i\alpha)}) = \sum_{j=1}^{s} \overline{m}_{ji} ZIC(\mathbf{G}_j; a_{d_{jk}}^{(i\alpha)}, b_{d_{jk}}^{(i\alpha)}, c_{d_{jk}}^{(i\alpha)})
\tag{19.55}
$$

for $i = 1, 2, \ldots, s$, where the SCI-CF (ZIC) in the right-hand side comes from Def. 19.3.

By starting from eq. 19.34, we have

$$\begin{aligned}
\sum_{[\theta]} B_{\theta i} W_\theta &= \sum_{[\theta]} \sum_{j=1}^{s} \rho_{\theta j} \overline{m}_{ji} W_\theta \\
&= \sum_{j=1}^{s} \overline{m}_{ji} \sum_{[\theta]} \rho_{\theta j} W_\theta \\
&= \sum_{j=1}^{s} \overline{m}_{ji} ZIC(\mathbf{G}_j; a_{d_{jk}}^{(i\alpha)}, b_{d_{jk}}^{(i\alpha)}, c_{d_{jk}}^{(i\alpha)}) \\
&= PCIC(\mathbf{G}_i; a_{d_{jk}}^{(i\alpha)}, b_{d_{jk}}^{(i\alpha)}, c_{d_{jk}}^{(i\alpha)}),
\end{aligned} \tag{19.56}$$

where we use eq. 19.27 and Def. 19.6. This derivation is summarized in the form of a theorem.

Theorem 19.6 *A generating function for calculating $B_{\theta i}$ is represented by*

$$\sum_{[\theta]} B_{\theta i} W_\theta = PCIC(\mathbf{G}_i; a_{d_{jk}}^{(i\alpha)}, b_{d_{jk}}^{(i\alpha)}, c_{d_{jk}}^{(i\alpha)}) \ (i = 1, 2, \ldots, s), \tag{19.57}$$

wherein the variables in the right-hand side are substituted by

$$a_{d_{jk}}^{(i\alpha)} = \sum_{\ell=1}^{|\mathbf{X}|} w_{i\alpha}(X_\ell^{(a)})^{d_{jk}}, \tag{19.58}$$

$$b_{d_{jk}}^{(i\alpha)} = \sum_{\ell=1}^{|\mathbf{X}|} w_{i\alpha}(X_\ell)^{d_{jk}}, \tag{19.59}$$

$$c_{d_{jk}}^{(i\alpha)} = \sum_{\ell=1}^{|\mathbf{X}|} w_{i\alpha}(X_\ell^{(a)})^{d_{jk}} + 2 \sum_{\ell=1}^{|\mathbf{X}|} \left(w_{i\alpha}(X_\ell^{(c)}) w_{i\alpha}(\overline{X}_\ell^{(c)}) \right)^{d_{jk}/2} \tag{19.60}$$

This theorem corresponds to Theorem 16.3 (Chapter 16). Theorem 19.6 is a generating-function version of Theorem 19.4 and Lemma 19.2.

By using the PCI-CF, we have an alternative definition of the CI-CF as follows.

Definition 19.7 *A cycle index with chirality fittingness (CI-CF) is represented by*

$$CIC(\mathbf{G}; a_{d_{jk}}^{(i\alpha)}, b_{d_{jk}}^{(i\alpha)}, c_{d_{jk}}^{(i\alpha)}) = \sum_{i=1}^{s} PCIC(\mathbf{G}_i; a_{d_{jk}}^{(i\alpha)}, b_{d_{jk}}^{(i\alpha)}, c_{d_{jk}}^{(i\alpha)}). \tag{19.61}$$

When we take no account of OMVs, we have a definition of PCI-CF for this case by starting from Def. 19.6.

Definition 19.8 *Without taking account of OMVs, a partial cycle index with chirality fittingness (PCI-CF) is represented by*

$$PCIC'(\mathbf{G}_i; a_{d_{jk}}, b_{d_{jk}}, c_{d_{jk}}) = \sum_{j=1}^{s} \overline{m}_{ji} ZIC'(\mathbf{G}_j; a_{d_{jk}}, b_{d_{jk}}, c_{d_{jk}}) \tag{19.62}$$

for $i = 1, 2, \ldots, s$, where the SCI-CF (ZIC') in the right-hand side comes from Def. 19.5.

Theorem 19.6 is transformed to the following form that is suitable for the present case without OMV restriction.

Theorem 19.7 *A generating function for calculating $B_{\theta i}$ is represented by*

$$\sum_{[\theta]} B_{\theta i} W_\theta = PCIC'(\mathbf{G}_i; a_{d_{jk}}, b_{d_{jk}}, c_{d_{jk}}) \quad (i = 1, 2, \ldots, s), \tag{19.63}$$

wherein the variables in the right-hand side are substituted by

$$a_{d_{jk}} = \sum_{\ell=1}^{|\mathbf{X}|} (X_\ell^{(a)})^{d_{jk}}, \tag{19.64}$$

$$b_{d_{jk}} = \sum_{\ell=1}^{|\mathbf{X}|} (X_\ell)^{d_{jk}}, \tag{19.65}$$

$$c_{d_{jk}} = \sum_{\ell=1}^{|\mathbf{X}|} (X_\ell^{(a)})^{d_{jk}} + 2 \sum_{\ell=1}^{|\mathbf{X}|} \left(X_\ell^{(c)} \overline{X}_\ell^{(c)} \right)^{d_{jk}/2} \tag{19.66}$$

We have described the definitions of several indices by starting from USCI-CFs. Their relationship is represented as follows. All of these indices are effective to solving enumeration problems.

$$\begin{array}{ccccc}
\text{USCI(-CF)} & \longrightarrow & \text{SCI(-CF)} & \longrightarrow & \text{PCI(-CF)} \\
& & & & \downarrow \\
& & \longrightarrow & & \text{CI(-CF)}
\end{array}$$

Chapter 20

Compounds with Rotatable Ligands [1]

Pólya[1] has already mentioned the enumeration of non-rigid molecules in terms of "coronas", which are equivalent to wreath products. The counting of non-rigid cyclohexane isomers was discussed by introducing a ring-flip-rotation operator along with usual symmetry operations.[2] Isomers derived from a non-rigid ethane were enumerated by means of the concept of covering groups.[3] A generalized wreath product method has been presented for enumerating stereo- and positional isomers.[4] However, systematic enumeration of non-rigid isomers that takes into account the spacial symmetries of isomers is open to further investigation. As a continuation of our studies on USCIs,[5] the present chapter will deal with this type of enumeration by using 2,2-diphenyl- and 2,2-dimethylpropane as examples. In particular, we discuss a general case in which a non-rigid molecule involves chiral ligands as well as achiral ones.

20.1 Rigid Skeleton and Rotatable Ligands

In order to enumerate isomers derived from a non-rigid parent molecule, the parent molecule is regarded as a three-dimensional object that has terminal substitution positions. These positions can be chemically classified into several equivalence classes. Although this classification should be discussed by using wreath products[1] and generalized wreath products,[4] we here adopt a more intuitive approach that is suitable for the application of USCIs.

A non-rigid molecule is considered to be divided into a rigid skeleton and rotatable ligands. For example, 2,2-diphenylpropane (1) is divided into three parts (1a, 1b, and 1c) as shown in Fig. 20.1. Note that free bond-rotations are considered to be allowed in this compound. The rigid skeleton (1a) of C_{2v} symmetry has four joints (• and ◐) to which the rotatable ligands attach. The rotatable ligand (1b) of C_{3v} symmetry consists of the joint (◐) and 3 substitution

[1]Reprinted with permission from S. Fujita, *Bull. Chem. Soc. Jpn.*, 63, 2033–2043 (1990). ©(1990) The Chemical Society of Japan. See also S. Fujita, *Theor. Chim. Acta*, 77, 307–321 (1990).

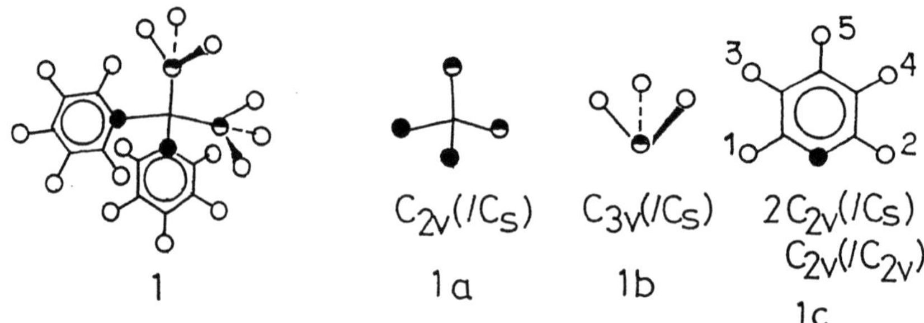

Figure 20.1: A $C_{2v}[(/C_s)[C_{3v}(/C_s)], (/C'_s)[C_{2v}(2/C_s, C_{2v})]]$ molecule

positions (o). The rotatable ligand (1c) of C_{2v} symmetry contains the joint (•) and 5 substitution positions (o). The other positions of these parts are not taken into consideration.

In the light of this formulation, the rigid skeleton (1a) is invariant on any chemical bond-rotations around the bonds contained in the skeleton. Moreover, the rotatable ligands (1b and 1c) are allowed to be regarded as rigid objects, since they are also invariant on any bond-rotations. Hence, the previously developed method (Chapter 15)[5] for enumerating rigid molecules are applicable to the present case. The joints (• and ◓) are thus considered to be substitution positions and divided into two equivalence classes, which are called *orbits* in terms of the terminology of permutation groups.[6, 7] The 3 positions of 1b and the 5 positions of 1c are also divided into several orbits.

Each of the orbits corresponds to a coset representation $G(/G_i)$ in one-to-one fashion, where G is a (point) group for characterizing the symmetry of such a rigid skeleton or a rotatable ligand and G_i is a subgroup of G.[5] A table of marks for the G-symmetry[8] is used for assigning an orbit to a CR ($G(/G_i)$).

The 4 joints of the rigid skeleton (1a) construct two orbits that are subject to CRs, $C_{2v}(/C_s)$ and $C_{2v}(/C'_s)$, respectively. This fact can be formulated as

$$\mathbf{P_{1a}} = C_{2v}(/C_s) + C_{2v}(/C'_s). \tag{20.1}$$

The 3 substitution positions (o) of the rotatable ligand (1b) belong to an orbit governed by a CR, $C_{3v}(/C_s)$. The rotatable ligand (1c) has 5 vertices to be considered (o). Since this unit belongs to C_{2v}-symmetry, they are classified into 3 orbits that are subject to $C_{2v}(/C_s)$, $C_{2v}(/C_s)$, and $C_{2v}(/C_{2v})$. This assignment is algebraically represented by

$$\mathbf{P_{1c}} = 2\mathbf{C}_{2v}(/\mathbf{C}_s) + \mathbf{C}_{2v}(/\mathbf{C}_{2v}), \tag{20.2}$$

the determination of which can be accomplished by using a table of marks,[5] where $\mathbf{P_{1c}}$ is a permutation representation of the $\mathbf{C}_{2v}$ unit (1c). As a result, the non-rigid molecule (1) is represented by $\mathbf{C}_{2v}[(/\mathbf{C}_s)[\mathbf{C}_{3v}(/\mathbf{C}_s)], (/\mathbf{C'}_s)[\mathbf{C}_{2v}(2/\mathbf{C}_s,\mathbf{C}_{2v})]]$, to which we refer as the *extended wreath product (EWP) notation* in the present chapter.

20.2 Enumeration of Rotatable Ligands

A mathematical treatment for enumeration of rotatable ligands is essentially the same as described for rigid molecules in Chapter 15. Suppose that a given ligand of **H**-symmetry consists of several atoms whose symmetrical properties are characterized by a permutation representation $(\mathbf{P_H})$. Then, $\mathbf{P_H}$ can be reduced into a sum of coset representations, *i.e.*,

$$\mathbf{P_H} = \sum_{p=1}^{C_\mathbf{H}} \gamma_p \mathbf{H}(/\mathbf{H}_p), \tag{20.3}$$

where the symbol $\mathbf{H}_p$ $(p = 1, 2, \ldots, C_\mathbf{H})$ denotes a representative of conjugate subgroups and the multiplicity, γ_p, is a non-negative integer. The multiplicity γ_p can be algebraically obtained by using a table of marks for **H** group (Chapters 5 and 6).[5] Equation 20.2 is a simple example of this formulation.

According to the subduction of the CRs $(\mathbf{H}(/\mathbf{H}_p) \downarrow \mathbf{H}_q)$, we have defined unit subduced cycle indices (USCIs) as

$$Z(\mathbf{H}(/\mathbf{H}_p) \downarrow \mathbf{H}_q; t_{d_{qr}}) = \prod_{r=1}^{v_q} (t_{d_{qr}})^{\delta_{pqr}} \tag{20.4}$$

for $p = 1, 2, \ldots, C_\mathbf{H}$ and $q = 1, 2, \ldots, C_\mathbf{H}$, where the symbol v_q denotes the number of representatives of conjugate subgroups of $\mathbf{H}_q$; and the power δ_{pqr} is determined by the subduction (Chapter 9).[5] Note that the t variables are just dummy symbols. The symbol d_{qr} denotes the length of the $\mathbf{H}_q(/\mathbf{H}_{qr})$ orbit which is in turn represented by

$$d_{qr} = |\, \mathbf{H}_q \,| \,/\, |\, \mathbf{H}_{qr} \,|, \tag{20.5}$$

where the symbol $\mathbf{H}_{qr}$ $(r = 1, 2, \ldots, v_q)$ denotes a representative of conjugate subgroups of $\mathbf{H}_q$. Tables 20.1 and 20.2 list USCIs for $\mathbf{C}_{2v}$ and $\mathbf{C}_{3v}$.

In terms of the USCIs (eq. 20.4), a subduced cycle index (SCI) for each subgroup $(\mathbf{H}_q)$ is defined as:

$$ZI(\mathbf{H}_q; t_{d_{qr}}) = \prod_{p=1}^{C_\mathbf{H}} [Z(\mathbf{H}(/\mathbf{H}_p) \downarrow \mathbf{H}_q; t_{d_{qr}})]^{\gamma_p}$$

Table 20.1: USCIs for C_{2v} point group

	unit subduced cycle index[a] for				
	$\downarrow C_1$	$\downarrow C_2$	$\downarrow C_s$	$\downarrow C'_s$	$\downarrow C_{3v}$
$C_{2v}(/C_1)$	$s_1^4\ (b_1^4)$	$s_2^2\ (b_2^2)$	$s_2^2\ (c_2^2)$	$s_2^2\ (c_2^2)$	$s_4\ (c_4)$
$C_{2v}(/C_2)$	$s_1^2\ (b_1^2)$	$s_1^2\ (b_1^2)$	$s_2\ (c_2)$	$s_2\ (c_2)$	$s_2\ (c_2)$
$C_{2v}(/C_s)$	$s_1^2\ (b_1^2)$	$s_2\ (b_2)$	$s_1^2\ (a_1^2)$	$s_2\ (c_2)$	$s_2\ (a_2)$
$C_{2v}(/C'_s)$	$s_1^2\ (b_1^2)$	$s_2\ (b_2)$	$s_2\ (c_2)$	$s_1^2\ (a_1^2)$	$s_2\ (a_2)$
$C_{2v}(/C_{2v})$	$s_1\ (b_1)$	$s_1\ (b_1)$	$s_1\ (a_1)$	$s_1\ (a_1)$	$s_1\ (a_1)$

[a] A variable in the parentheses is a USCI with chirality fittingness.

Table 20.2: USCIs for C_{3v} point group

	unit subduced cycle index[a] for			
	$\downarrow C_1$	$\downarrow C_s$	$\downarrow C_3$	$\downarrow C_{3v}$
$C_{3v}(/C_1)$	$s_1^6\ (b_1^6)$	$s_2^3\ (c_2^3)$	$s_3^2\ (b_3^2)$	$s_6\ (c_6)$
$C_{3v}(/C_s)$	$s_1^3\ (b_1^3)$	$s_1 s_2\ (a_1 c_2)$	$s_3\ (b_3)$	$s_3\ (a_3)$
$C_{3v}(/C_3)$	$s_1^2\ (b_1^2)$	$s_2\ (c_2)$	$s_1^2\ (b_1^2)$	$s_2\ (c_2)$
$C_{3v}(/C_{3v})$	$s_1\ (b_1)$	$s_1\ (a_1)$	$s_1\ (b_1)$	$s_1\ (a_1)$

[a] A variable in the parentheses is a USCI with chirality fittingness.

$$= \prod_{p=1}^{C_{\mathbf{H}}} [\prod_{r=1}^{v_p} (t_{d_{qr}})^{\delta_{pqr}}]^{\gamma_p}$$

$$= \prod_{r=1}^{v_p} (t_{d_{qr}})^{\sum_{p=1}^{C_{\mathbf{H}}} \gamma_p \delta_{pqr}} \tag{20.6}$$

for $q = 1, 2, \ldots, C_{\mathbf{H}}$ (Chapter 9).[5]

Suppose that rotatable ligands belonging to an equivalence class have the same molecular formula, $w_\xi = X_1^{n_1} X_2^{n_2} \cdots X_{|\mathbf{X}|}^{n_{|\mathbf{X}|}}$, where $| \Psi |$ is the number of positions in a rotatable ligand and $n_1 + n_2 + \ldots + n_{|\mathbf{X}|} = | \Psi |$. The number of isomeric ligands is regarded as the number of such equivalence classes with the same molecular formula. Let $\rho_{\xi q}$ be the number of such isomeric ligands with the weight w_ξ and $\mathbf{H}_q$. Then, $\rho_{\xi q}$ is obtained by means of generating functions (Chaper 15),[5] *i.e.*,

Lemma 20.1 . (Calculation of $\rho_{\xi q}$)

$$\sum_{[\xi]} \rho_{\xi q} w_\xi = ZI(\mathbf{H}_q; t_{d_{qr}}) \tag{20.7}$$

for $q = 1, 2, \ldots, C_{\mathbf{H}}$, in which every term of the right-hand side is substituted by

$$t_{d_{qr}} = \sum_{r=1}^{|\mathbf{X}|} X_r^{d_{qr}}. \tag{20.8}$$

In the light of this lemma, the number of non-equivalent rotatable ligands $(B_{\xi p})$ is obtained by the following lemma.

Lemma 20.2 . (Enumeration of rotatable ligands)
When $\overline{m}_{qp}$ is an element of the inverse of the mark table, the number of non-equivalent rotatable ligands $(B_{\xi p})$ is obtained by

$$B_{\xi p} = \sum_{q=1}^{C_{\mathbf{H}}} \rho_{\xi q} \overline{m}_{qp}. \tag{20.9}$$

Note that we count every pair of antipodes ($y_{\xi q}$ and $\tilde{y}_{\xi q}$).

The following example illustrates the enumeraton of rotatable ligands.

Example **20.1** . A $\mathbf{C}_{2v}(/\mathbf{C}_s)$ rotatable ligand (**1c**).

Suppose that the 5 positions of **1c** are occupied by either X or Y. Then, the codomain of the present case is $\mathbf{X} = \{ X, Y \}$. The domain containing the 5 positions is designated by $\Psi = \{ 1, 2, 3, 4, 5 \}$, which is divided into three orbits,

$$\begin{aligned}
\Psi_1 &= \{1,2\} \quad \text{subject to } \mathbf{C}_{2v}(/\mathbf{C}_s), \\
\Psi_2 &= \{3,4\} \quad \text{subject to } \mathbf{C}_{2v}(/\mathbf{C}_s), \text{ and} \\
\Psi_3 &= \{5\} \quad \text{subject to } \mathbf{C}_{2v}(/\mathbf{C}_{2v}),
\end{aligned}$$

according to eq. 20.2. Hence, the SCIs for this case are obtained from the $C_{2v}(/C_s)$ (twice) and $C_{2v}(/C_{2v})$ rows of Table 20.1. The introduction of a ligand-inventory, $s_d = X^d + Y^d$, into these SCIs provides generating functions,

$$
\begin{aligned}
(s_1^2)^2 s_1 &= (X+Y)^5 \quad \text{for } C_1, &(20.10)\\
(s_2)^2 s_1 &= (X+Y)(X^2+Y^2)^2 \quad \text{for } C_2, &(20.11)\\
(s_1^2)^2 s_1 &= (X+Y)^5 \quad \text{for } C_s, &(20.12)\\
(s_2)^2 s_1 &= (X+Y)(X^2+Y^2)^2 \quad \text{for } C'_s, &(20.13)\\
&\text{and}\\
(s_2)^2 s_1 &= (X+Y)(X^2+Y^2)^2 \quad \text{for } C_{2v}. &(20.14)
\end{aligned}
$$

These equations are expanded to give a matrix $(\rho_{\xi q})$, in which each column consists of the coefficients of respective terms. This matrix is in turn multiplied by the inverse of a mark table of C_{2v}. Thereby, we arrive at

$$
\begin{array}{c}
X^5 \\ X^4Y \\ X^3Y^2 \\ X^2Y^3 \\ XY^4 \\ Y^5
\end{array}
\left(
\begin{array}{ccccc}
1 & 1 & 1 & 1 & 1 \\
5 & 1 & 5 & 1 & 1 \\
10 & 2 & 10 & 2 & 2 \\
10 & 2 & 10 & 2 & 2 \\
5 & 1 & 5 & 1 & 1 \\
1 & 1 & 1 & 1 & 1
\end{array}
\right)
\left(
\begin{array}{ccccc}
\frac{1}{4} & 0 & 0 & 0 & 0 \\
-\frac{1}{4} & \frac{1}{2} & 0 & 0 & 0 \\
-\frac{1}{4} & 0 & \frac{1}{2} & 0 & 0 \\
-\frac{1}{4} & 0 & 0 & \frac{1}{2} & 0 \\
\frac{1}{2} & -\frac{1}{2} & -\frac{1}{2} & -\frac{1}{2} & 1
\end{array}
\right)
$$

$$
(\rho_{\xi q}) \qquad \text{the inverse}
$$

$$
=
\begin{array}{ccccc}
C_1 & C_2 & C_s & C'_s & C_{2v}
\end{array}
\left(
\begin{array}{ccccc}
0 & 0 & 0 & 0 & 1 \\
0 & 0 & 2 & 0 & 1 \\
0 & 0 & 4 & 0 & 2 \\
0 & 0 & 4 & 0 & 2 \\
0 & 0 & 2 & 0 & 1 \\
0 & 0 & 0 & 0 & 1
\end{array}
\right) \qquad (20.15)
$$

$$
(B_{\xi p})
$$

The resulting matrix affords the numbers of respective ligands, which are depicted in Fig. 20.2. The set of symbols (m of • and n of ○) in Fig. 20.2 corresponds to $X^m Y^n$ and $X^n Y^m$. All of the rotatable ligands collected in Fig. 20.2 are achiral.

20.3 Enumeration of Non-Rigid Isomers

Suppose that the rigid skeleton of a non-rigid molecule of G-symmetry has $|\Phi|$ substitution positions. If a permutation group P_G acts on the positions, the action

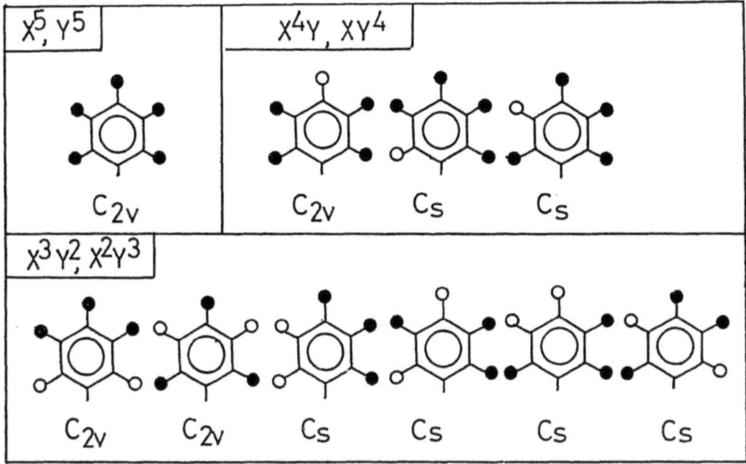

Figure 20.2: Rotatable ligands based on 1c

is expressed by

$$\mathbf{P_G} = \sum_{i=1}^{C_G} \alpha_i \mathbf{G}(/\mathbf{G}_i), \tag{20.16}$$

wherein α_i is the multiplicity of the CR, $\mathbf{G}(/\mathbf{G}_i)$. According to the subduction of the CR,

$$\mathbf{G}(/\mathbf{G}_i) \downarrow \mathbf{G}_j = \sum_{k=1}^{u_j} \beta_{ijk} \mathbf{G}_j(/\mathbf{G}_{jk}), \tag{20.17}$$

we define a USCI with chirality fittingness as

$$ZC(\mathbf{G}(/\mathbf{G}_i) \downarrow \mathbf{G}_j; \$_{d_{jk}}) = \prod_{k=1}^{u_j} (\$_{d_{jk}})^{\beta_{ijk}}, \tag{20.18}$$

for $i = 1, 2, \ldots, C_G$ and $j = 1, 2, \ldots, C_G$ (Chapter 9). This USCI is essentially equivalent to that described for a rotatable ligand. In a similar way, the subscript for \$ is calculated by

$$d_{jk} = |\mathbf{G}_j| \, / \, |\mathbf{G}_{jk}|. \tag{20.19}$$

The dummy variable \$ is replaced by the variable (a, b, or c) in accord with the chirality fittingness of the participating CR. The variable (a) is selected when both $\mathbf{G}_j$ and $\mathbf{G}_{jk}$ are achiral. The variable (b) is adopted, if both $\mathbf{G}_j$ and $\mathbf{G}_{jk}$ are chiral. The variable (c) is for the case in which $\mathbf{G}_j$ is achiral and $\mathbf{G}_{jk}$ is chiral. Since the dummy variable depends upon the $i\alpha$-th orbit through the corresponding CR, this dependence is designated by the superscript of the symbol $\$_{d_{jk}}^{(i\alpha)}$. The SCI for this case is obtained in a similar way described in the preceding sections, *i.e.*,

$$\begin{aligned} ZIC(\mathbf{G}_j; \$_{d_{qr}}^{(i\alpha)}) &= \prod_{\substack{i=1 \\ (\alpha_i \neq 0)}}^{C_G} \prod_{\alpha=1}^{\alpha_i} \mathbf{Z}(\mathbf{G}(/\mathbf{G}_i) \downarrow \mathbf{G}_j; \$_{d_{qr}}^{(i\alpha)}) \\ &= \prod_{\substack{i=1 \\ (\alpha_i \neq 0)}}^{C_G} \prod_{\alpha=1}^{\alpha_i} [\prod_{k=1}^{u_j} (\$_{d_{qr}}^{i\alpha})^{\beta_{ijk}}] \end{aligned} \tag{20.20}$$

$$(j = 1, 2, \ldots, C_G).$$

This equation is identical with Def. 19.3 (Chapter 19) except notation.

A non-rigid compound is regarded as a derivative of a $\mathbf{G}$ rigid skeleton, in which the α-th $\mathbf{G}(/\mathbf{G}_i)$ orbit is substituted by the ligands of $\mathbf{H}_p^{(i\alpha)}$ that have been enumerated in the preceding sections.

Suppose that non-rigid molecules belonging to an equivalence class have the same molecular formula, $W_\theta = X_1^{n_1} X_2^{n_2} \cdots X_{|\mathbf{X}|}^{n_{|\mathbf{X}|}}$, where $|\Phi|$ is the number of positions (joints) in the rigid skeleton and $n_1 + n_2 + \ldots + n_{|\mathbf{X}|} = |\Phi|$.

If all the rotatable ligands ($y_{\xi p}^{(i\alpha)}$'s) and their antipodes ($\tilde{y}_{\xi p}^{(i\alpha)}$'s) ($p = 1, 2, \ldots, C_{\mathbf{H}^{(i\alpha)}}$) have the same weight ($w_\xi^{(i\alpha)}$), the following lemma can be obtained by a slight modification of the method described in Chapter 9.[9] We arrive at the following lemma, which is a modification of Lemma 19.2 (Chapter 19).

Lemma 20.3 (Calculation of $\sigma_{\theta j}$)

Let $\sigma_{\theta j}$ be the number of such derivatives with the weight (W_θ) that are invariant (or fixed) on the operation of $\mathbf{G}_j$. We calculate $\sigma_{\theta j}$ by means of generating functions,[5]

$$\sum_{[\theta]} \sigma_{\theta j} W_\theta = ZIC(\mathbf{G}_j; \$_{d_{jk}}^{(i\alpha)}) \tag{20.21}$$

for $j = 1, 2, \ldots, C_{\mathbf{G}}$, in which every variable of the right-hand side is substituted by ligand-inventories,

$$a_{d_{jk}}^{(i\alpha)} = \sum_{[\xi]} \kappa_{\xi a}(w_\xi^{(i\alpha)})^{d_{jk}} \; for \; \$ = a \tag{20.22}$$

$$b_{d_{jk}}^{(i\alpha)} = \sum_{[\xi]} \kappa_{\xi a}(w_\xi^{(i\alpha)})^{d_{jk}} + 2\sum_{[\xi]} \kappa_{\xi c}(w_\xi^{(i\alpha)})^{d_{jk}} \; for \; \$ = b \; and \tag{20.23}$$

$$c_{d_{jk}}^{(i\alpha)} = \sum_{[\xi]} \kappa_{\xi a}(w_\xi^{(i\alpha)})^{d_{jk}} + 2\sum_{[\xi]} \kappa_{\xi c}(w_\xi^{(i\alpha)})^{d_{jk}} \; for \; \$ = c, \tag{20.24}$$

where $\kappa_{\xi a}$ and $\kappa_{\xi c}$ are represented by

$$\kappa_{\xi a} = \sum_{\substack{p=1 \\ achiral}}^{C_{\mathbf{H}^{(i\alpha)}}} B_{\xi p}^{(i\alpha)} \tag{20.25}$$

and

$$\kappa_{\xi c} = \sum_{\substack{p=1 \\ chiral}}^{C_{\mathbf{H}^{(i\alpha)}}} B_{\xi p}^{(i\alpha)}. \tag{20.26}$$

By means of this lemma, we end up with a theorem.

Theorem 20.1 . (Enumeration of $W_\theta,\mathbf{G}_i$-isomers)

Let $A_{\theta i}$ be the number of $W_\theta,\mathbf{G}_i$-isomers. This is calculated by using $\sigma_{\theta j}$ (Lemma 20.3) by means of

$$\sigma_{\theta j} = \sum_{i=1}^{C_{\mathbf{G}}} A_{\theta i} m_{ij} \tag{20.27}$$
$$for \; j = 1, 2, \ldots, C_{\mathbf{G}},$$

or inversely,

$$A_{\theta i} = \sum_{j=1}^{C_{\mathbf{G}}} \sigma_{\theta j} \overline{m}_{ji} \tag{20.28}$$
$$for \; i = 1, 2, \ldots, C_{\mathbf{G}},$$

wherein m_{ij} is the ij-element of a mark table (M) for the $\mathbf{G}$-group and $(\overline{m}_{ji})$ denotes the inverse of the matrix (M^{-1}).

If we thus restrict the rotatable ligands within achiral ones, we arrive at the following lemma.

Lemma 20.4 (Calculation of $\hat{\sigma}_{\theta j}$)
Let $\hat{\sigma}_{\theta j}$ be the number of such derivatives with the weight (W_θ) that are invariant (or fixed) on the operations of $\mathbf{G}_j$, when only achiral rotatable ligands are allowed. We obtain $\hat{\sigma}_{\theta j}$ by means of generating functions,[5]

$$\sum_{[\theta]} \hat{\sigma}_{\theta j} W_\theta = ZI(\mathbf{G}_j; s_{d_{jk}}^{(i\alpha)}) \tag{20.29}$$

for $j = 1, 2, \ldots, C_\mathbf{G}$, in which every variable of the right-hand side is substituted by a ligand-inventory,

$$s_{d_{jk}}^{(i\alpha)} = \sum_{[\xi]} \kappa_{\xi a} (w_\xi^{(i\alpha)})^{d_{jk}}. \tag{20.30}$$

Using Lemma 20.4, we can enumerate non-rigid molecules without any chiral rotatable ligands. For this purpose, we obtain

Corollary 20.1 . (Enumeration of non-rigid isomers with achiral rotatable ligands)
Let $\widehat{A}_{\theta i}$ be the number of $W_\theta, \mathbf{G}_i$-isomers without any chiral rotatable ligands. This is calculated from $\hat{\sigma}_{\theta j}$ (Lemma 20.4) by means of

$$\hat{\sigma}_{\theta j} = \sum_{i=1}^{C_\mathbf{G}} \widehat{A}_{\theta i} M_{ij} \tag{20.31}$$
$$\text{for } j = 1, 2, \ldots, C_\mathbf{G},$$

or inversely,

$$\widehat{A}_{\theta i} = \sum_{j=1}^{C_\mathbf{G}} \hat{\sigma}_{\theta j} \overline{m}_{ji} \tag{20.32}$$
$$\text{for } i = 1, 2, \ldots, C_\mathbf{G},$$

wherein m_{ij} is the ij-element of a mark table (M) for the $\mathbf{G}$-group and $(\overline{m}_{ji})$ denotes the inverse of the matrix (M^{-1}).

Lemmas 20.3 and 20.4 afford a foundation to the enumeration of the case concerning derivatives with at least one chiral rotatable ligand. That is to say, we have a useful lemma.

Lemma 20.5 (Calculation of $\tilde{\sigma}_{\theta j}$)
Let $\tilde{\sigma}_{\theta j}$ be the number of such derivatives with the weight (W_θ) that are invariant (or fixed) on the operations of $\mathbf{G}_j$, when permitting achiral ligands in addition to at least one chiral rotatable ligand. Then,

$$\sum_{[\theta]} \tilde{\sigma}_{\theta j} W_\theta = ZIC(\mathbf{G}_j; \$_{d_{jk}}^{(i\alpha)}) - ZI(\mathbf{G}_j; s_{d_{jk}}^{(i\alpha)}) \tag{20.33}$$

for $j = 1, 2, \ldots, C_\mathbf{G}$, in which every variable of the right-hand side is given by Lemmas 20.3 and 20.4.

In the light of Lemma 20.5, we can enumerate non-rigid molecules having at least one chiral rotatable ligand. We arrive at a corollary.

Corollary 20.2 . (Enumeration of non-rigid isomers with at least one chiral rotatable ligand)

Let $\widetilde{A}_{\theta i}$ be the number of W_θ, G_i-isomers with at least one chiral rotatable ligand. This is calculated from $\tilde{\sigma}_{\theta j}$ (Lemma 20.5) by means of

$$\tilde{\sigma}_{\theta j} = \sum_{i=1}^{C_{\mathbf{G}}} \widetilde{A}_{\theta i} m_{ij} \tag{20.34}$$
$$for\ j = 1, 2, \ldots, C_{\mathbf{G}},$$

or inversely,

$$\widetilde{A}_{\theta i} = \sum_{j=1}^{C_{\mathbf{G}}} \tilde{\sigma}_{\theta j} \overline{m}_{ji} \tag{20.35}$$
$$for\ i = 1, 2, \ldots, C_{\mathbf{G}},$$

wherein m_{ij} is the ij-element of a mark table (M) for the G-group and $(\overline{m}_{ji})$ denotes the inverse of the matrix (M^{-1}).

Example **20.2** . (Enumeration based on 2,2-diphenylpropane **1**).

Let us examine the parent **1** with substituents selected from $\{X, Y\}$. Since this case produces no chiral rotatable ligands, we can obtain the same result, whether we use Lemma 20.3 or Lemma 20.4. A ligand-inventory for **1c** is obtained in the light of Example 20.1. Thus,

$$\kappa_{\xi a} \begin{cases} = 1 & \text{for } w_\xi = X^5 \text{and } Y^5 \\ = 3 & \text{for } w_\xi = X^4Y \text{and } XY^4 \\ = 6 & \text{for } w_\xi = X^3Y^2 \text{and } X^2Y^3 \end{cases} \tag{20.36}$$

According to Lemma 20.4, we obtain a ligand-inventory,

$$s_d^{(2)} = X^{5d} + 3(X^4Y)^d + 6(X^3Y^2)^d + 6(X^2Y^3)^d + 3(XY^4)^d + Y^{5d} \tag{20.37}$$
$$\text{for } \mathbf{1c}.$$

A ligand-inventory for **1b** can be obtained by a similar enumeration as described in Example 20.1. Alternatively, we can easily obtain the same result, because there are four patterns of substitutions (X^3, X^2Y, XY^2, and Y^3) on the basis of the rotatable ligand (**1b**).

$$s_d^{(1)} = X^{3d} + (X^2Y)^d + (XY^2)^d + Y^{3d} \tag{20.38}$$
$$\text{for } \mathbf{1b}$$

The four vertices of the rigid skeleton (**1a**) are divided into two categories which are subject to $C_{2v}(/C_s)$ and $C_{2v}(/C_s')$. Hence, the above ligand-inventories are introduced into SCIs which are derived from the $\mathbf{C}_{2v}(/\mathbf{C}_s)$ and $\mathbf{C}_{2v}(/\mathbf{C}'_s)$ rows of Table 20.1. Thereby, we obtain

$$
\begin{aligned}
(s_1^2)^{(1)}(s_1^2)^{(2)} &= (X^3 + X^2Y + XY^2 + Y^3)^2 \times \\
&\quad (X^5 + 3X^4Y + 6X^3Y^2 + 6X^2Y^3 + 3XY^4 + Y^5)^2 \quad (20.39) \\
&\quad \text{for } \mathbf{C}_1, \\
(s_2)^{(1)}(s_2)^{(2)} &= (X^6 + X^4Y^2 + X^2Y^4 + Y^6) \times \\
&\quad (X^{10} + 3X^8Y^2 + 6X^6Y^4 + 6X^4Y^6 + 3X^2Y^8 + Y^{10}) (20.40) \\
&\quad \text{for } \mathbf{C}_2, \\
(s_1^2)^{(1)}(s_2)^{(2)} &= (X^3 + X^2Y + XY^2 + Y^3)^2 \times \\
&\quad (X^{10} + 3X^8Y^2 + 6X^6Y^4 + 6X^4Y^6 + 3X^2Y^8 + Y^{10}) (20.41) \\
&\quad \text{for } \mathbf{C}_s, \\
(s_2)^{(1)}(s_1^2)^{(2)} &= (X^6 + X^4Y^2 + X^2Y^4 + Y^6) \times \\
&\quad (X^5 + 3X^4Y + 6X^3Y^2 + 6X^2Y^3 + 3XY^4 + Y^5)^2 \quad (20.42) \\
&\quad \text{for } \mathbf{C}'_s, \text{ and} \\
(s_2)^{(1)}(s_2)^{(2)} &= (X^6 + X^4Y^2 + X^2Y^4 + Y^6) \times \\
&\quad (X^{10} + 3X^8Y^2 + 6X^6Y^4 + 6X^4Y^6 + 3X^2Y^8 + Y^{10}) (20.43) \\
&\quad \text{for } \mathbf{C}_{2v},
\end{aligned}
$$

in which superscripts (1) and (2) denote the $\mathbf{C}_{2v}(/\mathbf{C}_s)$ and $\mathbf{C}_{2v}(/\mathbf{C}'_s)$ orbits, respectively. The expansion of these generating functions and the collection of the terms of the same power give a matrix involving the values of $\sigma_{\xi j}$. This matrix is multiplied by the inverse of the mark table of $\mathbf{C}_{2v}$, $i.e.$,

$$
\begin{array}{c}
\begin{array}{r}
X^{16} \\
X^{15}Y \\
X^{14}Y^2 \\
X^{13}Y^3 \\
X^{12}Y^4 \\
X^{11}Y^5 \\
X^{10}Y^6 \\
X^9Y^7 \\
X^8Y^8
\end{array}
\left(
\begin{array}{ccccc}
1 & 1 & 1 & 1 & 1 \\
8 & 0 & 2 & 6 & 0 \\
36 & 4 & 6 & 22 & 4 \\
112 & 0 & 10 & 54 & 0 \\
264 & 10 & 18 & 100 & 10 \\
496 & 0 & 26 & 146 & 0 \\
764 & 16 & 34 & 178 & 16 \\
986 & 0 & 42 & 194 & 0 \\
1070 & 18 & 42 & 198 & 18
\end{array}
\right) \\
(\sigma_{\xi j})
\end{array}
\quad
\begin{array}{c}
\left(
\begin{array}{ccccc}
\frac{1}{4} & 0 & 0 & 0 & 0 \\
-\frac{1}{4} & \frac{1}{2} & 0 & 0 & 0 \\
-\frac{1}{4} & 0 & \frac{1}{2} & 0 & 0 \\
-\frac{1}{4} & 0 & 0 & \frac{1}{2} & 0 \\
\frac{1}{2} & -\frac{1}{2} & -\frac{1}{2} & -\frac{1}{2} & 1
\end{array}
\right) \\
the\ inverse
\end{array}
$$

$$
= \begin{pmatrix}
\mathbf{C_1} & \mathbf{C_2} & \mathbf{C_s} & \mathbf{C'_s} & \mathbf{C_{2v}} \\
0 & 0 & 0 & 0 & 1 \\
0 & 0 & 1 & 3 & 1 \\
3 & 0 & 1 & 9 & 4 \\
12 & 0 & 5 & 27 & 0 \\
39 & 0 & 4 & 45 & 10 \\
81 & 0 & 13 & 73 & 0 \\
142 & 0 & 9 & 81 & 16 \\
187 & 0 & 21 & 97 & 0 \\
212 & 0 & 12 & 90 & 18
\end{pmatrix}. \qquad (20.44)
$$

$$(A_{\xi p})$$

Since the $X^m Y^n$ and $X^n Y^m$ terms afford the same results, eq. 20.44 lists the values of the $X^m Y^n$ term ($m \geq n$). Figure 20.3 depicts 17 $X^{14}Y^2$-isomers having the subsymmetries of the $\mathbf{C}_{2v}$-symmetry, in which the symbol ($\circ$) denotes a Y-substitution and the remaining 14 positions are occupied by X. Thus, there emerge 3 $\mathbf{C}_1$ (asymmetric) molecules, one $\mathbf{C}_s$-molecule, 9 $\mathbf{C}'_s$-molecules and 4 $\mathbf{C}_{2v}$-molecules in accord with the data listed at the 3rd row of the rightmost matrix in eq. 20.44. Note that these symmetries are concerned with the subduction of the rigid skeleton (**1a**).

20.4 Total Numbers

When we sum up all $A_{\theta i}$ over $\mathbf{G}_i$, we obtain the total number (A_θ) of isomers with W_θ-weight. That is to say,

$$
A_\theta = \sum_{i=1}^{C_\mathbf{G}} A_{\theta i} = \sum_{i=1}^{C_\mathbf{G}} \sum_{j=1}^{C_\mathbf{G}} \sigma_{\theta j} \overline{m}_{ji}. \qquad (20.45)
$$

The enumeration of A_θ can be accomplished by using a generating function represented by $\sum_{[\theta]} A_\theta W_\theta$. This term is converted as follows:

$$
\begin{aligned}
\sum_{[\theta]} A_\theta W_\theta &= \sum_{[\theta]} \left(\sum_{i=1}^{C_\mathbf{G}} \sum_{j=1}^{C_\mathbf{G}} \sigma_{\theta j} \overline{m}_{ji} \right) W_\theta \\
&= \sum_{[\theta]} \sum_{j=1}^{C_\mathbf{G}} \left(\sum_{i=1}^{C_\mathbf{G}} \overline{m}_{ji} \right) \sigma_{\theta j} W_\theta \\
&= \sum_{j=1}^{C_\mathbf{G}} \left(\sum_{i=1}^{C_\mathbf{G}} \overline{m}_{ji} \right) \sum_{[\theta]} \sigma_{\theta j} W_\theta.
\end{aligned}
$$

Since the last sum of the last side has been given by eq. 20.21, we obtain a theorem as follows.

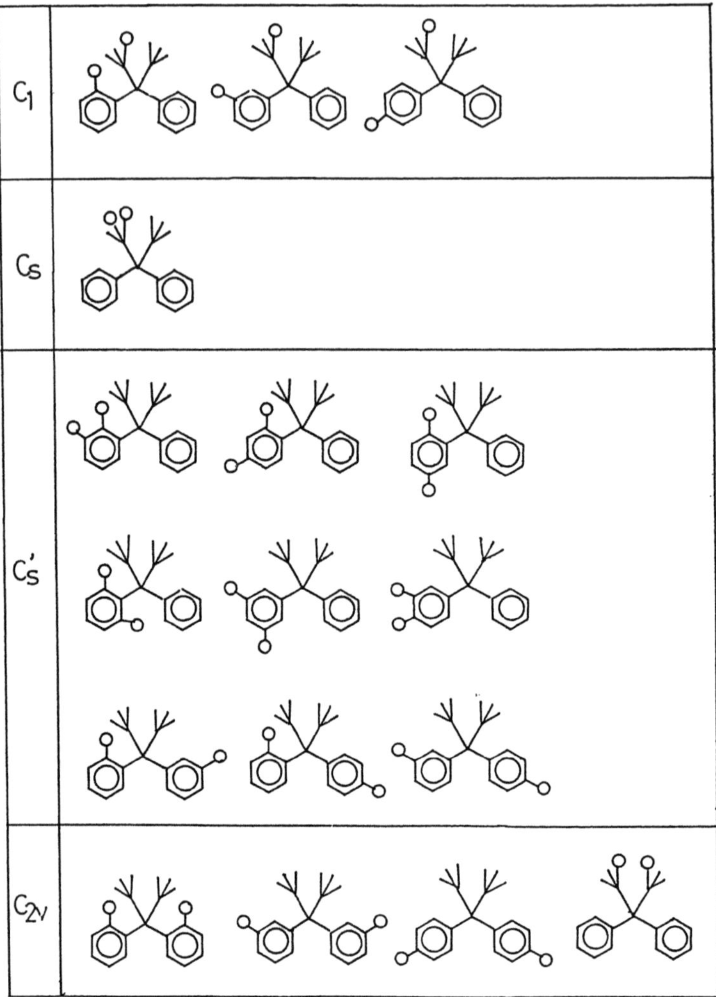

Figure 20.3: $X^{14}Y^2$-isomers based on 1

Theorem 20.2 . (The total number of non-rigid isomers for every weight)

$$\sum_{[\theta]} A_\theta W_\theta = \sum_{j=1}^{C_{\mathbf{G}}} (\sum_{i=1}^{C_{\mathbf{G}}} \overline{m}_{ji}) ZIC(\mathbf{G}_j; \$_{d_{jk}}^{(i\alpha)}), \tag{20.46}$$

in which the SCIs are generated as shown in Lemma 20.3.

It should be noted that the term, $\sum_{i=1}^{C_{\mathbf{G}}} \overline{m}_{ji}$, is calculated by summing up each row of the inverse of a mark table for $\mathbf{G}$. This term is zero if $\mathbf{G}$ is not a cyclic group. The right-hand side of eq. 20.46 is a cycle index (CI) that is an alternative form to that described by Pólya.[1] It should be emphasized that the present procedure generating USCIs $\longrightarrow$ SCIs $\longrightarrow$ CI provides us with a systematic tool for enumeration.[9]

If we sum up all $A_{\theta i}$ over W_θ, we are able to calculate the total number (A_i) of non-rigid isomers wtih $\mathbf{G}_i$-symmetry. This is expressed by

$$A_i = \sum_{[\theta]} A_{\theta i} = \sum_{[\theta]} \sum_{j=1}^{C_{\mathbf{G}}} \sigma_{\theta j} \overline{m}_{ji\cdot} = \sum_{j=1}^{C_{\mathbf{G}}} (\sum_{[\theta]} \sigma_{\theta j}) \overline{m}_{ji\cdot} \tag{20.47}$$

The term in the inner parentheses of the rightmost side is obtained by introducing

$$w_\xi^{(i\alpha)} = 1 \text{ for all } \xi \text{ and all } i\alpha; W_\theta = 1 \tag{20.48}$$

into the equations of Lemma 20.3. This operation affords a lemma as follows.

Lemma 20.6 (Calculation of σ_j)
Let σ_j be the number of derivatives that are invariant (or fixed) on the operation of $\mathbf{G}_j$. We calculate σ_j in the light of

$$\sigma_j \equiv \sum_{[\theta]} \sigma_{\theta j} = ZIC(\mathbf{G}_j; \$_{d_{jk}}^{(i\alpha)}) \tag{20.49}$$

for $j = 1, 2, \ldots, C_{\mathbf{G}}$, in which every variable of the right-hand side is substituted by

$$a_{d_{jk}}^{(i\alpha)} = \sum_{[\xi]} \kappa_{\xi a} = \sum_{[\xi]} \sum_{\substack{p=1 \\ achiral}}^{C_{\mathbf{H}^{(i\alpha)}}} B_{\xi p}^{(i\alpha)} \text{ for } \$ = a \tag{20.50}$$

and

$$b_{d_{jk}}^{(i\alpha)} = c_{d_{jk}}^{(i\alpha)} = \sum_{[\xi]} (\kappa_{\xi a} + 2\kappa_{\xi c}) = \sum_{[\xi]} (\sum_{\substack{p=1 \\ achiral}}^{C_{\mathbf{H}^{(i\alpha)}}} B_{\xi p}^{(i\alpha)} + 2 \sum_{\substack{p=1 \\ chiral}}^{C_{\mathbf{H}^{(i\alpha)}}} B_{\xi p}^{(i\alpha)}) \tag{20.51}$$

for $\$ = b$ and $\$ = c$.

By means of this lemma, we end up with the following corollary.

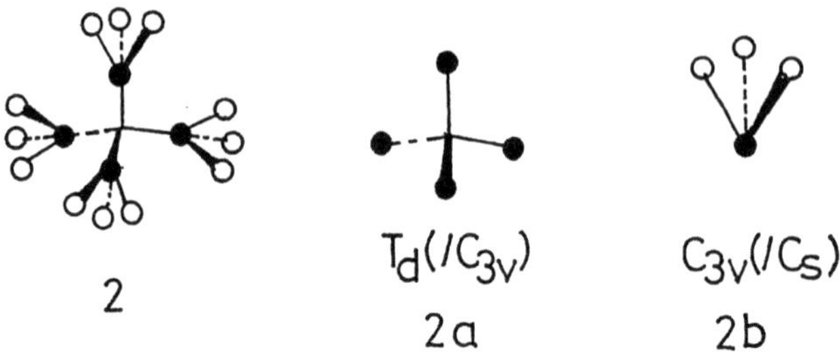

Figure 20.4: Enumeration based on 2,2-dimethylpropane (**2**)

Corollary 20.3 (The total number of non-rigid isomers for every subsymmetry)
Let A_i be the total number of G_i-isomers. This is calculated by using σ_j (Lemma 20.6) by means of

$$A_i = \sum_{j=1}^{c_G} \sigma_j \overline{m}_{ji} \qquad (20.52)$$
$$for\ i = 1, 2, \ldots, C_G,$$

wherein the symbol $(\overline{m}_{ji})$ denotes the ji-element of the inverse matrix (M^{-1}).

20.5   Typical Procedure for Enumeration

This section is devoted to giving a typical procedure for utilizing the propositions described above. The following example involves a participation of a chiral rotatable ligand that appears during an enumeration of non-rigid molecules. We consider 12 substituents selected from a domain, $\mathbf{X} = \{X,Y,Z\}$ to occupy the 12 positions of 2,2-dimethypropane (**2**).

Step 1 is the decomposition of **2** into a rigid skeleton (**2a**) and a rotatable ligand (**2b**) as shown in Fig. 20.4.

Step 2 is the classification of the vertices (substitution positions) of each rotatable ligand (*i.e.*, **2b** in this case) into orbits and the assignment of coset representations (CRs) to the orbits. This assignment has been described in Example 20.2. Thus, the vertices of **2b** are subject to $C_{3v}(/C_s)$.

In Step 3, we construct the SCIs for this case, using the $C_{3v}(/C_s)$ row of Table 20.2. When we introduce a ligand-inventory, $s_d = X^d + Y^d + Z^d$, into these SCIs, we have generating functions for $\rho_{\xi q}$, *i.e.,*

$$s_1^3 \;=\; (X+Y+Z)^3 \quad \text{for } C_1, \tag{20.53}$$
$$s_1 s_2 \;=\; (X+Y+Z)(X^2+Y^2+Z^2) \quad \text{for } C_s, \tag{20.54}$$
$$s_3 \;=\; X^3+Y^3+Z^3 \quad \text{for } C_3, \tag{20.55}$$

and

$$s_3 \;=\; X^3+Y^3+Z^3 \quad \text{for } C_{3v}. \tag{20.56}$$

These equations are expanded to afford a matrix $(\rho_{\xi q})$.

In Step 4, the matrix $(\rho_{\xi q})$ is multiplied by the inverse of a mark table of C_{3v}, *i.e.,*

$$
\begin{array}{c}
\\
X^3 \\
X^2Y \\
XY^2 \\
Y^3 \\
X^2Z \\
XYZ \\
Y^2Z \\
XZ^2 \\
YZ^2 \\
Z^3 \\
\end{array}
\begin{array}{cccc}
C_1 & C_s & C_3 & C_{3v} \\
\left(\begin{array}{cccc}
1 & 1 & 1 & 1 \\
3 & 1 & 0 & 0 \\
3 & 1 & 0 & 0 \\
1 & 1 & 1 & 1 \\
3 & 1 & 0 & 0 \\
6 & 0 & 0 & 0 \\
3 & 1 & 0 & 0 \\
3 & 1 & 0 & 0 \\
3 & 1 & 0 & 0 \\
1 & 1 & 1 & 1 \\
\end{array}\right)
\end{array}
\begin{array}{cccc}
\left(\begin{array}{cccc}
\frac{1}{6} & 0 & 0 & 0 \\
-\frac{1}{2} & 1 & 0 & 0 \\
\frac{1}{6} & 0 & \frac{1}{2} & 0 \\
\frac{1}{2} & -1 & -\frac{1}{2} & 1 \\
\end{array}\right)
\end{array}
$$

$$(\rho_{\xi q}) \qquad\qquad \textit{the inverse}$$

$$
=\;
\begin{array}{cccc}
C_1 & C_s & C_3 & C_{3v} \\
\left(\begin{array}{cccc}
0 & 0 & 0 & 1 \\
0 & 1 & 0 & 0 \\
0 & 1 & 0 & 0 \\
0 & 0 & 0 & 1 \\
0 & 1 & 0 & 0 \\
1 & 0 & 0 & 0 \\
0 & 1 & 0 & 0 \\
0 & 1 & 0 & 0 \\
0 & 1 & 0 & 0 \\
0 & 0 & 0 & 1 \\
\end{array}\right)
\end{array}
\tag{20.57}
$$

$$(B_{\xi p})$$

The matrix obtained indicates that there appear three types of rotatable ligands shown in Fig. 20.5. Note that the XYZ-ligand ia chiral.

Step 5 is the construction of ligand-inventories. The application of Lemma 20.3 to the data of eq. 20.57 yields the following inventories:

$$a_d \;=\; X^{3d} + (X^2Y)^d + (XY^2)^d + Y^{3d}$$

Table 20.3: Coefficients for enumeration

index	C_1	C_2	C_s	C_3	S_4	D_2	C_{2v}	C_{3v}	D_{2d}	T	T_d
[12,0,0]	1	1	1	1	1	1	1	1	1	1	1
[11,1,0]	4	0	2	1	0	0	0	1	0	0	0
[10,2,0]	10	2	4	1	0	0	2	1	0	0	0
[10,1,1]	20	0	2	2	0	0	0	0	0	0	0
[9,3.0]	20	0	6	2	0	0	0	2	0	0	0
[9,2,1]	52	0	6	1	0	0	0	1	0	0	0
[8,4,0]	31	3	7	1	1	1	3	1	1	1	1
[8,3,1]	100	0	6	1	0	0	0	1	0	0	0
[8,2,2]	138	6	14	0	0	0	2	0	0	0	0
[7,5,0]	40	0	8	1	0	0	0	1	0	0	0
[7,4,1]	152	0	8	2	0	0	0	0	0	0	0
[7,3,2]	256	0	16	1	0	0	0	1	0	0	0
[6,6,0]	44	4	8	2	0	0	4	2	0	0	0
[6,5,1]	184	0	8	1	0	0	0	1	0	0	0
[6,4,2]	364	8	22	1	0	0	4	1	0	0	0
[6,3,3]	448	0	16	4	0	0	0	2	0	0	0
[5,5,2]	408	0	20	0	0	0	0	0	0	0	0
[5,4,3]	584	0	20	2	0	0	0	0	0	0	0
[4,4,4]	682	10	30	4	2	2	6	0	0	2	0

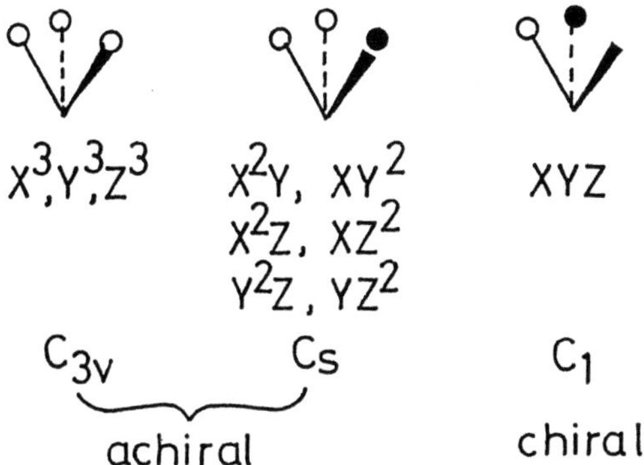

Figure 20.5: Rotatable ligands derived from **2b**

$$+(X^2Z)^d + (Y^2Z)^d + (XZ^2)^d + (YZ^2)^d + Z^{3d}, \tag{20.58}$$

$$b_d = X^{3d} + (X^2Y)^d + (XY^2)^d + Y^{3d}$$
$$+(X^2Z)^d + (Y^2Z)^d + (XZ^2)^d + (YZ^2)^d + Z^{3d} + 2(XYZ)^d, \tag{20.59}$$

and

$$c_d = X^{3d} + (X^2Y)^d + (XY^2)^d + Y^{3d}$$
$$+(X^2Z)^d + (Y^2Z)^d + (XZ^2)^d + (YZ^2)^d + Z^{3d} + 2(XYZ)^d. \tag{20.60}$$

Step 6 is the construction of SCIs for a rigid skeleton (*i.e.*, **2a** in this case). Since the four vertices (joints) are subject to $\mathbf{T}_d(/\mathbf{C}_{3v})$, we select the USCIs with chirality fittingness appearing in the $\mathbf{T}_d(/\mathbf{C}_{3v})$ row of a table of USCIs.[5] After intoducing the above ligand-inventories into the SCIs, we obtain generating functions:

$$b_1^4 = (X^3 + X^2Y + XY^2 + Y^3$$
$$+X^2Z + Y^2Z + XZ^2 + YZ^2 + Z^3 + 2XYZ)^4 \tag{20.61}$$
$$\text{for } \mathbf{C}_1,$$

$$b_2^2 = (X^6 + X^4Y^2 + X^2Y^4 + Y^6$$
$$+X^4Z^2 + Y^4Z^2 + X^2Z^4 + Y^2Z^4 + Z^6 + 2X^2Y^2Z^2)^2 \tag{20.62}$$
$$\text{for } \mathbf{C}_2,$$

$$a_1^2c_2 = (X^3 + X^2Y + XY^2 + Y^3 + X^2Z + Y^2Z + XZ^2 + YZ^2 + Z^3)^2$$
$$\times(X^6 + X^4Y^2 + X^2Y^4 + Y^6$$
$$+X^4Z^2 + Y^4Z^2 + X^2Z^4 + Y^2Z^4 + Z^6 + 2X^2Y^2Z^2) \tag{20.63}$$
$$\text{for } \mathbf{C}_s,$$

$$
\begin{aligned}
b_1 b_3 = \;& (X^3 + X^2 Y + XY^2 + Y^3 \\
& + X^2 Z + Y^2 Z + X Z^2 + Y Z^2 + Z^3 + 2XYZ) \\
& \times (X^9 + X^6 Y^3 + X^3 Y^6 + Y^9 \\
& + X^6 Z^3 + Y^6 Z^3 + X^3 Z^6 + Y^3 Z^6 + Z^9 + 2X^3 Y^3 Z^3) \\
& \text{for } \mathbf{C_3},
\end{aligned}
\tag{20.64}
$$

$$
\begin{aligned}
c_4 = \;& X^{12} + X^8 Y^4 + X^4 Y^8 + Y^{12} \\
& + X^8 Z^4 + Y^8 Z^4 + X^4 Z^8 + Y^4 Z^8 + Z^{12} + 2X^4 Y^4 Z^4 \\
& \text{for } \mathbf{S_4},
\end{aligned}
\tag{20.65}
$$

$$
\begin{aligned}
b_4 = \;& X^{12} + X^8 Y^4 + X^4 Y^8 + Y^{12} \\
& + X^8 Z^4 + Y^8 Z^4 + X^4 Z^8 + Y^4 Z^8 + Z^{12} + 2X^4 Y^4 Z^4 \\
& \text{for } \mathbf{D_2} \text{ and for } \mathbf{T},
\end{aligned}
\tag{20.66}
$$

$$
\begin{aligned}
a_2^2 = \;& (X^6 + X^4 Y^2 + X^2 Y^4 + Y^6 \\
& + X^4 Z^2 + Y^4 Z^2 + X^2 Z^4 + Y^2 Z^4 + Z^6)^2 \\
& \text{for } \mathbf{C_{2v}},
\end{aligned}
\tag{20.67}
$$

$$
\begin{aligned}
a_1 a_3 = \;& (X^3 + X^2 Y + XY^2 + Y^3 + X^2 Z + Y^2 Z + X Z^2 + Y Z^2 + Z^3) \\
& \times (X^9 + X^6 Y^3 + X^3 Y^6 + Y^9 \\
& + X^6 Z^3 + Y^6 Z^3 + X^3 Z^6 + Y^3 Z^6 + Z^9) \\
& \text{for } \mathbf{C_{3v}},
\end{aligned}
\tag{20.68}
$$

and

$$
\begin{aligned}
a_4 = \;& X^{12} + X^8 Y^4 + X^4 Y^8 + Y^{12} \\
& + X^8 Z^4 + Y^8 Z^4 + X^4 Z^8 + Y^4 Z^8 + Z^{12} \\
& \text{for } \mathbf{D_{2d}} \text{ and for } \mathbf{T_d},
\end{aligned}
\tag{20.69}
$$

according to eqs. 20.20 and 20.21. When these equations are expanded and the terms of the same powers are collected, the corresponding fixed-point matrix (FPM) is obtained (Table 20.3). In the present case, the terms are classified into 19 types, which are designated by a *type index* [l,m,n] for $X^l Y^m Z^n$, $X^m Y^n Z^l$ and so on. The terms of the same type have an equal coefficient. Hence, the FPM is simplified by collecting coefficients with respect to these type indices.

Step 7 is the multiplication of the FPM by the inverse of the mark table for $\mathbf{T_d}$ (Appendix B).[5] The resulting matrix (Table 20.4) shows the number of isomers of each type and each subsymmetry.

Figure 20.6 depicts [9,2,1]-isomers, each of which is denoted by the symmetry of the corresponding rigid skeleton.

The total values at the bottom of Table 20.4 are alternatively calculated in the light of Lemma 20.6 and Corollary 20.3. Thus Lemma 20.6 for this case yields a row vector $(11^4, 11^2, 9^2 \times 11, 11^2, 11, 11, 9^2, 9^2, 9, 11, 9)$. According to Corollary

Table 20.4: Number of non-rigid isomers derived from **2**

type index	C_1	C_2	C_s	C_3	S_4	D_2	C_{2v}	C_{3v}	D_{2d}	T	T_d	total
[12,0,0]	0	0	0	0	0	0	0	0	0	0	1	1
[11,1,0]	0	0	0	0	0	0	0	1	0	0	0	1
[10,2,0]	0	0	0	0	0	0	1	1	0	0	0	2
[10,1,1]	0	0	1	1	0	0	0	0	0	0	0	2
[9,3.0]	0	0	1	0	0	0	0	2	0	0	0	3
[9,2,1]	1	0	2	0	0	0	0	1	0	0	0	4
[8,4,0]	0	0	2	0	0	0	1	0	0	0	1	4
[8,3,1]	3	0	2	0	0	0	0	1	0	0	0	6
[8,2,2]	2	1	6	0	0	0	1	0	0	0	0	10
[7,5,0]	0	0	3	0	0	0	0	1	0	0	0	4
[7,4,1]	4	0	4	1	0	0	0	0	0	0	0	9
[7,3,2]	7	0	7	0	0	0	0	1	0	0	0	15
[6,6,0]	1	0	0	0	0	0	2	2	0	0	0	5
[6,5,1]	6	0	3	0	0	0	0	1	0	0	0	10
[6,4,2]	10	1	8	0	0	0	2	1	0	0	0	22
[6,3,3]	15	0	6	1	0	0	0	2	0	0	0	24
[5,5,2]	12	0	10	0	0	0	0	0	0	0	0	22
[5,4,3]	19	0	10	1	0	0	0	0	0	0	0	30
[4,4,4]	21	0	12	1	1	0	3	0	0	1	0	39
total	411	9	333	19	1	0	36	72	0	1	19	891

Figure 20.6: Isomers with the index [9,2,1] based on **2**

20.3, this vector is multiplied by the inverse of the mark table for $\mathbf{T}_d$ to afford (411,9,333,19,1,0,36,72,0,1,9), each value of which is identical with that obtained by the summation of each column of Table 20.4.

The total value of each row of Table 20.4 can be alternatively obtained in the form of a generating function (a cycle index) if we apply Theorem 20.2 to the present case. The cycle index (CI) of this case is given to be

$$\sum_{[\theta]} A_\theta W_\theta = (1/24)b_1^4 + (1/8)b_2^2 + (1/4)a_1^2 c_2 + (1/3)b_1 b_3 + (1/4)c_4, \qquad (20.70)$$

into which eqs. 20.58 to 20.60 are introduced.

For calculating non-rigid isomers containing at least one chiral ligand, Corollary 20.2 should be applied to this case. A similar procedure as above affords Table 20.5.

Table 20.5: Number of non-rigid isomers with chiral ligands based on 2

type index	C_1	C_2	C_s	C_3	S_4	D_2	C_{2v}	C_{3v}	D_{2d}	T	T_d	total
[12,0,0]	0	0	0	0	0	0	0	0	0	0	0	0
[11,1,0]	0	0	0	0	0	0	0	0	0	0	0	0
[10,2,0]	0	0	0	0	0	0	0	0	0	0	0	0
[10,1,1]	0	0	0	1	0	0	0	0	0	0	0	1
[9,3.0]	0	0	0	0	0	0	0	0	0	0	0	0
[9,2,1]	1	0	0	0	0	0	0	0	0	0	0	1
[8,4,0]	0	0	0	0	0	0	0	0	0	0	0	0
[8,3,1]	2	0	0	0	0	0	0	0	0	0	0	2
[8,2,2]	2	1	1	0	0	0	0	0	0	0	0	4
[7,5,0]	0	0	0	0	0	0	0	0	0	0	0	0
[7,4,1]	3	0	0	1	0	0	0	0	0	0	0	4
[7,3,2]	5	0	2	0	0	0	0	0	0	0	0	7
[6,6,0]	0	0	0	0	0	0	0	0	0	0	0	0
[6,5,1]	4	0	0	0	0	0	0	0	0	0	0	4
[6,4,2]	7	1	3	0	0	0	0	0	0	0	0	11
[6,3,3]	10	0	2	1	0	0	0	0	0	0	0	13
[5,5,2]	8	0	4	0	0	0	0	0	0	0	0	12
[5,4,3]	13	0	4	1	0	0	0	0	0	0	0	18
[4,4,4]	15	0	6	1	1	0	0	0	0	1	0	24·

For illustrating the result of Table 20.5, several selected examples are depicted in Fig 20.7. Note that the point groups assigned in Fig. 20.7 are concerned with the

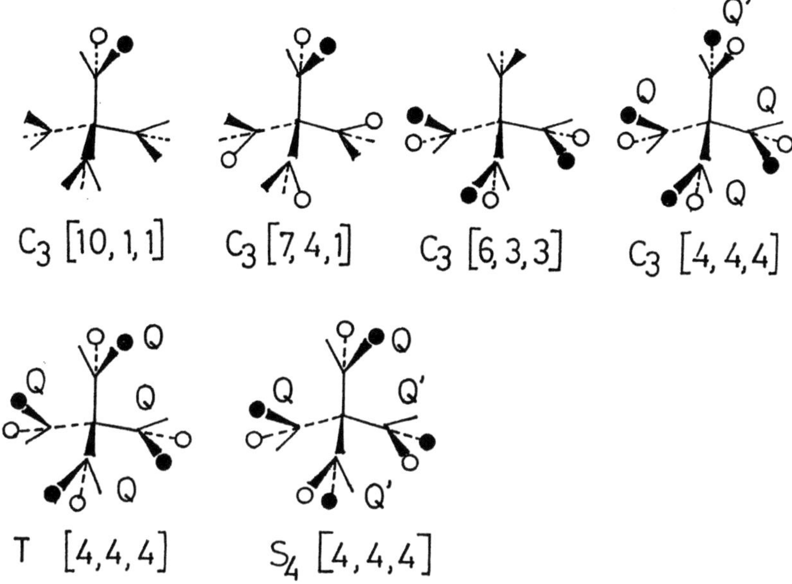

Figure 20.7: Selected isomers with chiral ligands based on 2

rigid skeletons, not with the molecules themselves. For example, the $[6,3,3],C_3$-isomer in Fig. 20.7 belongs to C_3-point group. On the other hand, the other isomers are restricted within C_1 point group, even if they have conformations of the highest symmetries. The $[4,4,4],T$-molecule belongs to D_2 symmetry, strictly speaking. In contrast to this, a full symmetry is realized in the $[4,4,4],S_4$-molecule. Several remarks have been made on such problems.[10]

Unit subduced cycle indices (USCIs) are applied to the enumeration of non-rigid molecules such as 2,2-diphenyl- and 2,2-dimethylpropane. Non-rigid molecules are regarded as a combination of a rigid skeleton and rotatable ligands. This formulation allows us to treat non-rigidity in the same line as rigid molecules, since both the rigid skeleton and the rotatable ligand can be treated as *rigid*.

Bibliography

[1] a) G. Pólya, *Acta Math.*, **68**, 145 (1937). b) G. Pólya and R. C. Read, *Combinatorial Enumeration of Groups, Graphs, and Chemical Compounds*, Springer-Verlag, New York-Berlin-Heidelberg (1987).

[2] a) J. E. Leonard, G. S. Hammond and H. E. Simmons, *J. Am. Chem. Soc.*, **97**, 5052 (1975). b) J. E. Leonard, *J. Phys. Chem.*, **81**, 2212 (1977).

[3] R. L. Flurry, Jr., *J. Chem. Ed.*, **61**, 663 (1984).

[4] K. Balasubramanian, *Theor. Chim. Acta*, **51**, 37 (1979).

[5] a) S. Fujita, *Theor. Chim. Acta*, **76**, 247 (1989). b) S. Fujita, *Bull. Chem. Soc. Jpn.*, **63**, 203 (1990). c) S. Fujita, *Tetrahedron*, **46**, 365 (1990).

[6] B. Baumslag and B. Chandler, *Theory and Problems of Group Theory*, McGraw-Hill, New York (1968).

[7] M. A. Armstrong, *Groups and Symmetry*, Springer-Verlag, New York-Berlin-Heidelberg (1988).

[8] W. Burnside, *Theory of Groups of Finite Order (2nd ed.)*, Cambridge Univ. Press, Cambridge (1911).

[9] S. Fujita, *J. Math. Chem.*, **5**, 121 (1990).

[10] a) K. Mislow, *J. Chem. Soc., Chem. Commun.*, 234 (1981). b) M. Farina and C. Morandi, *Tetrahedron*, **30**, 1819 (1974).

Chapter 21

Promolecules [1]

We have discussed the usefulness of coset representations (CRs) in enumerating various molecules (Chapters 15–20),[1]–[6] in classifying molecular symmetry (Chapter 6),[7] and in specifying stereochemical equivalency (Chapters 11 and 12).[8] These applications are based on the fact that each CR governs an orbit that consists of equivalent objects (atoms, bonds, faces and so on). We have also discussed the stereochemical relationship between global symmetry and local ones in a rigid molecule (Chapters 7–9).[9] This treatment affords a foundation for deciding how to obtain a molecule of a given symmetry. The present chapter deals with global-local relationships in *non-rigid* molecules, where several new concepts such as promolecules and proligands are proposed.

21.1   Molecular Models

Organic stereochemistry has been discussed on the basis of three-dimensional molecular models, which vary with purposes of discussions.[10]–[12] The formulation of a molecule as an achiral or chiral skeleton with several ligands is one of the dominant methodologies for discussing molecular symmetry.[13]–[15] This model directly succeeds to that of van't Hoff.[16] According to this model, Farina and Morandi[17] have proposed principles for the design of high symmetry chiral molecules. This approach has provided us with a useful method of determining how to realize a molecule of a desired symmetry. However, the highest attainable symmetry of a molecule is often difficult to determine, especially when the molecule contains several ligands that are mobile through bond rotations. For example, a methane molecule (**1a**) belongs to T_d symmetry, pentaerythritol (**1b**) has D_{2d} symmetry, and pentaerythritol tetra($-$)-menthyloxyacetate (**1c**)[18] belongs to D_2 symmetry. Although such highest attainable symmetries can be assigned by careful inspection, there have been no systematic methods of doing this task.

---

[1]Reprinted in part with permission from S. Fujita, *Tetrahedron*, **47**, 31-46 (1991). ©(1991) Pergamon Press PLC.

Nakazaki *et al.*[19] synthesized (−)-1,3,5,7-tetrakis[2-(1S,3S,5R,6S,8R,10R)-D_3-trishomocubanylacetoxymethyl]adamantane (**2a**), which they once claimed to belong to **T** symmetry. Mislow[20] pointed out that the highest attainable symmetry of this compound was D_2 in the same line as the McCasland compound (**1c**). Later, Nakazaki *et al.*[21] synthesized (+)-1,3,5,7-tetrakis[2-(1S,3S,5R,6S,8R,10R)-D_3-trishomocubanylbuta-1,3-diynyl]adamantane (**2b**) as a true **T**-molecule.

This short history indicates the importance of the effect of ligand symmetries upon the whole symmetry of such a molecule. This effect is closely related to the relationship between the global symmetry and the local symmetry in a molecule, which has been discussed by Mislow.[22]

Mislow[23] dealt with compound (**3**) as a *meso*-compound containing only C_1-conformers. The biphenyl part of this molecule has D_{2d} symmetry, which is incompatible with the symmetries of terminal chiral groups. Hence, this compound (**3**) has a C_1-conformation as the highest attainable symmetry at any time; however, this is an achiral compound because of internal rotations. Although this phenomenon is not so strange but rather common as pointed out in a critical review,[24] it also indicates the importance of the global-local relationship in a molecule.

1a: R = H

1b: R = CH$_2$OH

1c: R = CH$_2$OCCH$_2$O-

2a: R = CH$_2$OCCH$_2$-

2b: R = C≡C-C≡C-

3

21.2 Proligands and Promolecules

In this chapter, a molecule is regarded as a three-dimensional (3D) object that consists of a skeleton and ligands, where the skeleton has joint positions that are occupied by the ligands. The skeleton is considered to be a rigid 3D object belonging to **G** point group which is a supergroup of the molecular symmetry (**M**).[2] Each of the ligands has its own joint, which is identical with one of the

---

[2]This condition is important to avoid erroneous interpretations.

joint positions if the ligand is built into the molecule. This means that ligands in a molecule are capable of internal rotation. Such a ligand-in-molecule is called a *segment*. If a ligand is assumed to be free from the symmetry environment of the molecule, it is called a *fragment*. The intrinsic symmetry ($\mathbf{F}$) of a ligand appears in full in an isolated state (*i.e.* as a fragment), but becomes restricted when incorporated in a molecule (*i.e.* as a segmant). When the symmetry of the fragment is $\mathbf{F}$, we call it an $\mathbf{F}$-fragment. Such a fragment is defined as *chiral* if it cannot be superimposed onto its antipode (mirror image); and otherwise, as *achiral*.

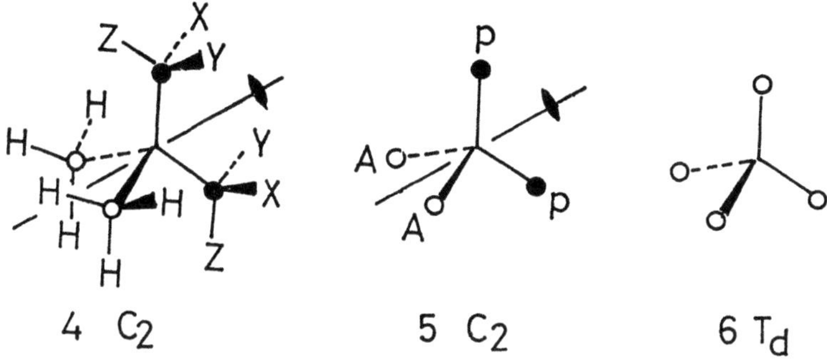

Figure 21.1: Molecule, promolecule and skeleton

Figure 21.1 illustrates the present molecular model for describing molecular symmetry. Consider a molecule (**4**) that belongs to $\mathbf{C_2}$ symmetry in its highest attainable state. This molecule contains a skeleton (**6**) having four joint positions. Two methyl groups and two CXYZ groups in this molecule are regarded as segments. According to the above definition, the methyl group is a $\mathbf{C_{3v}}$-fragment (an achiral fragment) and the CXYZ group is a $\mathbf{C_1}$-fragment (or an asymmetric and chiral fragment). The skeleton (**6**) has $\mathbf{T_d}$ symmetry in isolation; but a restricted $\mathbf{C_2}$ symmetry in the molecular environment. Thus, the $\mathbf{C_2}$ symmetry is generated by replacing the four positions of **6** by the two $\mathbf{C_{3v}}$-fragments and the two $\mathbf{C_1}$-fragments.

This molecular model has several conformers becasue of internal bond rotations. Such non-rigidity can be treated by several approaches.[25]–[27] In order to discuss organic stereochemistry, however, a more simplified approach is desirable. For this purpose, we introduce the concepts of proligands and promolecules. A *proligand* [3] [4] is defined as a 3D object that is structureless but has chirality. The

---

[3]The prefix *pro-* has been used in chemistry in the sense "before", *e.g.*, pro-γ-carotene, provitamin and prochirality.

[4]For similar formulation, see refs. [14] and [28]. In contrast with these predecessors, the present term *proligand* is used together with the concepts of segments and fragments.

term "having chirality" means that there are two types of proligands as fragments: achiral (A, B, C, D, $\cdots$) and chiral (p, q, r, s, $\cdots$, and their antipodes, $\overline{p}$, $\overline{q}$, $\overline{r}$, $\overline{s}$, $\cdots$). A *promolecule* is then defined as a 3D object that consists of a skeleton and such proligands. With respect to **4**, we replace the methyl group by A (achiral) and the CXYZ group by p (chiral). This replacement gives the corresponding promolecule (**5**), which retains the $\mathbf{C}_2$ symmetry. We conceptually construct a molecule by consecutive processes: (1) starting from an appropriate skeleton, (2) substituting proligands for joint positions to produce a promolecule, and (3) further replacing the proligands with ligands. In the light of this formulation, such promolecules are manipulated as hypothetically rigid 3D objects, which maintain some of the symmetrical properties of related molecules. Our targets are to enumerate such promolecules; to calrify their symmetrical properties which are perturbed by the symmetry of the skeleton and the chirality/achirality of the proligands; as well as to characterize relationships between symmetries of the promolecule and of the molecule.

21.3   Enumeration of Promolecules

Tetrahedral promolecules of various symmetries. Promolecules derived from the tetrahedral skeleton (**6**) can be enumerated by means of unit subduced cycle indices with chirality fittingness (USCI-CF).[3, 5] We use USCI-CFs for $\mathbf{T}_d$-($/\mathbf{C}_{3v}$), *i.e.*, b_1^4 for $\mathbf{C}_1$, b_2^2 for $\mathbf{C}_2$, $a_1^2 c_2$ for $\mathbf{C}_s$, $b_1 b_3$ for $\mathbf{C}_3$, c_4 for $\mathbf{S}_4$, b_4 for $\mathbf{D}_2$, a_2^2 for $\mathbf{C}_{2v}$, $a_1 a_3$ for $\mathbf{C}_{3v}$, a_4 for $\mathbf{D}_{2d}$, b_4 for $\mathbf{T}$, and a_4 for $\mathbf{T}_d$, which themselves constitute respective subduced cycle indices with chirality fittingness (SCI-CFs). In this case, we take account of the following ligand-inventories:

$$a_d = A^d + B^d + C^d + D^d, \tag{21.1}$$
$$b_d = A^d + B^d + C^d + D^d + p^d + q^d + r^d + s^d + \overline{p}^d + \overline{q}^d + \overline{r}^d + \overline{s}^d, \tag{21.2}$$
and
$$c_d = A^d + B^d + C^d + D^d + 2[(p\overline{p})^{d/2} + (q\overline{q})^{d/2} + (r\overline{r})^{d/2} + (s\overline{s})^{d/2}]. \tag{21.3}$$

These ligand-inventories are introduced into the SCI-CFs. We expand the resulting generating functions and collect coefficients of terms for every racemic pair. Among the terms representing equivalent proligand partitions (*e.g.* A_3B and AB_3), we can select an arbitrary term as a representative without losing generality. Then, using the inverse of a mark table for $\mathbf{T}_d$, we calculate the number of isomeric promolecules (Table 21.1). Thus, we have one $\mathbf{T}_d$, one $\mathbf{T}$, one $\mathbf{C}_{3v}$, one $\mathbf{C}_{2v}$, one $\mathbf{S}_4$, four $\mathbf{C}_3$, four $\mathbf{C}_s$, two $\mathbf{C}_2$, and 22 $\mathbf{C}_1$ promolecules. These values are itemized with respect to the corresponding terms (proligand partitions) in the present enumeration. It should be noted that there exist no $\mathbf{D}_{2d}$ and no $\mathbf{D}_2$ promolecules in this enumeration.

Table 21.1: Number of promolecules derived from a tetrahederal skeleton (6)

Proligand	Number of promolecules										
partition	C_1	C_2	C_s	C_3	S_4	D_2	C_{2v}	C_{3v}	D_{2d}	T	T_d
A^4	0	0	0	0	0	0	0	0	0	0	1
A^3B	0	0	0	0	0	0	0	1	0	0	0
A^3p	0	0	0	1	0	0	0	0	0	0	0
A^2B^2	0	0	0	0	0	0	1	0	0	0	0
A^2BC	0	0	1	0	0	0	0	0	0	0	0
A^2Bp	1	0	0	0	0	0	0	0	0	0	0
A^2p^2	0	1	0	0	0	0	0	0	0	0	0
$A^2p\overline{p}$	0	0	1	0	0	0	0	0	0	0	0
A^2pq	1	0	0	0	0	0	0	0	0	0	0
$ABCD$	1	0	0	0	0	0	0	0	0	0	0
$ABCp$	2	0	0	0	0	0	0	0	0	0	0
ABp^2	1	0	0	0	0	0	0	0	0	0	0
$ABp\overline{p}$	0	0	2	0	0	0	0	0	0	0	0
$ABpq$	2	0	0	0	0	0	0	0	0	0	0
Ap^3	0	0	0	1	0	0	0	0	0	0	0
$Ap^2\overline{p}$	1	0	0	0	0	0	0	0	0	0	0
Ap^2q	1	0	0	0	0	0	0	0	0	0	0
$Ap\overline{p}r$	2	0	0	0	0	0	0	0	0	0	0
$Apqr$	2	0	0	0	0	0	0	0	0	0	0
p^4	0	0	0	0	0	0	0	0	0	1	0
$p^3\overline{p}$	0	0	0	1	0	0	0	0	0	0	0
p^3q	0	0	0	1	0	0	0	0	0	0	0
$p^2\overline{p}^2$	0	0	0	0	1	0	0	0	0	0	0
$p^2\overline{p}r$	1	0	0	0	0	0	0	0	0	0	0
p^2q^2	0	1	0	0	0	0	0	0	0	0	0
$p^2q\overline{q}$	1	0	0	0	0	0	0	0	0	0	0
p^2qr	1	0	0	0	0	0	0	0	0	0	0
$p\overline{p}q\overline{q}$	1	0	0	0	0	0	0	0	0	0	0
$p\overline{p}qr$	2	0	0	0	0	0	0	0	0	0	0
$pqrs$	2	0	0	0	0	0	0	0	0	0	0

Figure 21.2 depicts all of these promolecules, where an arbitrary representative is selected from every racemic pair.[5]

Table 21.2 collects symmetrical properties of the promolecules (**7** to **21**). The four joint positions of the skeleton (**6**) construct a $\mathbf{T}_d(/\mathbf{C}_{3v})$ orbit, which remains unchanged in the promolecule (**7**). In the other promolecules, the orbit is divided into several orbits shown in Table 21.2.

These divisions are strictly controlled by a desymmetrization lattice (Fig. 21.3) which contains subductions of the CR $(\mathbf{T}_d(/\mathbf{C}_{3v}))$.[8] For example, the derivation of **9** from **6** is represented by

$$\mathbf{T}_d(/\mathbf{C}_{3v}) \downarrow \mathbf{C}_{3v} = \mathbf{C}_{3v}(/\mathbf{C}_{3v}) + \mathbf{C}_{3v}(/\mathbf{C}_s), \tag{21.4}$$

which is found in Fig. 21.3. Equation 21.4 is schematically represented by

$$\boxed{\quad}\boxed{\quad}\boxed{\quad}\boxed{\quad} \quad \xrightarrow{\downarrow \mathbf{C}_{3v}} \quad \boxed{A}\boxed{A}\boxed{A} \cdots \;\; \mathbf{C}_{3v}(/\mathbf{C}_s)$$
$$\mathbf{T}_d(/\mathbf{C}_{3v}). \qquad\qquad\qquad \boxed{B} \qquad \cdots \;\; \mathbf{C}_{3v}(/\mathbf{C}_{3v})$$

Each CR has its chirality fittingness, which determines a mode of substitution of proligands. Thus, a homospheric orbit takes only achiral proligands of the same kind. An enantiospheric orbit takes (1) achiral proligands of the same kind or (2) one half of chiral proligands and the other half of their antipodes. A hemispheric orbit permits (1) achiral proligands of the same kind and (2) chiral proligands of the same chirality. These criteria hold true in the data of Table 21.2. For example, two orbits generated by eq. 21.4 are both homospheric; in addition, the length of the $\mathbf{C}_{3v}(/\mathbf{C}_{3v})$ orbit is equal to 1 and that of $\mathbf{C}_{3v}(/\mathbf{C}_s)$ orbit is euqal to 3. Hence, we have an A_3B-promolecule (**9**).

Examination of the $\mathbf{C}_s$ promolecules (**16** to **19**) is instructive, since there emerge several modes of occupation. The above criteria indicate that, in anyone of these promolecules, the two $\mathbf{C}_s(/\mathbf{C}_s)$ orbits require distinct achiral proligands. In addition, the $\mathbf{C}_s(/\mathbf{C}_1)$ orbit takes two achiral proligands or a pair of antipodal chiral proligands. Thus, **16** is an example of the former occupation of the $\mathbf{C}_s(/\mathbf{C}_1)$ orbit; and **18** and **19** are examples of the latter. The promolecule (**17**) shows that the two $\mathbf{C}_s(/\mathbf{C}_s)$ can take proligands of the same kind; however, these two proligands belong to distinct orbits.

It is worthwhile comparing the present enumeration with Prelog's one[13] as well as with our previous one.[15] Prelog has used Young's diagrams that designate ligand partitions. His method takes no account of symmetrical properties in the process of enumeration. After the enumeration, the symmetry of each molecule is

---

[5]Prelog[13] reported an essentially equivalent enumeration, although his enumeration did not contain the present concept of promolcules.

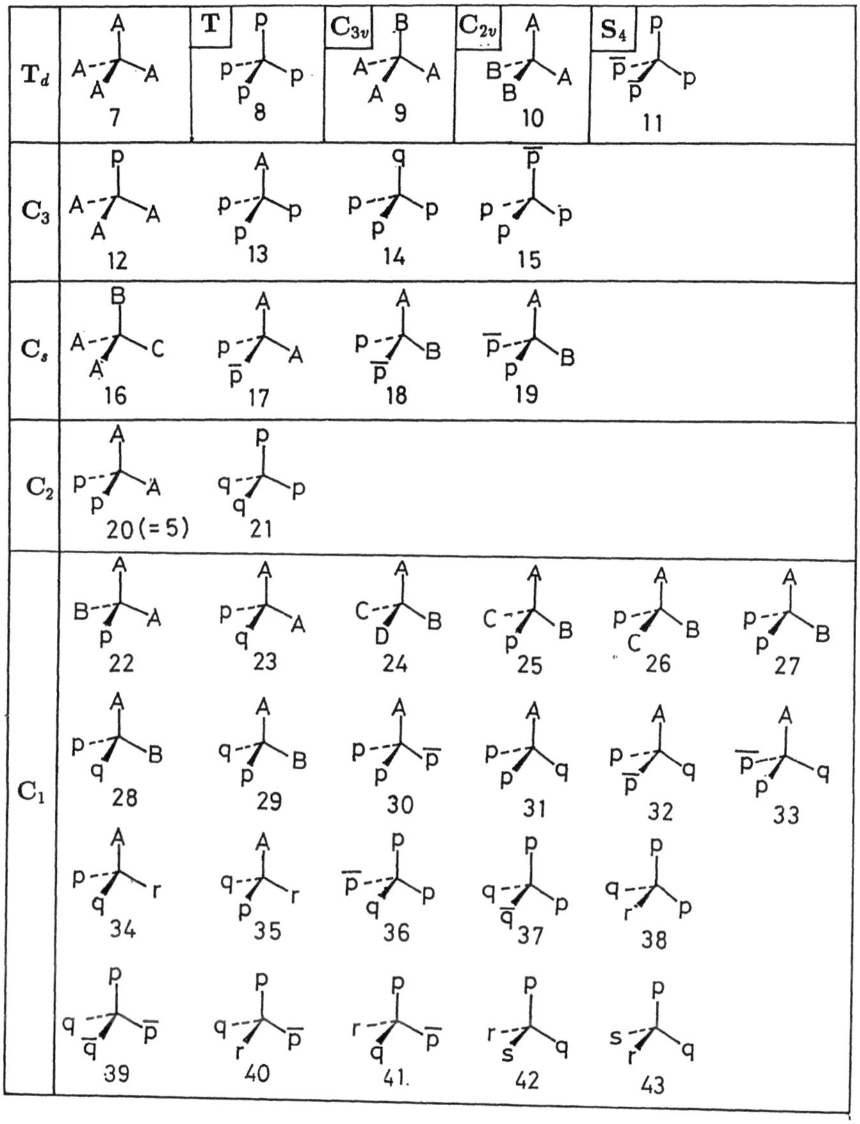

Figure 21.2: Tetrahedral promolecules of various symmetries

Table 21.2: Orbits and coset representations in promolecules (7-21)

Pro-molecule	Symmetry	Orbit	Members of an orbit	Coset representation	Chirality fittingness (Sphericity)
7	T_d	Δ	A_4	$T_d(/C_{3v})$	homospheric
8	T	Δ	p_4	$T(/C_3)$	hemispheric
9	C_{3v}	Δ_1	B	$C_{3v}(/C_{3v})$	homospheric
		Δ_2	A_3	$C_{3v}(/C_s)$	homospheric
10	C_{2v}	Δ_1	A_2	$C_{2v}(/C_s)$	homospheric
		Δ_2	B_2	$C_{2v}(/C_s')$	homospheric
11	S_4	Δ	$p_2\bar{p}_2$	$S_4(/C_1)$	enantiospheric
12	C_3	Δ_1	p	$C_3(/C_3)$	homospheric
		Δ_2	A_3	$C_3(/C_1)$	hemispheric
13	C_3	Δ_1	A	$C_3(/C_3)$	homospheric
		Δ_2	p_3	$C_3(/C_1)$	hemispheric
14	C_3	Δ_1	q	$C_3(/C_3)$	homospheric
		Δ_2	p_3	$C_3(/C_1)$	hemispheric
15	C_3	Δ_1	$\bar{p}$	$C_3(/C_3)$	homospheric
		Δ_2	p_3	$C_3(/C_1)$	hemispheric
16	C_s	Δ_1	B	$C_s(/C_s)$	homospheric
		Δ_2	C	$C_s(/C_s)$	homospheric
		Δ_3	A_2	$C_s(/C_1)$	enantiospheric
17	C_s	Δ_1	A	$C_s(/C_s)$	homospheric
		Δ_2	A	$C_s(/C_s)$	homospheric
		Δ_3	$p\bar{p}$	$C_s(/C_1)$	enantiospheric
18	C_s	Δ_1	A	$C_s(/C_s)$	homospheric
		Δ_2	B	$C_s(/C_s)$	homospheric
		Δ_3	$p\bar{p}$	$C_s(/C_1)$	enantiospheric
19	C_s	Δ_1	A	$C_s(/C_s)$	homospheric
		Δ_2	B	$C_s(/C_s)$	homospheric
		Δ_3	$\bar{p}p$	$C_s(/C_1)$	enantiospheric
20 (5)	C_2	Δ_1	A_2	$C_2(/C_1)$	hemispheric
		Δ_2	p_2	$C_2(/C_1)$	hemispheric
21	C_2	Δ_1	p_2	$C_2(/C_1)$	hemispheric
		Δ_2	q_2	$C_2(/C_1)$	hemispheric

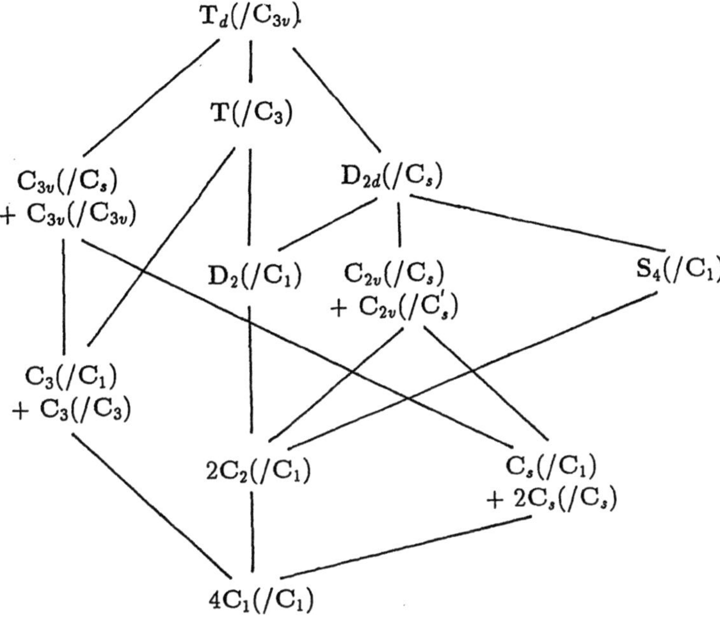

Figure 21.3: Desymmetralization lattice for $T_d(/C_{3v})$

determined by finding its symmetry elements. There is hence no systematic item-ization concerning molecular symmetries in a strict sense. Our previous method has adopted an analogous procedure.

On the other hand, we now use a term such as $A_2p\bar{p}$ for denoting ligand par-titions; and a symbol of a CR or a Young's diagram for designating a site partition (*i.e.* division of an orbit into suborbits). Our method first determines a symmetry at issue, into which a given orbit is subduced. We then fill the resulting suborbits in the light of their chirality fittingnesses. This process is closely related to enu-meration by USCI-CFs, which algebraically provides the numbers of promolecules in an itemized form concerning proligand partitions and symmetries (Chapter 19).

Prelog's[13] and our previous enumeration[15] may afford some confusions in specifying stereochemical equivalency. For example, compare promolecule **44** (equivalent to **17**) with **16** (Fig. 21.4). By means of these treatments, one may take no account of the difference between the As of **16** and those of **44**. Thus, they may be erroneously considered to be placed in the same symmetrical environment, although this mistake can be, of course, avoided by careful inspection. On the other hand, the present method discriminates them in terms of orbits. Thus, the two As of **16** construct a two-membered $C_s(/C_1)$ orbit, whereas the two As of **44** construct two distinct one-membered $C_s(/C_s)$ orbits.

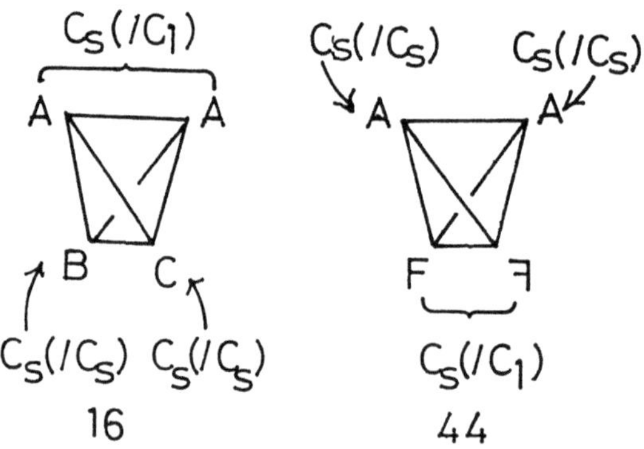

Figure 21.4: Orbits of C_s promolecules (**16** and **44**)

Promolecules derived from an allene skeleton. Consider an allene skeleton (**45**). The four joint positions of the skeleton (**45**) construct a $D_{2d}(/C_s)$ orbit.

$$\text{45} \quad D_{2d}$$

Promolecules derived from an allene skeleton (**45**) are enumerated by means of USCI-CFs. The SCI-CFs used are equal to the USCI-CFs. Equations 21.1–21.3 are used as ligand-inventories. By the same procedure as above, we have the results collected in Table 21.3. See also Example 15.1 (Chapter 15).

The promolecules are depicted in Figure 21.5, which contains an arbitrary representative selected from every racemic pair. Comparison of Fig. 21.2 with Fig. 21.5 are useful to understand the difference between the skeletons **6** and **45**. For example, the ligand partition (A^2B^2) produces one molecule (**10**) of C_{2v} symmetry (Fig. 21.2). On the other hand, it produces **48** (C_{2v}) and **57** (C_2') by starting from **45** (Fig. 21.5). The ligand partition (A^2p^2) produces one promolecule (**20**) in the series of **6**, whereas it produces three promolecules (**55**, **58**, and **59**) on the basis of **45**.

Table 21.4 collects symmetrical properties of the promolecules (**46** to **62**). The $D_{2d}(/C_s)$ orbit of **45** is divided into several orbits according to a desymmetrization lattice (Fig. 9.3 in Chapter 9), which contains subductions of the CR ($D_{2d}(/C_s)$).

21.4 Molecules Based on Promolecules

Matched molecules. Suppose that a set of equivalent proligands in a promolecule belong to an $H(/H_i)$ orbit, where H is the point group of the promolecule and H_i is its subgroup indicating the local symmetry of the orbit.[29] Any proligand agrees with the local symmetry (H_i), because it is considered to be structureless. However, a ligands that has a 3D structure is not always compatible with such local symmetry.

Consider that a proligand in a promolecule is replaced by a ligand, where their achiral/chiral characters are presumed to be equal. If the symmetry (F) of the ligand as a fragment is a supergroup of the local symmetry (H_i), in other words,

Table 21.3: Number of promolecules derived from an allene skeleton (45)

Proligand	Number of promolecules							
partition	C_1	C_2	C_2'	C_s	S_4	C_{2v}	D_2	D_{2d}
A^4	0	0	0	0	0	0	0	1
A^3B	0	0	0	1	0	0	0	0
A^3p	1	0	0	0	0	0	0	0
A^2B^2	0	0	1	0	0	1	0	0
A^2BC	1	0	0	1	0	0	0	0
A^2Bp	3	0	0	0	0	0	0	0
A^2p^2	0	1	2	0	0	0	0	0
$A^2p\overline{p}$	1	0	0	1	0	0	0	0
A^2pq	3	0	0	0	0	0	0	0
ABCD	3	0	0	0	0	0	0	0
ABCp	6	0	0	0	0	0	0	0
ABp^2	3	0	0	0	0	0	0	0
$ABp\overline{p}$	2	0	0	0	0	0	0	0
ABpq	6	0	0	0	0	0	0	0
Ap^3	1	0	0	0	0	0	0	0
$Ap^2\overline{p}$	3	0	0	0	0	0	0	0
Ap^2q	3	0	0	0	0	0	0	0
$Ap\overline{p}r$	6	0	0	0	0	0	0	0
Apqr	6	0	0	0	0	0	0	0
p^4	0	0	0	0	0	0	1	0
$p^3\overline{p}$	1	0	0	0	0	0	0	0
p^3q	1	0	0	0	0	0	0	0
$p^2\overline{p}^2$	0	0	1	0	1	0	0	0
$p^2\overline{p}r$	3	0	0	0	0	0	0	0
p^2q^2	0	1	2	0	0	0	0	0
$p^2q\overline{q}$	3	0	0	0	0	0	0	0
p^2qr	3	0	0	0	0	0	0	0
$p\overline{p}q\overline{q}$	3	0	0	0	0	0	0	0
$p\overline{p}qr$	6	0	0	0	0	0	0	0
pqrs	6	0	0	0	0	0	0	0

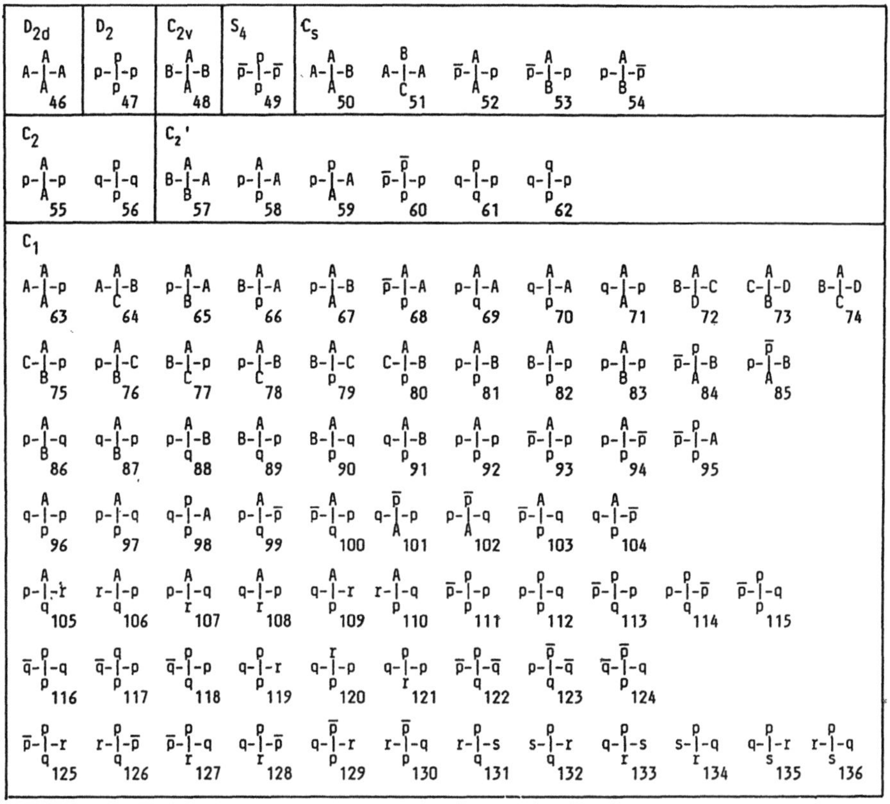

Figure 21.5 presents promolecules derived from an allene skeleton, arranged by symmetry group. Each structure is drawn with a central axis bearing a substituent above and below the horizontal axis (written here as top / X–|–Y / bottom), and is labelled with a number.

D_{2d}
- 46: A / A–|–A / A

D_2
- 47: p / p–|–p / p

C_{2v}
- 48: A / B–|–B / A

S_4
- 49: p / $\bar{p}$–|–$\bar{p}$ / p

C_s
- 50: A / A–|–B / A
- 51: B / A–|–A / C
- 52: A / $\bar{p}$–|–p / A
- 53: A / $\bar{p}$–|–p / B
- 54: A / p–|–$\bar{p}$ / B

C_2
- 55: A / p–|–p / A
- 56: p / q–|–q / p

$C_2{}'$
- 57: A / B–|–A / B
- 58: A / p–|–A / p
- 59: p / p–|–A / A
- 60: $\bar{p}$ / $\bar{p}$–|–p / p
- 61: p / q–|–p / q
- 62: q / q–|–p / p

C_1
- 63: A / A–|–p / A
- 64: A / A–|–B / C
- 65: A / p–|–A / B
- 66: A / B–|–A / p
- 67: A / p–|–B / A
- 68: A / $\bar{p}$–|–A / p
- 69: A / p–|–A / q
- 70: A / q–|–A / p
- 71: A / q–|–p / A
- 72: A / B–|–C / D
- 73: A / C–|–D / B
- 74: A / B–|–D / C
- 75: A / C–|–p / B
- 76: A / p–|–C / B
- 77: A / B–|–p / C
- 78: A / p–|–B / C
- 79: A / B–|–C / p
- 80: A / C–|–B / p
- 81: A / p–|–B / p
- 82: A / B–|–p / p
- 83: A / p–|–p / B
- 84: $\bar{p}$ / $\bar{p}$–|–B / A
- 85: $\bar{p}$ / p–|–B / A
- 86: A / p–|–q / B
- 87: A / q–|–p / B
- 88: A / p–|–B / q
- 89: A / B–|–p / q
- 90: A / B–|–q / p
- 91: A / q–|–B / p
- 92: A / p–|–p / p
- 93: A / $\bar{p}$–|–p / p
- 94: A / p–|–$\bar{p}$ / p
- 95: p / $\bar{p}$–|–A / p
- 96: A / q–|–p / p
- 97: A / p–|–q / p
- 98: p / q–|–A / p
- 99: A / p–|–$\bar{p}$ / q
- 100: A / $\bar{p}$–|–p / q
- 101: $\bar{p}$ / q–|–p / A
- 102: $\bar{p}$ / p–|–q / A
- 103: A / $\bar{p}$–|–q / p
- 104: A / q–|–$\bar{p}$ / p
- 105: A / p–|–r / q
- 106: A / r–|–p / q
- 107: A / p–|–q / r
- 108: A / q–|–p / r
- 109: A / q–|–r / p
- 110: A / r–|–q / p
- 111: p / $\bar{p}$–|–p / p
- 112: p / p–|–q / p
- 113: p / $\bar{p}$–|–p / q
- 114: p / p–|–$\bar{p}$ / q
- 115: p / $\bar{p}$–|–q / p
- 116: p / $\bar{q}$–|–q / p
- 117: q / $\bar{q}$–|–p / p
- 118: p / $\bar{q}$–|–p / q
- 119: p / q–|–r / p
- 120: r / q–|–p / p
- 121: p / q–|–p / r
- 122: p / $\bar{p}$–|–$\bar{q}$ / q
- 123: $\bar{p}$ / p–|–$\bar{q}$ / q
- 124: $\bar{p}$ / $\bar{q}$–|–q / p
- 125: p / $\bar{p}$–|–r / q
- 126: p / r–|–$\bar{p}$ / q
- 127: p / $\bar{p}$–|–q / r
- 128: p / q–|–$\bar{p}$ / r
- 129: $\bar{p}$ / q–|–r / p
- 130: $\bar{p}$ / r–|–q / p
- 131: p / r–|–s / p
- 132: p / s–|–r / q
- 133: p / q–|–s / r
- 134: p / s–|–q / r
- 135: p / q–|–r / s
- 136: p / r–|–q / s

Figure 21.5: Promolecules derived from an allene skeleton (45)

Table 21.4: Orbits and coset representations in promolecules (**46-62**)

Pro-molecule	Symmetry	Orbit	Members of an orbit	Coset representation	Chirality fittingness (Sphericity)
46	$\mathbf{D}_{2d}$	Δ	A_4	$\mathbf{D}_{2d}(/\mathbf{C}_s)$	homospheric
47	$\mathbf{D}_2$	Δ	p_4	$\mathbf{D}_2(/\mathbf{C}_1)$	hemispheric
48	$\mathbf{C}_{2v}$	Δ_1	A_2	$\mathbf{C}_{2v}(/\mathbf{C}_s)$	homospheric
		Δ_2	B_2	$\mathbf{C}_{2v}(/\mathbf{C}'_s)$	homospheric
49	$\mathbf{S}_4$	Δ	$p_2\overline{p}_2$	$\mathbf{S}_4(/\mathbf{C}_1)$	enantiospheric
50	$\mathbf{C}_s$	Δ_1	A	$\mathbf{C}_s(/\mathbf{C}_s)$	homospheric
		Δ_2	B	$\mathbf{C}_s(/\mathbf{C}_s)$	homospheric
		Δ_3	A_2	$\mathbf{C}_s(/\mathbf{C}_1)$	enantiospheric
51	$\mathbf{C}_s$	Δ_1	B	$\mathbf{C}_s(/\mathbf{C}_s)$	homospheric
		Δ_2	C	$\mathbf{C}_s(/\mathbf{C}_s)$	homospheric
		Δ_3	A_2	$\mathbf{C}_s(/\mathbf{C}_1)$	enantiospheric
52	$\mathbf{C}_s$	Δ_1	A	$\mathbf{C}_s(/\mathbf{C}_s)$	homospheric
		Δ_2	A	$\mathbf{C}_s(/\mathbf{C}_s)$	homospheric
		Δ_3	$p\overline{p}$	$\mathbf{C}_s(/\mathbf{C}_1)$	enantiospheric
53	$\mathbf{C}_s$	Δ_1	A	$\mathbf{C}_s(/\mathbf{C}_s)$	homospheric
		Δ_2	B	$\mathbf{C}_s(/\mathbf{C}_s)$	homospheric
		Δ_3	$p\overline{p}$	$\mathbf{C}_s(/\mathbf{C}_1)$	enantiospheric
54	$\mathbf{C}_s$	Δ_1	A	$\mathbf{C}_s(/\mathbf{C}_s)$	homospheric
		Δ_2	B	$\mathbf{C}_s(/\mathbf{C}_s)$	homospheric
		Δ_3	$\overline{p}p$	$\mathbf{C}_s(/\mathbf{C}_1)$	enantiospheric
55	$\mathbf{C}_2$	Δ_1	A_2	$\mathbf{C}_2(/\mathbf{C}_1)$	hemispheric
		Δ_2	p_2	$\mathbf{C}_2(/\mathbf{C}_1)$	hemispheric
56	$\mathbf{C}_2$	Δ_1	p_2	$\mathbf{C}_2(/\mathbf{C}_1)$	hemispheric
		Δ_2	q_2	$\mathbf{C}_2(/\mathbf{C}_1)$	hemispheric

Table 21.4: Continued

Pro-molecule	Symmetry	Orbit	Members of an orbit	Coset representation	Chirality fittingness (Sphericity)
57	C_2'	Δ_1	A_2	$C_2'(/C_1)$	hemispheric
		Δ_2	B_2	$C_2'(/C_1)$	hemispheric
58	C_2'	Δ_1	A_2	$C_2'(/C_1)$	hemispheric
		Δ_2	p_2	$C_2'(/C_1)$	hemispheric
59	C_2'	Δ_1	A_2	$C_2'(/C_1)$	hemispheric
		Δ_2	p_2	$C_2'(/C_1)$	hemispheric
60	C_2'	Δ_1	p_2	$C_2'(/C_1)$	hemispheric
		Δ_2	$\overline{p}_2$	$C_2'(/C_1)$	hemispheric
61	C_2'	Δ_1	p_2	$C_2'(/C_1)$	hemispheric
		Δ_2	q_2	$C_2'(/C_1)$	hemispheric
62	C_2'	Δ_1	p_2	$C_2'(/C_1)$	hemispheric
		Δ_2	q_2	$C_2'(/C_1)$	hemispheric

if the ligand matches the local symmetry of the promolecule, the symmetry of the promolecule remains unchanged in the resulting molecule. We call the resulting molecule a *matched molecule*.[6]

Such matched molecules can be classified into two categories ($F = H_i$ and $F > H_i$). The first case is that F is equal to H_i. There appears no restriction in this type of construction of matched molecules. For example, the promolecule (7) can be converted into neopentane and like, when each A is replaced by a methyl group. The methyl group as a fragment has C_{3v} symmetry, which is compatible with the local symmetry of the $T_d(/C_{3v})$ orbit of 7. Hence, this molecule ideally has the same T_d symmetry as 7, where the global symmetry of 7 retains in the resulting neopentane.[7] Moreover, there emerges no restriction concerning the C_{3v}-fragment. The Nakazaki compound (2b) is another example of this case. Obviously, the D_3-trishomocubanyl group belongs to C_3, which is compatible with the $T(/C_3)$ orbit of an adamantane promolecule (analogous to 8 collected in Table 21.2).

If $F > H_i$, the symmetry (F) of a ligand is restricted to H_i as a segment, whereas global symmetry is not affected. Consider the case of 46 derived from the allene skeleton (45). Since the four joints of 46 construct a $D_{2d}(/C_s)$ orbit, this orbit has C_s local symmetry. A methyl group (a C_{3v}-fragment) matches this local

---

[6]The existence and non-existence of promolecules (and matched molecules) can be qualitatively predicted by using unit subduced cycle indices with chirality fittingness. See ref. [7]. The enumeration described in ref. [3] is concerned with promolecules (and matched molecules) in the present sense.

[7]It should be noted that the present method predicts the highest attainable symmetry; however, steric hindrance may possibly restrict such a molecule into a lower symmetry, as reported by Mislow.[30]

symmetry. This means that the resulting tetramethylallene retains the same D_{2d} symmetry as the promolecule (46). However, the C_{3v} symmetry of the methyl fragment is restricted to C_s according to the following expression:

$$C_{3v}(/C_s) \downarrow C_s = C_s(/C_1) + C_s(/C_s) \quad \text{for three hydrogens} \quad (21.5)$$
$$\text{and}$$
$$C_{3v}(/C_{3v}) \downarrow C_s = C_s(/C_s) \quad \text{for the joint carbon.} \quad (21.6)$$

Equation 21.5 indicates that the three hydrogens are split into a $C_s(/C_1)$ orbit (two hydrogens) and a $C_s(/C_s)$ orbit (one hydrogen), if there appears an appropriate energy barrier.

Figure 21.1 illustrates an example that involves the two cases, *i.e.*, $F = H_i$ for the CXYZ groups and $F > H_i$ for the methyl groups, because H_i is equal to C_1 in 5 ($= 20$). Obviously, if H_i is equal to C_1 (an identity group), the corresponding orbit (governed by a regular representation) accepts any ligands, since the relation $F \geq H_i$ ($= C_1$) holds for any F.

Mismatched molecules. Suppose that $F < H_i$. Then, the symmetry (F) of a ligand is incompatible with the local symmetry (H_i) of an $H(/H_i)$ orbit in a promolecule. The resulting molecule (called a *mismatched molecule*) no longer retains the original symmetry (H) of the promolecule.

Let M be the symmetry of the molecule ($M < H \leq G$). If H is chiral, M can be selected as chiral; if H is achiral, M can be selected as achiral. If F is a subgroup of M and if $|M| / |F|$ is equal to $|H| / |H_i|$, the $H(/H_i)$ orbit in the promolecule restricted into an $M(/F)$ orbit in the resulting molecule. Such an M group can be obtained by the inspection of a desymmetrization lattice for H.

Consider the promolecule (7), in which the four As are replaced by four hydroxymethyl ligands (CH_2OH). This ligand in isolation has C_s symmetry, which is incompatible with the C_{3v} local symmetry. When we examine the desymmetrization lattice (Fig. 21.3), we find $D_{2d}(/C_s)$, where $|T_d| / |C_{3v}| = |D_{2d}| / |C_s| = 2$. Hence, the resulting pentaerythritol belongs to D_{2d} symmetry; and the four CH_2OH ligands construct a $D_{2d}(/C_s)$ orbit.

The McCasland D_2 molecule ($1c$) is a mismatched molecule derived from the promolecule (8); *i.e.*, $T(/C_3) \rightarrow D_2(/C_1)$. This desymmetrization is rationalized by Fig. 21.3. That is to say, the symmetry (C_1) of the ligand mismatches the C_3 local symmetry of $T(/C_3)$ and lowers the symmetry of the molecule ($1c$) to have D_2. Thereby, the C_1 symmetry becomes compatible to the CR ($D_2(/C_1)$). The Nakazaki D_2 molecule ($2a$) can be explained in the same line.

The Mislow molecule (3) can be derived from a promolecule (137), where $p\overline{p}$ proligands construct an $S_4(/C_2)$ orbit. Since CXYZ ligands (C_1) are incompatible with this CR, the resulting molecule (3) no longer belongs to S_4 but to C_1. It should be noted that *the chirality/achirality of such a molecule is determined by the chirality/achirality of the corresponding promolecule.*

$$S_4(/C_2)$$

137

There is another type of mismatched molecules. Suppose that the symmetry $\mathbf{F}'$ of a ligand is neither a subgroup nor a supergroup of $\mathbf{H}_i$. If $\mathbf{F}$ is selected as being equal to $\mathbf{F}' \cap \mathbf{H}_i$, the above treatment holds for this case. There exists such an $\mathbf{F}$ group; in the lowest cases, $\mathbf{F}$ may be $\mathbf{C}_s$ for an achiral $\mathbf{F}'$ and $\mathbf{C}_1$ for a chiral $\mathbf{F}'$.

Let us examine the promolecule (7), in which As are replaced by phenyl groups. The symmetry of the phenyl group is $\mathbf{C}_{2v}$, whereas the local symmetry is $\mathbf{C}_{3v}$. Hence, we select $\mathbf{F} = \mathbf{C}_{2v} \cap \mathbf{C}_{3v} = \mathbf{C}_s$. We look up a $(/\mathbf{C}_s)$ term in Fig. 21.3; in a similar way as pentaerythritol, we are able to find $\mathbf{D}_{2d}(/\mathbf{C}_s)$. Hence, we conclude that the resulting tetraphenylmethane has $\mathbf{D}_{2d}$ symmetry.

Highest attainable symmetry. The above discussions provide us with a general approach to the judgement of the highest attainable symmetry of a given molecule. The procedure is summarized as follows. (1) The molecule is converted into the corresponding promolecule by replacing ligands by proligands. (2) The symmetry and orbits (CRs) of the promolecule are determined. (3) The chirality/achirality of the molecule is determined by the symmetry of the promolecule. (4) From each CR, we have a local symmetry for each proligand. (5) We examine whether the symmetry of a ligand is compatible with such a local symmetry or not. (6) A matched (compatible) molecule retains the symmetry of the promolecule in the highest attainable state. (7) A mismatched (incompatible) molecule lowers the symmetry of the promolecule. The resulting symmetry is judged in terms of a desymmetrization lattice.

21.5 Prochiralities of Promolecules and Molecules

In Chapter 10,[8] we have proposed a novel definition of *prochirality*. Although this definition can be applied to all types of molecules, the previous discussion mainly took rigid molecules into consideration. We here extend this definition in order to treat promolecules as well as non-rigid molecules.

Definition of prochirality. The concept of prochirality is ascribed to the presence of an enantiospheric orbit. This is based on a theorem:[8] *an enantiospheric orbit is capable of separating into two hemispheric orbits of the same length under a chiral environment, whether the change is reversible or irreversible.* As exemplified in Fig. 21.1, a promolecule (*e.g.* **5**) has joint positions which are occupied by proligands. A set of equivalent proligands (as segments) constructs an orbit, which corresponds to an orbit of the joints in one-to-one fashion. Hence, we can regard the chirality fittingness (sphericity) for the orbit of proligands as being the same as that for the orbit of joints. A *prochiral promolecule* is then defined as a promolecule that has at least one enantiospheric orbit of proligands.[8]

Tables 21.2 and 21.4 contain such chirality fittingnesses for promolecules. Among them, enantiospheric orbits are concerned with prochirality. For example, the promolecule (**11**) has four proligands that are the members of an enantiospheric ($S_4(/C_1)$) orbit. Under a chiral environment, the orbit is split into two hemispheric orbits (p_2 and $\bar{p}_2$ perturbed), which are energetically different. In other words, the original enantiospheric orbit is degenerated under an achiral environment.

The prochiralities of the C_s-promolecules (**16-19**) come from their respective enantiospheric orbits (Δ_3) governed by $C_s(/C_1)$. The $C_s(/C_1)$ orbit (Δ_3) of **16** takes two achiral proligands (A_2); on the other hand, the Δ_3 orbit of **17**, **18**, or **19** is occupied by a pair of antipodal proligands ($p\bar{p}$). These two modes of occupation are characteristic of such enantiospheric orbits.

It should be noted that p and $\bar{p}$ of an enantiospheric orbit (*e.g.* in **17**) are equivalent to each other as segments. This fact may be strange to organic chemists, since their convention discriminates a chiral moiety from its mirror image. On the other hand, it is natural for organic chemists to recognize that the two A's of an enantiospheric orbit (*e.g.* in **16**) are equivalent. In spite of such opposite recognitions, the relationship between p and $\bar{p}$ has the same effects as does the relationship between the two A's. Mathematically speaking, such an enantiospheric orbit is an equivalence class, the two halves of which are equivalent in the sense that they coincide with each other by an improper rotation, but not by a proper rotation. Thus, the two A's coincide with each other by an improper rotation only in the same manner as does the $p\bar{p}$. That is to say, each of the two A's is restricted to C_1 (asymmetric) in such a promolecule as **16**. In addition, the one A is the mirror image of the other A under the restricted condition (as segments). Misunderstanding regarding these facts has created a vast number of confusions

---

[8] Strictly speaking, the prochirality is concerned with a given orbit (of sites, faces, bonds *etc.*). In this chapter, we take orbits of ligands and of proligands into account.

concerning prochirality and related concepts, as discussed in the following section.

Matched molecules can be treated in the same line as promolecules. Fig. 21.1 shows that a matched molecule (*e.g.* **4**) has joint positions which are occupied by ligands. A set of equivalent ligands (as segments) constructs an orbit which is equivalent to that of proligands. This corresponds to an orbit of the joints in one-to-one fashion. Hence, we regard the chirality fittingness (sphericity) for the orbit of ligands as being the same as that for the orbit of joints. A *prochiral matched molecule* is then defined as a matched molecule that has at least one enantiospheric orbit of ligands.

Orbits of ligands in mismatched molecules are also considered to be orbits of such joints. Since the symmetry of such a mismatched molecule differs from that of the corresponding promolecule, two types of prochiralities should be taken into account. The one is concerned with such a mismatched molecule and the other with a parent promolecule. A *prochiral mismatched molecule* in the former sense (*i.e.*, as an appropriate conformer of the highest attainable symmetry) is defined as a mismatched molecule that has at least one enantiospheric orbit of ligands, where the symmetry at issue is that of the mismatched molecule. A *prochiral mismatched molecule* in the latter sense (*i.e.*, as an average conformation) is defined as a mismatched molecule that corresponds to a prochiral promolecule. Which prochiralities are effective would be variable and should be determined experimentally. For example, the Mislow molecule (**3**) is chiral in the highest attainable symmetry; but prochiral in the latter sense, because the corresponding promolecule (**137**) has an enantiospheric orbit $S_4(/C_2)$. As a result, the replacement of p or $\bar{\text{p}}$ by an appropriate achiral ligand can produce a chiral molecule.

Comments on conventional definitions. In connection with the preceding analysis, we should re-examine previous definitions of the term "prochirality". The original version of Hanson[28] was as follows: "If a chiral assembly is obtained when a point ligand in a finite nonchiral assembly of point ligands is replaced by a new point ligand, the original assembly is prochiral." This definition is obviously a special case of the present one if the terms "assembly" and "point ligand" are replaced by the terms *promolecule* and *achiral proligand*. The original definition was revised by Hirschmann and Hanson so that "prochirality" was used with reference to prostereoisomerism[31] and further with reference to the terms "graphochiral" and "pherochiral".[32] This revision, however, changed the meanings of the original terms so as to aim at preparing sequence rules in a formal fashion. In the light of this revision, Cg^+g^-hi (equivalent to **18**) and $Cg^+g^-h^+i^-$ (equivalent to **40** or **41**) are "achiral" centers of stereoisomerism with a chiral configuration; and $Cg^+g^-h^+h^-$ (equivalent to **39**) is a "chiral" center of setereoisomerism with a chiral configuration. Contrary to this, **18** is achiral; and **40** (or **41**) and **39** are chiral in the present criterion. Since the achiral **18** (Cg^+g^-hi) can produce the chiral

22 (Cg^+jhi), it seems hard to understand why **18** (Cg^+g^-hi) should not be called "prochiral".

Prelog and Helmchen[14] differentiated centers of "prochirality" (*e.g.*, for **16**), those of "pseudoasymmetry" (*e.g.*, for **18** and **19**) and those of "propseudoasymmetry" (*e.g.*, for **17** or **44**) from each other. The term "propseudoasymmetry" would produce a serious confusion, even though their statament on stereochemistry is to the point. Using this term, they focussed their attention on the two A's of **17** (equivalently **44** in Fig. 21.4). The propseudoasymmetry indicates that one of the two A's is replaced by B producing a *meso* promolecule (**18** or **19**) having a pseudoasymmetric center. However, this process is chemoselective and by no means stereoselective.[8] There exists a stereoselective process for the promolecule (**17**), which is the replacement of either one of $p\overline{p}$ to create a chiral promolecule (*e.g.* **22**). Thus, the term "propseudoasymmetry" contradicts itself, because a promolecule with a center of propseudoasymmetry can produce a chiral promolecule in addition to a promolecule having a center of pseudoasymmetry.

The latter example also indicates that the term "centers" of prochirality *etc.* has no sound basis. Thus, the conversion of **17** into **22** does not stem from the handedness of such an center, but from a stereoselective distinction regarding the halves of the enantiospheric orbit (Δ_3) of **17**. Such an orbit by no means depends upon any of such centers. Hence, stereochemical phenomena must be discussed on the basis of orbits governed by coset representations. They should not be ascribed to such "centers", as pointed out by Mislow and Siegel.[22]

The Mislow-Siegel defintion of prochirality[22] can give results equivalent to ours by careful examination, although it does not contain the present concepts such as sphericities of orbits. We have recently discussed their concept of $(\text{pro})^p$-chirality (Chapter 11).[8]

The preceding analysis implies that there are two types of representations for stereochemistry: (1) a representation for reproducing or rewriting a stereostructure and (2) a representation for discussing stereochemical relationships. In the begining and as for rather simple molecules, the two representations were identical with each other in all of the proposed rules.[33, 28, 34] However, every revision for aiming at a more complicated cases has more and more separated the two representations and has attached greater importance to (1) than to (2). Conventional confusions stem from such approaches that substitute (1) for (2) without recognizing this fact. The present change of viewpoints thinks much of (2) and thereby provides us with a deeper insight on stereochemistry.

21.6 Concluding Remarks

In this chapter, we have introduced several concepts in order to characterize symmetrical properties of a non-rigid molecule, where the non-rigidity stems from internal bond rotations. One of such concepts is a proligand, which is defined as a hypothetical ligand being structureless but having chirality. A promolecule is an abstract three-dimensional object that consists of a skeleton and such proligands. According to this formulation, promolecules can be manipulated as rigid objects. Thus, promolecules based on a methane and an allene skeleton are enumerated systematically by using unit subduced cycle indices with chirality fittingness. By starting from one of these promolecules, a molecule can be constructed, where proligands contained are replaced by approriate ligands that have three-dimensional structures. This construction is controlled by a coset representation that governs an orbit of such proligands or equivalently by the corresponding local symmetry. The resulting molecules are classified into matched molecule of which symmetries retain the symmetries of the starting promolecules; or into mismatched molecules of which symmetries become lower. Modes of such desymmetrizations are explained by a desymmetrization lattice indicating subduction of coset representations.

Bibliography

[1] S. Fujita, *Theor. Chim. Acta*, **76**, 247 (1989).

[2] S. Fujita, *Bull. Chem. Soc. Jpn.*, **62**, 3771 (1989).

[3] S. Fujita, *Bull. Chem. Soc. Jpn.*, **63**, 203 (1990).

[4] S. Fujita, *Tetrahedron*, **46**, 365 (1990).

[5] S. Fujita, *J. Math. Chem.*, **5**, 99, 121 (1990).

[6] S. Fujita, *Bull. Chem. Soc. Jpn.*, **63**, 1876 (1990).

[7] S. Fujita, *Bull. Chem. Soc. Jpn.*, **63**, 315 (1990).

[8] S. Fujita, *J. Am. Chem. Soc.*, **112**, 3390 (1990).

[9] S. Fujia, *Theor. Chim. Acta*, **78**, 45 (1990).

[10] J. K. O'loane, *Chem. Rev.*, **80**, 41 (1980).

[11] E. Ruch, *Angew. Chem. Int. Ed. Engl.*, **16**, 65 (1977).

[12] I. Ugi, J. Dugundji, R. Kopp, D. Marquarding, *Perspective in Theoretical Stereochemistry (Lecture Notes in Chemistry, Vol. 36)*, Springer-Verlag, Berlin-Heidelberg (1984).

[13] V. Prelog, *Science*, **193**, 17 (1976).

[14] a) V. Prelog, G. Helmchen, *Helv. Chim. Acta*, **55**, 2581 (1972). b) V. Prelog, J. Thix, *Helv. Chim. Acta*, **65**, 2622 (1982).

[15] S. Fujita, *J. Chem. Educ.*, **63**, 744 (1986).

[16] J. H. van't Hoff, *Arch. Néer.*, **9**, 445 (1874); *Bull. Soc. Chim. Fr.*, **23**, 295 (1875).

[17] M. Farina, C. Morandi, *Tetrahedron*, **30**, 1819 (1974).

[18] G. E. McCasland, R. Horvat, M. R. Roth, *J. Am. Chem. Soc.*, **81**, 2399 (1959).

[19] M. Nakazaki, K. Naemura, *J. Chem. Soc., Chem. Commun.*, 911 (1980); *J. Org. Chem.*, **46**, 106 (1981).

[20] K. Mislow, *J. Chem. Soc., Chem. Commun.*, 234 (1981).

[21] M. Nakazaki, K. Naemura, Y. Hokura, *J. Chem. Soc., Chem. Commun.*, 1245 (1982).

[22] K. Mislow, J. Siegel, *J. Am. Chem. Soc.*, **106**, 3319 (1984).

[23] a) K. Mislow, *Science*, **120**, 232 (1954). b) K. Mislow, *Introduction to Stereochemistry*, Benjamin, New York, p. 93 (1965).

[24] M. Nakazaki, *Bunshi no Katachi to Taisho–Sono Hyoji Hou (Shapes and Symmetry of Molecules–Their Notations)*, Nankodo, Tokyo, Chapter 8 (1969).

[25] H. C. Longuet-Higgins, *Mol. Phys.*, **6**, 445 (1963).

[26] C. M. Woodmann, *Mol. Phys.*, **19**, 753 (1970).

[27] K. Balasbramanian, *Theor. Chim. Acta*, **51**, 37 (1979).

[28] K. R. Hanson, *J. Am. Chem. Soc.*, **88**, 2731 (1966).

[29] a) S. Fujita, *Theor. Chim. Acta*, **77**, 307 (1990). b) S. Fujita, *Bull. Chem. Soc. Jpn.*, **63**, 2033 (1990).

[30] L. D. Iroff, K. Mislow, *J. Am. Chem. Soc.*, **100**, 2121 (1978).

[31] H. Hirschmann, K. R. Hanson, *J. Org. Chem.*, **36**, 3293 (1971).

[32] H. Hirschmann, K. R. Hanson, *J. Org. Chem.*, **37**, 2984 (1972); *Top. Stereochem.*, **14**, 183 (1983).

[33] a) R. S. Chan, C. K. Ingold, V. Prelog, *Angew. Chem. Int. Ed. Engl.*, **5**, 385 (1966). b) V. Prelog, G. Helmchen, *Angew. Chem. Int. Ed. Engl.*, **21**, 567 (1982).

[34] IUPAC Recomendations. Section E (Stereochemistry), *J. Org. Chem.*, **35**, 2849 (1970); *Pure Appl. Chem.*, **45**, 11 (1976).

Appendix A

Mark Tables

This is a list of mark tables of several point groups.

A.1 T_d Point Group and Its Subgroups

Table A.1: Mark table of C_2

	C_1	C_2
$C_2(/C_1)$	2	0
$C_2(/C_2)$	1	1

Table A.2: Mark table of C_s

	C_1	C_s
$C_s(/C_1)$	2	0
$C_s(/C_s)$	1	1

Table A.3: Mark table of C_3

	C_1	C_3
$C_3(/C_1)$	3	0
$C_3(/C_3)$	1	1

Table A.4: Mark table of S_4

	C_1	C_2	S_4
$S_4(/C_1)$	4	0	0
$S_4(/C_2)$	2	2	0
$S_4(/S_4)$	1	1	1

Table A.5: Mark table of C_{2v}

	C_1	C_2	C_s	C_s'	C_{2v}
$C_{2v}(/C_1)$	4	0	0	0	0
$C_{2v}(/C_2)$	2	2	0	0	0
$C_{2v}(/C_s)$	2	0	2	0	0
$C_{2v}(/C_s')$	2	0	0	2	0
$C_{2v}(/C_{2v})$	1	1	1	1	1

Table A.6: Mark table of D_2

	C_1	C_2	C_2'	C_2''	D_2
$D_2(/C_1)$	4	0	0	0	0
$D_2(/C_2)$	2	2	0	0	0
$D_2(/C_2')$	2	0	2	0	0
$D_2(/C_2'')$	2	0	0	2	0
$D_2(/D_2)$	1	1	1	1	1

Table A.7: Mark table of C_{3v}

	C_1	C_s	C_3	C_{3v}
$C_{3v}(/C_1)$	6	0	0	0
$C_{3v}(/C_s)$	3	1	0	0
$C_{3v}(/C_3)$	2	0	2	0
$C_{3v}(/C_{3v})$	1	1	1	1

Table A.8: Mark table of D_{2d}

	C_1	C_2	C_2'	C_s	S_4	C_{2v}	D_2	D_{2d}
$D_{2d}(/C_1)$	8	0	0	0	0	0	0	0
$D_{2d}(/C_2)$	4	4	0	0	0	0	0	0
$D_{2d}(/C_2')$	4	0	2	0	0	0	0	0
$D_{2d}(/C_s)$	4	0	0	2	0	0	0	0
$D_{2d}(/S_4)$	2	2	0	0	2	0	0	0
$D_{2d}(/C_{2v})$	2	2	0	2	0	2	0	0
$D_{2d}(/D_2)$	2	2	2	0	0	0	2	0
$D_{2d}(/D_{2d})$	1	1	1	1	1	1	1	1

Table A.9: Mark table of T

	C_1	C_2	C_3	D_2	T
$T(/C_1)$	12	0	0	0	0
$T(/C_2)$	6	2	0	0	0
$T(/C_3)$	4	0	1	0	0
$T(/D_2)$	3	3	0	3	0
$T(/T)$	1	1	1	1	1

Table A.10: Mark table of $\mathbf{T}_d$

	$\mathbf{C}_1$	$\mathbf{C}_2$	$\mathbf{C}_s$	$\mathbf{C}_3$	$\mathbf{S}_4$	$\mathbf{D}_2$	$\mathbf{C}_{2v}$	$\mathbf{C}_{3v}$	$\mathbf{D}_{2d}$	$\mathbf{T}$	$\mathbf{T}_d$
$\mathbf{T}_d(/\mathbf{C}_1)$	24	0	0	0	0	0	0	0	0	0	0
$\mathbf{T}_d(/\mathbf{C}_2)$	12	4	0	0	0	0	0	0	0	0	0
$\mathbf{T}_d(/\mathbf{C}_s)$	12	0	2	0	0	0	0	0	0	0	0
$\mathbf{T}_d(/\mathbf{C}_3)$	8	0	0	2	0	0	0	0	0	0	0
$\mathbf{T}_d(/\mathbf{S}_4)$	6	2	0	0	2	0	0	0	0	0	0
$\mathbf{T}_d(/\mathbf{D}_2)$	6	6	0	0	0	6	0	0	0	0	0
$\mathbf{T}_d(/\mathbf{C}_{2v})$	6	2	2	0	0	0	2	0	0	0	0
$\mathbf{T}_d(/\mathbf{C}_{3v})$	4	0	2	1	0	0	0	1	0	0	0
$\mathbf{T}_d(/\mathbf{D}_{2d})$	3	3	1	0	1	3	1	0	1	0	0
$\mathbf{T}_d(/\mathbf{T})$	2	2	0	2	0	2	0	0	0	2	0
$\mathbf{T}_d(/\mathbf{T}_d)$	1	1	1	1	1	1	1	1	1	1	1

A.2 D_{3h} Point Group and Its Subgroups

Table A.11: Mark table of C_{3h}

	C_1	C_s	C_3	C_{3h}
$C_{3h}(/C_1)$	6	0	0	0
$C_{3h}(/C_s)$	3	3	0	0
$C_{3h}(/C_3)$	2	0	2	0
$C_{3h}(/C_{3h})$	1	1	1	1

Table A.12: Mark table of D_3

	C_1	C_2	C_3	D_3
$D_3(/C_1)$	6	0	0	0
$D_3(/C_2)$	3	1	0	0
$D_3(/C_3)$	2	0	2	0
$D_3(/D_3)$	1	1	1	1

Table A.13: Mark Table for D_{3h} Point Group

	C_1	C_2	C_s	C'_s	C_3	C_{2v}	C_{3v}	C_{3h}	D_3	D_{3h}
$D_{3h}(/C_1)$	12	0	0	0	0	0	0	0	0	0
$D_{3h}(/C_2)$	6	2	0	0	0	0	0	0	0	0
$D_{3h}(/C_s)$	6	0	2	0	0	0	0	0	0	0
$D_{3h}(/C'_s)$	6	0	0	6	0	0	0	0	0	0
$D_{3h}(/C_3)$	4	0	0	0	4	0	0	0	0	0
$D_{3h}(/C_{2v})$	3	1	1	3	0	1	0	0	0	0
$D_{3h}(/C_{3v})$	2	0	2	0	2	0	2	0	0	0
$D_{3h}(/C_{3h})$	2	0	0	2	2	0	0	2	0	0
$D_{3h}(/D_3)$	2	2	0	0	2	0	0	0	2	0
$D_{3h}(/D_{3h})$	1	1	1	1	1	1	1	1	1	1

Appendix B

Inverses of Mark Tables

This is a list of the inverse matrices of mark tables for several point groups.

B.1 T_d Point Group and Its Subgroups

Table B.1: The Inverse of the Mark Table for C_2 Point Group

	$C_2(/C_1)$	$C_2(/C_2)$	sum[a]
C_1	1/2	0	1/2
C_2	$-1/2$	1	1/2

[a] sum $= \sum_{i=1}^{s} \overline{m}_{ji}$.

Table B.2: The Inverse of the Mark Table for C_s Point Group

	$C_s(/C_1)$	$C_s(/C_s)$	sum[a]
C_1	1/2	0	1/2
C_s	$-1/2$	1	1/2

[a] sum $= \sum_{i=1}^{s} \overline{m}_{ji}$.

Table B.3: The Inverse of the Mark Table for C_3 Point Group

	$C_3(/C_1)$	$C_3(/C_3)$	sum[a]
C_1	1/3	0	1/3
C_3	−1/3	1	2/3

[a] sum $= \sum_{i=1}^{s} \overline{m}_{ji}$.

Table B.4: The Inverse of the Mark Table for S_4 Point Group

	$S_4(/C_1)$	$S_4(/C_2)$	$S_4(/S_4)$	sum[a]
C_1	1/4	0	0	1/4
C_2	−1/4	1/2	0	1/4
S_4	0	−1/2	1	1/2

[a] sum $= \sum_{i=1}^{s} \overline{m}_{ji}$.

Table B.5: The Inverse of the Mark Table for C_{2v} Point Group

	$C_{2v}(/C_1)$	$C_{2v}(/C_2)$	$C_{2v}(/C_s)$	$C_{2v}(/C_s')$	$C_{2v}(/C_{2v})$	sum[a]
C_1	1/4	0	0	0	0	1/4
C_2	−1/4	1/2	0	0	0	1/4
C_s	−1/4	0	1/2	0	0	1/4
C_s'	−1/4	0	0	1/2	0	1/4
C_{2v}	1/2	−1/2	−1/2	−1/2	1	0

[a] sum $= \sum_{i=1}^{s} \overline{m}_{ji}$.

Table B.6: The Inverse of the Mark Table for D_2 Point Group

	$D_2(/C_1)$	$D_2(/C_2)$	$D_2(/C_2')$	$D_2(/C_2'')$	$D_2(/D_2)$	sum[a]
C_1	1/4	0	0	0	0	1/4
C_2	−1/4	1/2	0	0	0	1/4
C_2'	−1/4	0	1/2	0	0	1/4
C_2''	−1/4	0	0	1/2	0	1/4
D_2	1/2	−1/2	−1/2	−1/2	1	0

[a] sum $= \sum_{i=1}^{s} \overline{m}_{ji}$.

Table B.7: The Inverse of the Mark Table for C_{3v} Point Group

	$C_{3v}(/C_1)$	$C_{3v}(/C_s)$	$C_{3v}(/C_3)$	$C_{3v}(/C_{3v})$	sum[a]
C_1	1/6	0	0	0	1/6
C_s	−1/2	1	0	0	1/2
C_3	−1/6	0	1/2	0	1/3
C_{3v}	1/2	−1	−1/2	1	0

[a] sum $= \sum_{i=1}^{s} \overline{m}_{ji}$.

Table B.8: The Inverse of the Mark Table for D_{2d} Point Group

	D_{2d} $(/C_1)$	D_{2d} $(/C_2)$	D_{2d} $(/C_2')$	D_{2d} $(/C_s)$	D_{2d} $(/S_4)$	D_{2d} $(/C_{2v})$	D_{2d} $(/D_2)$	D_{2d} $(/D_{2d})$	sum[a]
C_1	1/8	0	0	0	0	0	0	0	1/8
C_2	−1/8	1/4	0	0	0	0	0	0	1/8
C_2'	−1/4	0	1/2	0	0	0	0	0	1/4
C_s	−1/4	0	0	1/2	0	0	0	0	1/4
S_4	0	−1/4	0	0	1/2	0	0	0	1/4
C_{2v}	1/4	−1/4	0	−1/2	0	1/2	0	0	0
D_2	1/4	−1/4	−1/2	0	0	0	1/2	0	0
D_{2d}	0	1/2	0	0	−1/2	−1/2	−1/2	1	0

[a] sum $= \sum_{i=1}^{s} \overline{m}_{ji}$.

Table B.9: The Inverse of the Mark Table for T Point Group

	T $(/C_1)$	T $(/C_2)$	T $(/C_3)$	T $(/D_2)$	T $(/T)$	sum[a]
C_1	1/12	0	0	0	0	1/12
C_2	−1/4	1/2	0	0	0	1/4
C_3	−1/3	0	1	0	0	2/3
D_2	1/6	−1/2	0	1/3	0	0
T	1/3	0	−1	−1/3	1	0

[a] sum $= \sum_{i=1}^{s} \overline{m}_{ji}$.

Table B.10: The inverse of the mark table of T_d

	T_d (/C_1)	T_d (/C_2)	T_d (/C_s)	T_d (/C_3)	T_d (/S_4)	T_d (/D_2)	T_d (/C_{2v})	T_d (/C_{3v})	T_d (/D_{2d})	T_d (/T)	T_d (/T_d)	sum[a]
C_1	$\frac{1}{24}$	0	0	0	0	0	0	0	0	0	0	$\frac{1}{24}$
C_2	$-\frac{1}{8}$	$\frac{1}{4}$	0	0	0	0	0	0	0	0	0	$\frac{1}{8}$
C_s	$-\frac{1}{4}$	0	$\frac{1}{2}$	0	0	0	0	0	0	0	0	$\frac{1}{4}$
C_3	$-\frac{1}{6}$	0	0	$\frac{1}{2}$	0	0	0	0	0	0	0	$\frac{1}{3}$
S_4	0	$-\frac{1}{4}$	0	0	$\frac{1}{2}$	0	0	0	0	0	0	$\frac{1}{4}$
D_2	$\frac{1}{12}$	$-\frac{1}{4}$	0	0	0	$\frac{1}{6}$	0	0	0	0	0	0
C_{2v}	$\frac{1}{4}$	$-\frac{1}{4}$	$-\frac{1}{2}$	0	0	0	$\frac{1}{2}$	0	0	0	0	0
C_{3v}	$\frac{1}{2}$	0	-1	$-\frac{1}{2}$	0	0	0	1	0	0	0	0
D_{2d}	0	$\frac{1}{2}$	0	0	$-\frac{1}{2}$	$-\frac{1}{2}$	$-\frac{1}{2}$	0	1	0	0	0
T	$\frac{1}{6}$	0	0	$-\frac{1}{2}$	0	$-\frac{1}{6}$	0	0	0	$\frac{1}{2}$	0	0
T_d	$-\frac{1}{2}$	0	1	$\frac{1}{2}$	0	$\frac{1}{2}$	0	-1	-1	$-\frac{1}{2}$	1	0

[a] $\mathrm{sum} = \sum_{i=1}^{s} \overline{m}_{ji}.$

B.2 C_{3h} Point Group and Its Subgroups

Table B.11: The Inverse of the Mark Table for C_{3h} Point Group

	$C_{3h}(/C_1)$	$C_{3h}(/C_s)$	$C_{3h}(/C_3)$	$C_{3h}(/C_{3h})$	sum[a]
C_1	1/6	0	0	0	1/6
C_s	−1/6	1/3	0	0	1/6
C_3	−1/6	0	1/2	0	1/3
C_{3h}	1/6	−1/3	−1/2	1	1/3

[a] sum $= \sum_{i=1}^{s} \overline{m}_{ji}$.

Table B.12: The Inverse of the Mark Table for D_3 Point Group

	$D_3(/C_1)$	$D_3(/C_2)$	$D_3(/C_3)$	$D_3(/D_3)$	sum[a]
C_1	1/6	0	0	0	1/6
C_2	−1/2	1	0	0	1/2
C_3	−1/6	0	1/2	0	1/3
D_3	1/2	−1	−1/2	1	0

[a] sum $= \sum_{i=1}^{s} \overline{m}_{ji}$.

Table B.13: The Inverse of the Mark Table for D_{3h} Point Group

	D_{3h} (/C_1)	D_{3h} (/C_2)	D_{3h} (/C_s)	D_{3h} (/C_s')	D_{3h} (/C_3)	D_{3h} (/C_{2v})	D_{3h} (/C_{3v})	D_{3h} (/C_{3h})	D_{3h} (/D_3)	D_{3h} (/D_{3h})	sum[a]
C_1	1/12	0	0	0	0	0	0	0	0	0	1/12
C_2	−1/4	1/2	0	0	0	0	0	0	0	0	1/4
C_s	−1/4	0	1/2	0	0	0	0	0	0	0	1/4
C_s'	−1/12	0	0	1/6	0	0	0	0	0	0	1/12
C_3	−1/12	0	0	0	1/4	0	0	0	0	0	1/6
C_{2v}	1/2	−1/2	−1/2	−1/2	0	1	0	0	0	0	0
C_{3v}	1/4	0	−1/2	0	−1/4	0	1/2	0	0	0	0
C_{3h}	1/12	0	0	−1/6	−1/4	0	0	1/2	0	0	1/6
D_3	1/4	−1/2	0	0	−1/4	0	0	0	1/2	0	0
D_{3h}	−1/2	1/2	1/2	1/2	1/2	−1	−1/2	−1/2	−1/2	1	0

[a] sum $= \sum_{i=1}^{s} \overline{m}_{ji}$.

Appendix C

Subduction Tables

This is a list of subduction tables of several point groups. A coset representation
with an asterisk is forbidden.

C.1 T_d Point Group and Its Subgroups

Table C.1: Subduction Table of C_2

	$\downarrow C_1$	$\downarrow C_2$
$C_2(/C_1)$	$2C_1(/C_1)$	$C_2(/C_1)$
$C_2(/C_2)$	$C_1(/C_1)$	$C_2(/C_2)$

Table C.2: Subduction Table of C_s

	$\downarrow C_1$	$\downarrow C_s$
$C_s(/C_1)$	$2C_1(/C_1)$	$C_s(/C_1)$
$C_s(/C_s)$	$C_1(/C_1)$	$C_s(/C_s)$

Table C.3: Subduction Table of C_3

	$\downarrow C_1$	$\downarrow C_3$
$C_3(/C_1)$	$3C_1(/C_1)$	$C_3(/C_1)$
$C_3(/C_3)$	$C_1(/C_1)$	$C_3(/C_3)$

Table C.4: Subduction Table of S_4

	$\downarrow C_1$	$\downarrow C_2$	$\downarrow S_4$
$S_4(/C_1)$	$4C_1(/C_1)$	$2C_2(/C_1)$	$S_4(/C_1)$
$S_4(/C_2)$	$2C_1(/C_1)$	$2C_2(/C_2)$	$S_4(/C_2)$
$S_4(/S_4)$	$C_1(/C_1)$	$C_2(/C_2)$	$S_4(/S_4)$

Table C.5: Subduction Table of C_{2v}

	$\downarrow C_1$	$\downarrow C_2$	$\downarrow C_s$	$\downarrow C_s'$	$\downarrow C_{2v}$
$C_{2v}(/C_1)$	$4C_1(/C_1)$	$2C_2(/C_1)$	$2C_s(/C_1)$	$2C_s'(/C_1)$	$C_{2v}(/C_1)$
$C_{2v}(/C_2)^*$	$2C_1(/C_1)$	$2C_2(/C_2)$	$C_s(/C_1)$	$C_s'(/C_1)$	$C_{2v}(/C_2)$
$C_{2v}(/C_s)$	$2C_1(/C_1)$	$C_2(/C_1)$	$2C_s(/C_s)$	$C_s'(/C_1)$	$C_{2v}(/C_s)$
$C_{2v}(/C_s')$	$2C_1(/C_1)$	$C_2(/C_1)$	$C_s(/C_1)$	$2C_s'(/C_s)$	$C_{2v}(/C_s')$
$C_{2v}(/C_{2v})$	$C_1(/C_1)$	$C_2(/C_2)$	$C_s(/C_s)$	$C_s'(/C_s)$	$C_{2v}(/C_{2v})$

* Forbidden CR. See Chapter 7.

Table C.6: Subduction Table of $\mathbf{D}_2$

	$\downarrow\mathbf{C}_1$	$\downarrow\mathbf{C}_2$	$\downarrow\mathbf{C}_2'$	$\downarrow\mathbf{C}_2''$	$\downarrow\mathbf{D}_2$
$\mathbf{D}_2(/\mathbf{C}_1)$	$4\mathbf{C}_1(/\mathbf{C}_1)$	$2\mathbf{C}_2(/\mathbf{C}_1)$	$2\mathbf{C}_2'(/\mathbf{C}_1)$	$2\mathbf{C}_2''(/\mathbf{C}_1)$	$\mathbf{D}_2(/\mathbf{C}_1)$
$\mathbf{D}_2(/\mathbf{C}_2)$	$2\mathbf{C}_1(/\mathbf{C}_1)$	$2\mathbf{C}_2(/\mathbf{C}_2)$	$\mathbf{C}_2'(/\mathbf{C}_1)$	$\mathbf{C}_2''(/\mathbf{C}_1)$	$\mathbf{D}_2(/\mathbf{C}_2)$
$\mathbf{D}_2(/\mathbf{C}_2')$	$2\mathbf{C}_1(/\mathbf{C}_1)$	$\mathbf{C}_2(/\mathbf{C}_1)$	$2\mathbf{C}_2'(/\mathbf{C}_2)$	$\mathbf{C}_2''(/\mathbf{C}_1)$	$\mathbf{D}_2(/\mathbf{C}_2')$
$\mathbf{D}_2(/\mathbf{C}_2'')$	$2\mathbf{C}_1(/\mathbf{C}_1)$	$\mathbf{C}_2(/\mathbf{C}_1)$	$\mathbf{C}_2'(/\mathbf{C}_1)$	$2\mathbf{C}_2''(/\mathbf{C}_2)$	$\mathbf{D}_2(/\mathbf{C}_2'')$
$\mathbf{D}_2(/\mathbf{D}_2)$	$\mathbf{C}_1(/\mathbf{C}_1)$	$\mathbf{C}_2(/\mathbf{C}_2)$	$\mathbf{C}_2'(/\mathbf{C}_2)$	$\mathbf{C}_2''(/\mathbf{C}_2)$	$\mathbf{D}_2(/\mathbf{D}_2)$

Table C.7: Subduction Table of $\mathbf{C}_{3v}$

	$\downarrow\mathbf{C}_1$	$\downarrow\mathbf{C}_s$	$\downarrow\mathbf{C}_3$	$\downarrow\mathbf{C}_{3v}$
$\mathbf{C}_{3v}(/\mathbf{C}_1)$	$6\mathbf{C}_1(/\mathbf{C}_1)$	$3\,\mathbf{C}_s(/\mathbf{C}_1)$	$2\mathbf{C}_3(/\mathbf{C}_1)$	$\mathbf{C}_{3v}(/\mathbf{C}_1)$
$\mathbf{C}_{3v}(/\mathbf{C}_s)$	$3\mathbf{C}_1(/\mathbf{C}_1)$	$\mathbf{C}_s(/\mathbf{C}_1) + \mathbf{C}_s(/\mathbf{C}_s)$	$\mathbf{C}_3(/\mathbf{C}_1)$	$\mathbf{C}_{3v}(/\mathbf{C}_s)$
$\mathbf{C}_{3v}(/\mathbf{C}_3)^*$	$2\mathbf{C}_1(/\mathbf{C}_1)$	$\mathbf{C}_s(/\mathbf{C}_1)$	$2\mathbf{C}_3(/\mathbf{C}_3)$	$\mathbf{C}_{3v}(/\mathbf{C}_3)$
$\mathbf{C}_{3v}(/\mathbf{C}_{3v})$	$\mathbf{C}_1(/\mathbf{C}_1)$	$\mathbf{C}_s(/\mathbf{C}_s)$	$\mathbf{C}_3(/\mathbf{C}_3)$	$\mathbf{C}_{3v}(/\mathbf{C}_{3v})$

* Forbidden CR. See Chapter 7.

Table C.8: Subduction of coset representations for D_{2d} Point Group

	$\downarrow C_1$	$\downarrow C_2$	$\downarrow C_2'$	$\downarrow C_s$	$\downarrow S_4$	$\downarrow C_{2v}$	$\downarrow D_2$	$\downarrow D_{2d}$
$D_{2d}(/C_1)$	$8C_1(/C_1)$	$4C_2(/C_1)$	$4C_2'(/C_1)$	$4C_s(/C_1)$	$2S_4(/C_1)$	$2C_{2v}(/C_1)$	$2D_2(/C_1)$	$D_{2d}(/C_1)$
$D_{2d}(/C_2)^*$	$4C_1(/C_1)$	$4C_2(/C_2)$	$2C_2'(/C_1)$	$2C_s(/C_1)$	$2S_4(/C_2)$	$2C_{2v}(/C_2)$	$2D_2(/C_2)$	$D_{2d}(/C_2)$
$D_{2d}(/C_2')$	$4C_1(/C_1)$	$2C_2(/C_1)$	$C_2'(/C_1)$ $+2\,C_2'(/C_2)$	$2C_s(/C_1)$	$S_4(/C_1)$	$C_{2v}(/C_1)$	$D_2(/C_2')$ $+D_2(/C_2'')$	$D_{2d}(/C_2')$
$D_{2d}(/C_s)$	$4C_1(/C_1)$	$2C_2(/C_1)$	$2C_2'(/C_1)$	$C_s(/C_1)$ $+2\,C_s(/C_s)$	$S_4(/C_1)$	$C_{2v}(/C_s)$ $+C_{2v}(/C_s')$	$D_2(/C_1)$	$D_{2d}(/C_s)$
$D_{2d}(/S_4)^*$	$2C_1(/C_1)$	$2C_2(/C_2)$	$C_2'(/C_1)$	$C_s(/C_1)$	$2S_4(/S_4)$	$C_{2v}(/C_2)$	$D_2(/C_2)$	$D_{2d}(/S_4)$
$D_{2d}(/C_{2v})$	$2C_1(/C_1)$	$2C_2(/C_2)$	$C_2'(/C_1)$	$2C_s(/C_s)$	$S_4(/C_2)$	$2C_{2v}(/C_{2v})$	$D_2(/C_2)$	$D_{2d}(/C_{2v})$
$D_{2d}(/D_2)^*$	$2C_1(/C_1)$	$2C_2(/C_2)$	$2C_2'(/C_2)$	$C_s(/C_1)$	$S_4(/C_2)$	$C_{2v}(/C_2)$	$2D_2(/D_2)$	$D_{2d}(/D_2)$
$D_{2d}(/D_{2d})$	$C_1(/C_1)$	$C_2(/C_2)$	$C_2'(/C_2)$	$C_s(/C_s)$	$S_4(/S_4)$	$C_{2v}(/C_{2v})$	$D_2(/D_2)$	$D_{2d}(/D_{2d})$

* Forbidden CR. See Chapter 7.

Table C.9: Subduction Table of T

	↓C$_1$	↓C$_2$	↓C$_3$	↓D$_2$	↓T
T(/C$_1$)	12C$_1$(/C$_1$)	6C$_2$(/C$_1$)	4C$_3$(/C$_1$)	3D$_2$(/C$_1$)	**T**(/C$_1$)
T(/C$_2$)	6C$_1$(/C$_1$)	2C$_2$(/C$_1$)	2C$_3$(/C$_1$)	**D**$_2$(/C$_2$)	**T**(/C$_2$)
		+ 2C$_2$(/C$_2$)		+ **D**$_2$(/C$_2'$)	
				+ **D**$_2$(/C$_2''$)	
T(/C$_3$)	4C$_1$(/C$_1$)	2C$_2$(/C$_1$)	C$_3$(/C$_1$)	**D**$_2$(/C$_1$)	**T**(/C$_3$)
			+ C$_3$(/C$_3$)		
T(/D$_2$)*	3C$_1$(/C$_1$)	3C$_2$(/C$_2$)	C$_3$(/C$_1$)	3D$_2$(/D$_2$)	**T**(/D$_2$)
T(/T)	C$_1$(/C$_1$)	C$_2$(/C$_2$)	C$_3$(/C$_3$)	**D**$_2$(/D$_2$)	**T**(/T)

* Forbidden CR. See Chapter 7.

Table C.10: Subduction Table of T_d

	$\downarrow C_1$	$\downarrow C_2$	$\downarrow C_s$	$\downarrow C_3$	$\downarrow S_4$	$\downarrow D_2$	$\downarrow C_{2v}$	$\downarrow C_{3v}$	$\downarrow D_{2d}$	$\downarrow T$	$\downarrow T_d$
$T_d(/C_1)$	$24C_1(/C_1)$	$12C_2(/C_1)$	$12C_s(/C_1)$	$8C_3(/C_1)$	$6S_4(/C_1)$	$6D_2(/C_1)$	$6C_{2v}(/C_1)$	$4C_{3v}(/C_1)$	$3D_{2d}(/C_1)$	$2T(/C_1)$	$T_d(/C_1)$
$T_d(/C_2)^*$	$12C_1(/C_1)$	$4C_2(/C_1)$ $+ 4C_2(/C_2)$	$6C_s(/C_1)$	$4C_3(/C_1)$	$2S_4(/C_1)$ $+ 2S_4(/C_2)$	$2D_2(/C_2)$ $+ 2D_2(/C_2')$ $+ 2D_2(/C_2'')$	$2C_{2v}(/C_1)$ $+ 2C_{2v}(/C_2)$	$2C_{3v}(/C_1)$	$D_{2d}(/C_2)$ $+ 2D_{2d}(/C_2')$	$2T(/C_2)$	$T_d(/C_2)$
$T_d(/C_s)$	$12C_1(/C_1)$	$6C_2(/C_1)$	$5C_s(/C_1)$ $+ 2C_s(/C_s)$	$4C_3(/C_1)$	$3S_4(/C_1)$	$3D_2(/C_1)$	$2C_{2v}(/C_1)$ $+ C_{2v}(/C_s)$ $+ C_{2v}(/C_s')$	$C_{3v}(/C_1)$ $+ 2C_{3v}(/C_s)$	$D_{2d}(/C_1)$ $+ D_{2d}(/C_s)$	$T(/C_1)$	$T_d(/C_2)$
$T_d(/C_3)^*$	$8C_1(/C_1)$	$4C_2(/C_1)$	$4C_s(/C_1)$	$2C_3(/C_1)$ $+ 2C_3(/C_3)$	$2S_4(/C_1)$	$2D_2(/C_1)$	$2C_{2v}(/C_1)$	$C_{3v}(/C_1)$ $+ C_{3v}(/C_3)$	$D_{2d}(/C_1)$	$2T(/C_3)$	$T_d(/C_3)$
$T_d(/S_4)^*$	$6C_1(/C_1)$	$2C_2(/C_1)$ $+ 2C_2(/C_2)$	$3C_s(/C_1)$	$2C_3(/C_1)$	$S_4(/C_1)$ $+ 2S_4(/S_4)$	$D_2(/C_2)$ $+ D_2(/C_2')$ $+ D_2(/C_2'')$	$C_{2v}(/C_1)$ $+ C_{2v}(/C_2)$	$C_{3v}(/C_1)$	$D_{2d}(/C_2')$ $+D_{2d}(/S_4)$	$T(/C_2)$	$T_d(/S_4)$
$T_d(/D_2)^*$	$6C_1(/C_1)$	$6C_2(/C_2)$	$3C_s(/C_1)$	$2C_3(/C_1)$	$3S_4(/C_2)$	$6D_2(/D_2)$	$3C_{2v}(/C_s)$	$C_{3v}(/C_1)$	$3D_{2d}(/D_2)$	$2T(/D_2)$	$T_d(/D_2)$
$T_d(/C_{2v})$	$6C_1(/C_1)$	$2C_2(/C_1)$ $+ 2C_2(/C_2)$	$2C_s(/C_1)$ $+ 2C_s(/C_s)$	$2C_3(/C_1)$	$S_4(/C_1)$ $+ S_4(/C_2)$	$D_2(/C_2)$ $+ D_2(/C_2')$ $+ D_2(/C_2'')$	$C_{2v}(/C_1)$ $+ 2C_{2v}(/C_{2v})$	$2C_{3v}(/C_s)$	$D_{2d}(/C_2')$ $+ D_{2d}(/C_{2v})$	$T(/C_2)$	$T_d(/C_{2v})$
$T_d(/C_{3v})$	$4C_1(/C_1)$	$2C_2(/C_1)$	$C_s(/C_1)$ $+ 2C_s(/C_s)$	$C_3(/C_1)$ $+ C_3(/C_3)$	$S_4(/C_1)$	$D_2(/C_1)$	$C_{2v}(/C_s)$ $+ C_{2v}(/C_s')$	$C_{3v}(/C_s)$ $+ C_{3v}(/C_{3v})$	$D_{2d}(/C_s)$	$T(/C_3)$	$T_d(/C_{3v})$
$T_d(/D_{2d})^*$	$3C_1(/C_1)$	$3C_2(/C_2)$	$C_s(/C_1)$ $+ C_s(/C_s)$	$C_3(/C_1)$	$S_4(/C_2)$ $+ S_4(/S_4)$	$3D_2(/D_2)$	$C_{2v}(/C_2)$ $+ C_{2v}(/C_{2v})$	$C_{3v}(/C_s)$	$D_{2d}(/D_2)$ $+ D_{2d}(/D_{2d})$	$T(/D_2)$	$T_d(/D_{2d})$
$T_d(/T)^*$	$2C_1(/C_1)$	$2C_2(/C_2)$	$C_s(/C_1)$	$2C_3(/C_3)$	$S_4(/C_2)$	$2D_2(/D_2)$	$C_{2v}(/C_2)$	$C_{3v}(/C_3)$	$D_{2d}(/D_2)$	$2T(/T)$	$T_d(/T)$
$T_d(/T_d)$	$C_1(/C_1)$	$C_2(/C_2)$	$C_s(/C_s)$	$C_3(/C_3)$	$S_4(/S_4)$	$D_2(/D_2)$	$C_{2v}(/C_{2v})$	$C_{3v}(/C_{3v})$	$D_{2d}(/D_{2d})$	$T(/T)$	$T_d(/T_d)$

* Forbidden CR. See Chapter 7.

C.2   D$_{3h}$ Point Group and Its Subgroups

Table C.11: Subduction table of C$_{3h}$

	$\downarrow$C$_1$	$\downarrow$C$_s$	$\downarrow$C$_3$	$\downarrow$C$_{3h}$
C$_{3h}$(/C$_1$)	6C$_1$(/C$_1$)	3C$_s$(/C$_1$)	2C$_3$(/C$_1$)	C$_{3h}$(/C$_1$)
C$_{3h}$(/C$_s$)	3C$_1$(/C$_1$)	3C$_s$(/C$_s$)	3C$_3$(/C$_3$)	C$_{3h}$(/C$_s$)
C$_{3h}$(/C$_3$)	2C$_1$(/C$_1$)	C$_s$(/C$_1$)	2C$_3$(/C$_3$)	C$_{3h}$(/C$_3$)
C$_{3h}$(/C$_{3h}$)	C$_1$(/C$_1$)	C$_s$(/C$_s$)	C$_3$(/C$_3$)	C$_{3h}$(/C$_{3h}$)

Table C.12: Subduction Table of D$_3$

	$\downarrow$C$_1$	$\downarrow$C$_2$	$\downarrow$C$_3$	$\downarrow$D$_3$
D$_3$(/C$_1$)	6C$_1$(/C$_1$)	3C$_2$(/C$_1$)	2C$_3$(/C$_1$)	D$_3$(/C$_1$)
D$_3$(/C$_2$)	3C$_1$(/C$_1$)	C$_2$(/C$_1$) + C$_2$(/C$_2$)	C$_3$(/C$_1$)	D$_3$(/C$_2$)
D$_3$(/C$_3$)	2C$_1$(/C$_1$)	C$_2$(/C$_1$)	2C$_3$(/C$_3$)	D$_3$(/C$_3$)
D$_3$(/D$_3$)	C$_1$(/C$_1$)	C$_2$(/C$_2$)	C$_3$(/C$_3$)	D$_3$(/D$_3$)

342

Table C.13: Subduction Table for D_{3h} Point Group

	$\downarrow C_1$	$\downarrow C_2$	$\downarrow C_s$	$\downarrow C_s'$	$\downarrow C_3$	$\downarrow C_{2v}$	$\downarrow C_{3v}$	$\downarrow C_{3h}$	$\downarrow D_3$	$\downarrow D_{3h}$
$D_{3h}(/C_1)$	$12C_1(/C_1)$	$6C_2(/C_1)$	$6C_s(/C_1)$	$6C_s'(/C_1)$	$4C_3(/C_1)$	$3C_{2v}(/C_1)$	$2C_{3v}(/C_1)$	$2C_{3h}(/C_1)$	$2D_3(/C_1)$	$D_{3h}(/C_1)$
$D_{3h}(/C_2)^*$	$6C_1(/C_1)$	$2C_2(/C_1)$ $+ 2C_2(/C_2)$	$3C_s(/C_1)$	$3C_s'(/C_1)$	$2C_3(/C_1)$	$C_{2v}(/C_1)$ $+ C_{2v}(/C_2)$	$C_{3v}(/C_1)$	$C_{3h}(/C_1)$	$2D_3(/C_2)$	$D_{3h}(/C_2)$
$D_{3h}(/C_s)$	$6C_1(/C_1)$	$3C_2(/C_1)$	$2C_s(/C_1)$ $+ 2C_s(/C_s)$	$3C_s'(/C_1)$	$2C_3(/C_1)$	$C_{2v}(/C_1)$ $+ C_{2v}(/C_2)$	$2C_{3v}(/C_s)$	$C_{3h}(/C_1)$	$D_3(/C_1)$	$D_{3h}(/C_s)$
$D_{3h}(/C_s')$	$6C_1(/C_1)$	$3C_2(/C_1)$	$3C_s(/C_1)$	$6C_s'(/C_s)$	$2C_3(/C_1)$	$3C_{2v}(/C_s')$	$C_{3v}(/C_1)$	$2C_{3h}(/C_s)$	$D_3(/C_1)$	$D_{3h}(/C_s')$
$D_{3h}(/C_3)^*$	$4C_1(/C_1)$	$2C_2(/C_1)$	$2C_s(/C_1)$	$2C_s'(/C_1)$	$4C_3(/C_3)$	$C_{2v}(/C_1)$	$2C_{3v}(/C_3)$	$2C_{3h}(/C_3)$	$2D_3(/C_3)$	$D_{3h}(/C_3)$
$D_{3h}(/C_{2v})$	$3C_1(/C_1)$	$C_2(/C_1)$ $+ C_2(/C_2)$	$C_s(/C_1)$ $+ C_s(/C_s)$	$3C_s'(/C_s)$	$C_3(/C_1)$	$C_{2v}(/C_s')$ $+ C_{2v}(/C_{2v})$	$C_{3v}(/C_s)$	$C_{3h}(/C_s)$	$D_3(/C_2)$	$D_{3h}(/C_{2v})$
$D_{3h}(/C_{3v})$	$2C_1(/C_1)$	$C_2(/C_1)$	$2C_s(/C_s)$	$C_s'(/C_1)$	$2C_3(/C_3)$	$C_{2v}(/C_s)$	$2C_{3v}(/C_{3v})$	$C_{3h}(/C_3)$	$D_3(/C_3)$	$D_{3h}(/C_{3v})$
$D_{3h}(/C_{3h})^*$	$2C_1(/C_1)$	$C_2(/C_1)$	$C_s(/C_1)$	$C_s'(/C_s)$	$2C_3(/C_3)$	$C_{2v}(/C_s')$	$C_{3v}(/C_3)$	$2C_{3h}(/C_{3h})$	$D_3(/C_3)$	$D_{3h}(/C_{3h})$
$D_{3h}(/D_3)^*$	$2C_1(/C_1)$	$2C_2(/C_2)$	$C_s(/C_1)$	$C_s'(/C_1)$	$2C_3(/C_3)$	$C_{2v}(/C_2)$	$C_{3v}(/C_3)$	$C_{3h}(/C_3)$	$2D_3(/D_3)$	$D_{3h}(/D_3)$
$D_{3h}(/D_{3h})$	$C_1(/C_1)$	$C_2(/C_2)$	$C_s(/C_s)$	$C_s'(/C_s)$	$C_3(/C_3)$	$C_{2v}(/C_{2v})$	$C_{3v}(/C_{3v})$	$C_{3h}(/C_{3h})$	$D_3(/D_3)$	$D_{3h}(/D_{3h})$

* Forbidden CR. See Chapter 7. Note that $C_{3h} = S_3$.

Appendix D

Tables of USCIs

This is a list of tables of unit subduced cycle indices for several point groups.

D.1 T_d Point Group and Its Subgroups

Table D.1: USCI Table of C_2

	$\downarrow C_1$	$\downarrow C_2$
$C_2(/C_1)$	s_1^2	s_2
$C_2(/C_2)$	s_1	s_1
sum	1/2	1/2

Table D.2: USCI Table of C_s

	$\downarrow C_1$	$\downarrow C_s$
$C_s(/C_1)$	s_1^2	s_2
$C_s(/C_s)$	s_1	s_1
sum	1/2	1/2

Table D.3: Subduction Table of C_3

	$\downarrow C_1$	$\downarrow C_3$
$C_3(/C_1)$	s_1^3	s_3
$C_3(/C_3)$	s_1	s_1
sum	1/3	2/3

Table D.4: USCI Table of S_4

	$\downarrow C_1$	$\downarrow C_2$	$\downarrow S_4$
$S_4(/C_1)$	s_1^4	s_2^2	s_4
$S_4(/C_2)$	s_1^2	s_1^2	s_2
$S_4(/S_4)$	s_1	s_1	s_1
sum	1/4	1/4	1/2

Table D.5: USCI Table of C_{2v} point group

	$\downarrow C_1$	$\downarrow C_2$	$\downarrow C_s$	$\downarrow C'_s$	$\downarrow C_{3v}$
$C_{2v}(/C_1)$	s_1^4	s_2^2	s_2^2	s_2^2	s_4
$C_{2v}(/C_2)$	s_1^2	s_1^2	s_2	s_2	s_2
$C_{2v}(/C_s)$	s_1^2	s_2	s_1^2	s_2	s_2
$C_{2v}(/C'_s)$	s_1^2	s_2	s_2	s_1^2	s_2
$C_{2v}(/C_{2v})$	s_1	s_1	s_1	s_1	s_1
sum	1/4	1/4	1/4	1/4	0

Table D.6: USCI Table of $\mathbf{D}_2$

	$\downarrow\mathbf{C}_1$	$\downarrow\mathbf{C}_2$	$\downarrow\mathbf{C}_2'$	$\downarrow\mathbf{C}_2''$	$\downarrow\mathbf{D}_2$
$\mathbf{D}_2(/\mathbf{C}_i)$	s_1^4	s_2^2	s_2^2	s_2^2	s_4
$\mathbf{D}_2(/\mathbf{C}_2)$	s_1^2	s_1^2	s_2	s_2	s_2
$\mathbf{D}_2(/\mathbf{C}_2')$	s_1^2	s_2	s_1^2	s_2	s_2
$\mathbf{D}_2(/\mathbf{C}_2'')$	s_1^2	s_2	s_2	s_1^2	s_2
$\mathbf{D}_2(/\mathbf{D}_2)$	s_1	s_1	s_1	s_1	s_1
sum	$1/4$	$1/4$	$1/4$	$1/4$	0

Table D.7: USCI Table of $\mathbf{C}_{3v}$ point group

	$\downarrow\mathbf{C}_1$	$\downarrow\mathbf{C}_s$	$\downarrow\mathbf{C}_3$	$\downarrow\mathbf{C}_{3v}$
$\mathbf{C}_{3v}(/\mathbf{C}_1)$	s_1^6	s_2^3	s_3^2	s_6
$\mathbf{C}_{3v}(/\mathbf{C}_s)$	s_1^3	$s_1 s_2$	s_3	s_3
$\mathbf{C}_{3v}(/\mathbf{C}_3)$	s_1^2	s_2	s_1^2	s_2
$\mathbf{C}_{3v}(/\mathbf{C}_{3v})$	s_1	s_1	s_1	s_1
sum	$1/6$	$1/2$	$1/3$	0

Table D.8: USCI Table of D_{2d} Point Group

	$\downarrow C_1$	$\downarrow C_2$	$\downarrow C_2'$	$\downarrow C_s$	$\downarrow S_4$	$\downarrow C_{2v}$	$\downarrow D_2$	$\downarrow D_{2d}$
$D_{2d}(/C_1)$	s_1^8	s_2^4	s_2^4	s_2^4	s_4^2	s_4^2	s_4^2	s_8
$D_{2d}(/C_2)$	s_1^4	s_1^4	s_2^2	s_2^2	s_2^2	s_2^2	s_2^2	s_4
$D_{2d}(/C_2')$	s_1^4	s_2^2	$s_1^2 s_2$	s_2^2	s_4	s_4	s_2^2	s_4
$D_{2d}(/C_s)$	s_1^4	s_2^2	s_2^2	$s_1^2 s_2$	s_4	s_2^2	s_4	s_4
$D_{2d}(/S_4)$	s_1^2	s_1^2	s_2	s_2	s_1^2	s_2	s_2	s_2
$D_{2d}(/C_{2v})$	s_1^2	s_1^2	s_2	s_1^2	s_2	s_1^2	s_2	s_2
$D_{2d}(/D_2)$	s_1^2	s_1^2	s_1^2	s_2	s_2	s_2	s_1^2	s_2
$D_{2d}(/D_{2d})$	s_1	s_1	s_1	s_1	s_1	s_1	s_1	s_1
$\sum_{i=1}^{s} \overline{m}_{ji}$	1/8	1/8	1/4	1/4	1/4	0	0	0

Table D.9: USCI Table of **T**

	$\downarrow C_1$	$\downarrow C_2$	$\downarrow C_3$	$\downarrow D_2$	$\downarrow T$
$T(/C_1)$	s_1^{12}	s_2^6	s_3^4	s_4^3	s_{12}
$T(/C_2)$	s_1^6	$s_1^2 s_2^2$	s_3^2	s_2^3	s_6
$T(/C_3)$	s_1^4	s_2^2	$s_1 s_3$	s_4	s_4
$T(/D_2)$	s_1^3	s_1^3	s_3	s_1^3	s_3
$T(/T)$	s_1	s_1	s_1	s_1	s_1
sum	$1/12$	$1/4$	$2/3$	0	0

Table D.10: USCI Table of T_d

	$\downarrow C_1$	$\downarrow C_2$	$\downarrow C_s$	$\downarrow C_3$	$\downarrow S_4$	$\downarrow D_2$	$\downarrow C_{2v}$	$\downarrow C_{3v}$	$\downarrow D_{2d}$	$\downarrow T$	$\downarrow T_d$
$T_d(/C_1)$	s_1^{24}	s_2^{12}	s_2^{12}	s_3^{8}	s_4^{6}	s_4^{6}	s_4^{6}	s_6^{4}	s_8^{3}	s_{12}^{2}	s_{24}
$T_d(/C_2)$	s_1^{12}	$s_1^{4}s_2^{4}$	s_2^{6}	s_3^{4}	$s_2^{2}s_4^{2}$	s_2^{6}	$s_2^{2}s_4^{2}$	s_6^{2}	s_4^{3}	s_6^{2}	s_{12}
$T_d(/C_s)$	s_1^{12}	s_2^{6}	$s_1^{2}s_2^{5}$	s_3^{4}	s_4^{3}	s_4^{3}	$s_2^{2}s_4^{2}$	$s_3^{2}s_6$	$s_4 s_8$	s_{12}	s_{12}
$T_d(/C_3)$	s_1^{8}	s_2^{4}	s_2^{4}	$s_1^{2}s_3^{2}$	s_4^{2}	s_4^{2}	s_4^{2}	$s_2 s_6$	s_8	s_4^{2}	s_8
$T_d(/S_4)$	s_1^{6}	$s_1^{2}s_2^{2}$	s_2^{3}	s_3^{2}	$s_1^{2}s_4$	s_2^{3}	$s_2 s_4$	s_6	$s_2 s_4$	s_6	s_6
$T_d(/D_2)$	s_1^{6}	s_1^{6}	s_2^{3}	s_3^{2}	s_2^{3}	s_1^{6}	s_2^{3}	s_6	s_2^{3}	s_3^{2}	s_6
$T_d(/C_{2v})$	s_1^{6}	$s_1^{2}s_2^{2}$	$s_1^{2}s_2^{2}$	s_3^{2}	$s_2 s_4$	s_2^{3}	$s_1^{2}s_4$	s_3^{2}	$s_2 s_4$	s_6	s_6
$T_d(/C_{3v})$	s_1^{4}	s_2^{2}	$s_1^{2}s_2$	$s_1 s_3$	s_4	s_4	s_2^{2}	$s_1 s_3$	s_4	s_4	s_4
$T_d(/D_{2d})$	s_1^{3}	s_1^{3}	$s_1 s_2$	s_3	$s_1 s_2$	s_1^{3}	$s_1 s_2$	s_3	$s_1 s_2$	s_3	s_3
$T_d(/T)$	s_1^{2}	s_1^{2}	s_2	s_1^{2}	s_2	s_1^{2}	s_2	s_2	s_2	s_1^{2}	s_2
$T_d(/T_d)$	s_1	s_1	s_1	s_1	s_1	s_1	s_1	s_1	s_1	s_1	s_1
sum	$1/24$	$1/8$	$1/4$	$1/3$	$1/4$	0	0	0	0	0	0

D.2 D$_{3h}$ Point Group and Its Subgroups

Table D.11: USCI Table of C$_{3h}$

	$\downarrow$C$_1$	$\downarrow$C$_s$	$\downarrow$C$_3$	$\downarrow$C$_{3h}$
C$_{3h}$(/C$_1$)	s_1^6	s_2^3	s_3^2	s_6
C$_{3h}$(/C$_s$)	s_1^3	s_1^3	s_1^3	s_3
C$_{3h}$(/C$_3$)	s_1^2	s_2	s_1^2	s_2
C$_{3h}$(/C$_{3h}$)	s_1	s_1	s_1	s_1
sum	1/6	1/6	1/3	1/3

Table D.12: USCI Table of D$_3$

	$\downarrow$C$_1$	$\downarrow$C$_2$	$\downarrow$C$_3$	$\downarrow$D$_3$
D$_3$(/C$_1$)	s_1^6	s_2^3	s_3^2	s_6
D$_3$(/C$_2$)	s_1^3	$s_1 s_2$	s_3	s_3
D$_3$(/C$_3$)	s_1^2	s_2	s_1^2	s_2
D$_3$(/D$_3$)	s_1	s_1	s_1	s_1
sum	1/6	1/2	1/3	0

Table D.13: USCI Table of D_{3h} Point Group

	$\downarrow C_1$	$\downarrow C_2$	$\downarrow C_s$	$\downarrow C_s'$	$\downarrow C_3$	$\downarrow C_{2v}$	$\downarrow C_{3v}$	$\downarrow C_{3h}$	$\downarrow D_3$	$\downarrow D_{3h}$
$D_{3h}(/C_1)$	s_1^{12}	s_2^6	s_2^6	s_2^6	s_3^4	s_4^3	s_6^2	s_6^2	s_6^2	s_{12}
$D_{3h}(/C_2)$	s_1^6	$s_1^2 s_2^2$	s_2^3	s_2^3	s_3^2	$s_2 s_4$	s_6	s_6	s_3^2	s_6
$D_{3h}(/C_s)$	s_1^6	s_2^3	$s_1^2 s_2^2$	s_2^3	s_3^2	$s_2 s_4$	s_3^2	s_6	s_6	s_6
$D_{3h}(/C_s')$	s_1^6	s_2^3	s_2^3	s_1^6	s_3^2	s_2^3	s_6	s_3^2	s_3^2	s_6
$D_{3h}(/C_3)$	s_1^4	s_2^2	s_2^2	s_2^2	s_1^4	s_4	s_2^2	s_2^2	s_2^2	s_4
$D_{3h}(/C_{2v})$	s_1^3	$s_1 s_2$	$s_1 s_2$	s_1^3	s_3	$s_1 s_2$	s_3	s_3	s_3	s_3
$D_{3h}(/C_{3v})$	s_1^2	s_2	s_1^2	s_2	s_1^2	s_2	s_1^2	s_2	s_2	s_2
$D_{3h}(/C_{3h})$	s_1^2	s_2	s_2	s_1^2	s_1^2	s_2	s_2	s_1^2	s_2	s_2
$D_{3h}(/D_3)$	s_1^2	s_1^2	s_2	s_2	s_1^2	s_2	s_2	s_2	s_1^2	s_2
$D_{3h}(/D_{3h})$	s_1	s_1	s_1	s_1	s_1	s_1	s_1	s_1	s_1	s_1
$\sum_{i=1}^{s} \overline{m}_{ji}$	1/12	1/4	1/4	1/12	1/6	0	0	1/6	0	0

Appendix E

Tables of USCI-CFs

This is a list of tables of unit subduced cycle indices with chirality fittingness for several point groups.

E.1 T_d Point Group and Its Subgroups

Table E.1: USCI-CF Table of C_2

	$\downarrow C_1$	$\downarrow C_2$
$C_2(/C_1)$	b_1^2	b_2
$C_2(/C_2)$	b_1	b_1
sum	1/2	1/2

Table E.2: USCI-CF Table of C_s

	$\downarrow C_1$	$\downarrow C_s$
$C_s(/C_1)$	b_1^2	c_2
$C_s(/C_s)$	b_1	a_1
sum	1/2	1/2

Table E.3: USCI-CF Table of C_3

	$\downarrow C_1$	$\downarrow C_3$
$C_3(/C_1)$	b_1^3	b_3
$C_3(/C_3)$	b_1	b_1
sum	$1/3$	$2/3$

Table E.4: USCI-CF Table of S_4

	$\downarrow C_1$	$\downarrow C_2$	$\downarrow S_4$
$S_4(/C_1)$	b_1^4	b_2^2	c_4
$S_4(/C_2)$	b_1^2	b_1^2	c_2
$S_4(/S_4)$	b_1	b_1	a_1
sum	$1/4$	$1/4$	$1/2$

Table E.5: USCI-CF Table of C_{2v} point group

	$\downarrow C_1$	$\downarrow C_2$	$\downarrow C_s$	$\downarrow C'_s$	$\downarrow C_{2v}$
$C_{2v}(/C_1)$	b_1^4	b_2^2	c_2^2	c_2^2	c_4
$C_{2v}(/C_2)$	b_1^2	b_1^2	c_2	c_2	c_2
$C_{2v}(/C_s)$	b_1^2	b_2	a_1^2	c_2	a_2
$C_{2v}(/C'_s)$	b_1^2	b_2	c_2	a_1^2	a_2
$C_{2v}(/C_{2v})$	b_1	b_1	a_1	a_1	a_1
sum	$1/4$	$1/4$	$1/4$	$1/4$	0

Table E.6: USCI-CF Table of $\mathbf{D}_2$

	$\downarrow\mathbf{C}_1$	$\downarrow\mathbf{C}_2$	$\downarrow\mathbf{C}_2'$	$\downarrow\mathbf{C}_2''$	$\downarrow\mathbf{D}_2$
$\mathbf{D}_2(/\mathbf{C}_1)$	b_1^4	b_2^2	b_2^2	b_2^2	b_4
$\mathbf{D}_2(/\mathbf{C}_2)$	b_1^2	b_1^2	b_2	b_2	b_2
$\mathbf{D}_2(/\mathbf{C}_2')$	b_1^2	b_2	b_1^2	b_2	b_2
$\mathbf{D}_2(/\mathbf{C}_2'')$	b_1^2	b_2	b_2	b_1^2	b_2
$\mathbf{D}_2(/\mathbf{D}_2)$	b_1	b_1	b_1	b_1	b_1
sum	1/4	1/4	1/4	1/4	0

Table E.7: USCI-CF Table of $\mathbf{C}_{3v}$ point group

	$\downarrow\mathbf{C}_1$	$\downarrow\mathbf{C}_s$	$\downarrow\mathbf{C}_3$	$\downarrow\mathbf{C}_{3v}$
$\mathbf{C}_{3v}(/\mathbf{C}_1)$	b_1^6	c_2^3	b_3^2	c_6
$\mathbf{C}_{3v}(/\mathbf{C}_s)$	b_1^3	a_1c_2	b_3	a_3
$\mathbf{C}_{3v}(/\mathbf{C}_3)$	b_1^2	c_2	b_1^2	c_2
$\mathbf{C}_{3v}(/\mathbf{C}_{3v})$	b_1	a_1	b_1	a_1
sum	1/6	1/2	1/3	0

Table E.8: USCI-CF Table of $\mathbf{D}_{2d}$ Point Group

	$\downarrow C_1$	$\downarrow C_2$	$\downarrow C_2'$	$\downarrow C_s$	$\downarrow S_4$	$\downarrow C_{2v}$	$\downarrow D_2$	$\downarrow D_{2d}$
$\mathbf{D}_{2d}(/C_1)$	b_1^8	b_2^4	b_2^4	c_2^4	c_4^2	c_4^2	b_4^2	c_8
$\mathbf{D}_{2d}(/C_2)$	b_1^4	b_1^4	b_2^2	c_2^2	c_2^2	c_2^2	b_2^2	c_4
$\mathbf{D}_{2d}(/C_2')$	b_1^4	b_2^2	$b_1^2 b_2$	c_2^2	c_4	c_4	b_2^2	c_4
$\mathbf{D}_{2d}(/C_s)$	b_1^4	b_2^2	b_2^2	$a_1^2 c_2$	c_4	a_2^2	b_4	a_4
$\mathbf{D}_{2d}(/S_4)$	b_1^2	b_1^2	b_2	c_2	a_1^2	c_2	b_2	a_2
$\mathbf{D}_{2d}(/C_{2v})$	b_1^2	b_1^2	b_2	a_1^2	c_2	a_1^2	b_2	a_2
$\mathbf{D}_{2d}(/D_2)$	b_1^2	b_1^2	b_1^2	c_2	c_2	c_2	b_1^2	c_2
$\mathbf{D}_{2d}(/D_{2d})$	b_1	b_1	b_1	a_1	a_1	a_1	b_1	a_1
$\sum_{i=1}^{s} \overline{m}_{ji}$	1/8	1/8	1/4	1/4	1/4	0	0	0

Table E.9: USCI-CF Table of $\mathbf{T}$

	$\downarrow\mathbf{C}_1$	$\downarrow\mathbf{C}_2$	$\downarrow\mathbf{C}_3$	$\downarrow\mathbf{D}_2$	$\downarrow\mathbf{T}$
$\mathbf{T}(/\mathbf{C}_1)$	b_1^{12}	b_2^6	b_3^4	b_4^3	b_{12}
$\mathbf{T}(/\mathbf{C}_2)$	b_1^6	$b_1^2b_2^2$	b_3^2	b_2^3	b_6
$\mathbf{T}(/\mathbf{C}_3)$	b_1^4	b_2^2	b_1b_3	b_4	b_4
$\mathbf{T}(/\mathbf{D}_2)$	b_1^3	b_1^3	b_3	b_1^3	b_3
$\mathbf{T}(/\mathbf{T})$	b_1	b_1	b_1	b_1	b_1
sum	$1/12$	$1/4$	$2/3$	0	0

Table E.10: USCI-CF Table of $\mathbf{T}_d$

	$\downarrow\mathbf{C}_1$	$\downarrow\mathbf{C}_2$	$\downarrow\mathbf{C}_s$	$\downarrow\mathbf{C}_3$	$\downarrow\mathbf{S}_4$	$\downarrow\mathbf{D}_2$	$\downarrow\mathbf{C}_{2v}$	$\downarrow\mathbf{C}_{3v}$	$\downarrow\mathbf{D}_{2d}$	$\downarrow\mathbf{T}$	$\downarrow\mathbf{T}_d$
$\mathbf{T}_d(/\mathbf{C}_1)$	b_1^{24}	b_2^{12}	c_2^{12}	b_3^{8}	c_4^{6}	b_4^{6}	c_4^{6}	c_6^{4}	c_8^{3}	b_{12}^{2}	c_{24}
$\mathbf{T}_d(/\mathbf{C}_2)$	b_1^{12}	$b_1^{4}b_2^{4}$	c_2^{6}	b_3^{4}	$c_2^{2}c_4^{2}$	b_2^{6}	$c_2^{2}c_4^{2}$	c_6^{2}	c_4^{3}	b_6^{2}	c_{12}
$\mathbf{T}_d(/\mathbf{C}_s)$	b_1^{12}	b_2^{6}	$a_1^{2}c_2^{5}$	b_3^{4}	c_4^{3}	b_4^{3}	$a_2^{2}c_4^{2}$	$a_3^{2}c_6$	a_4c_8	b_{12}	a_{12}
$\mathbf{T}_d(/\mathbf{C}_3)$	b_1^{8}	b_2^{4}	c_2^{4}	$b_1^{2}b_3^{2}$	c_4^{2}	b_4^{2}	c_4^{2}	c_2c_6	c_8	b_4^{2}	c_8
$\mathbf{T}_d(/\mathbf{S}_4)$	b_1^{6}	$b_1^{2}b_2^{2}$	c_2^{3}	b_3^{2}	$a_1^{2}c_4$	b_2^{3}	c_2c_4	c_6	a_2c_4	b_6	a_6
$\mathbf{T}_d(/\mathbf{D}_2)$	b_1^{6}	b_1^{6}	c_2^{3}	b_3^{2}	c_2^{3}	b_1^{6}	c_2^{3}	c_6	c_2^{3}	b_3^{2}	c_6
$\mathbf{T}_d(/\mathbf{C}_{2v})$	b_1^{6}	$b_1^{2}b_2^{2}$	$a_1^{2}c_2^{2}$	b_3^{2}	c_2c_4	b_2^{3}	$a_1^{2}c_4$	a_3^{2}	a_2c_4	b_6	a_6
$\mathbf{T}_d(/\mathbf{C}_{3v})$	b_1^{4}	b_2^{2}	$a_1^{2}c_2$	b_1b_3	c_4	b_4	a_2^{2}	a_1a_3	a_4	b_4	a_4
$\mathbf{T}_d(/\mathbf{D}_{2d})$	b_1^{3}	b_1^{3}	a_1c_2	b_3	a_1c_2	b_1^{3}	a_1c_2	a_3	a_1c_2	b_3	a_3
$\mathbf{T}_d(/\mathbf{T})$	b_1^{2}	b_1^{2}	c_2	b_1^{2}	a_2	b_1^{2}	c_2	c_2	c_2	b_1^{2}	c_2
$\mathbf{T}_d(/\mathbf{T}_d)$	b_1	b_1	a_1	b_1	a_1	b_1	a_1	a_1	a_1	b_1	a_1
sum	1/24	1/8	1/4	1/3	1/4	0	0	0	0	0	0

E.2 D_{3h} Point Group and Its Subgroups

Table E.11: USCI-CF Table of C_{3h}

	$\downarrow C_1$	$\downarrow C_s$	$\downarrow C_3$	$\downarrow C_{3h}$
$C_{3h}(/C_1)$	b_1^6	c_2^3	b_3^2	c_6
$C_{3h}(/C_s)$	b_1^3	a_1^3	b_1^3	a_3
$C_{3h}(/C_3)$	b_1^2	c_2	b_1^2	a_2
$C_{3h}(/C_{3h})$	b_1	a_1	b_1	a_1
sum	1/6	1/6	1/3	1/3

Table E.12: USCI-CF Table of D_3

	$\downarrow C_1$	$\downarrow C_2$	$\downarrow C_3$	$\downarrow D_3$
$D_3(/C_1)$	b_1^6	b_2^3	b_3^2	b_6
$D_3(/C_2)$	b_1^3	$b_1 b_2$	b_3	b_3
$D_3(/C_3)$	b_1^2	b_2	b_1^2	b_2
$D_3(/D_3)$	b_1	b_1	b_1	b_1
sum	1/6	1/2	1/3	0

Table E.13: USCI-CF Table of D_{3h} Point Group

	$\downarrow C_1$	$\downarrow C_2$	$\downarrow C_s$	$\downarrow C_s'$	$\downarrow C_3$	$\downarrow C_{2v}$	$\downarrow C_{3v}$	$\downarrow C_{3h}$	$\downarrow D_3$	$\downarrow D_{3h}$
$D_{3h}(/C_1)$	b_1^{12}	b_2^6	c_2^6	c_2^6	b_3^4	c_4^3	c_6^2	c_6^2	b_6^2	c_{12}
$D_{3h}(/C_2)$	b_1^6	$b_1^2 b_2^2$	c_2^3	c_2^3	b_3^2	$c_2 c_4$	c_6	c_6	b_3^2	c_6
$D_{3h}(/C_s)$	b_1^6	b_2^3	$a_1^2 c_2^2$	c_2^3	b_3^2	$a_2 c_4$	a_3^2	c_6	b_6	a_6
$D_{3h}(/C_s')$	b_1^6	b_2^3	c_2^3	a_1^6	b_3^2	a_2^3	c_6	a_3^2	b_3^2	a_6
$D_{3h}(/C_3)$	b_1^4	b_2^2	c_2^2	c_2^2	b_1^4	c_4	c_2^2	c_2^2	b_2^2	c_4
$D_{3h}(/C_{2v})$	b_1^3	$b_1 b_2$	$a_1 c_2$	a_1^3	b_3	$a_1 a_2$	a_3	a_3	b_3	a_3
$D_{3h}(/C_{3v})$	b_1^2	b_2	a_1^2	c_2	b_1^2	a_2	a_1^2	c_2	b_2	a_2
$D_{3h}(/C_{3h})$	b_1^2	b_2	c_2	a_1^2	b_1^2	a_2	c_2	a_1^2	b_2	a_2
$D_{3h}(/D_3)$	b_1^2	b_1^2	c_2	c_2	b_1^2	c_2	c_2	c_2	b_1^2	c_2
$D_{3h}(/D_{3h})$	b_1	b_1	a_1	a_1	b_1	a_1	a_1	a_1	b_1	a_1
$\sum_{i=1}^s \overline{m}_{ji}$	1/12	1/4	1/4	1/12	1/6	0	0	1/6	0	0

Index

D

W